N.B. Delone   V.P. Krainov

Atoms in Strong Light Fields

With 49 Figures

W0225767

Springer-Verlag
Berlin Heidelberg New York Tokyo

Professor Dr. Nikolai B. Delone

General Physics Institute, Academy of Sciences of the USSR, Vavilov Street 38
SU-117924 Moscow, USSR

Professor Dr. Vladimir P. Krainov

Moscow Engineering Physics Institute, 31 Kashirskoe shosse, SU-115409 Moscow, USSR

Translator
Dr. Evgeny M. Yankovsky

Series Editors

Professor Vitalii I. Goldanskii

Institute of Chemical Physics
Academy of Sciences
Kosygin Street 4
Moscow V-334, USSR

Professor Dr. Fritz Peter Schäfer

Max-Planck-Institut für
Biophysikalische Chemie
D-3400 Göttingen-Nikolausberg
Fed. Rep. of Germany

Professor Robert Gomer

The James Franck Institute
The University of Chicago
5640 Ellis Avenue
Chicago, IL 60637, USA

Professor Dr. J. Peter Toennies

Max-Planck-Institut für Strömungsforschung
Böttingerstraße 6-8
D-3400 Göttingen
Fed. Rep. of Germany

Title of the original Russian edition: *Atom v sil'nom svetovom pole*
© Atomizdat, 1978; Energoatomizdat, 1984

ISBN-13: 978-3-642-85693-8   e-ISBN-13: 978-3-642-85691-4
DOI: 10.1007/978-3-642-85691-4

Preface

The monograph is devoted to phenomena of nonlinear optics appearing on a macroscopic level in the interaction of intense light with an isolated atom. It is a first attempt to summarize the elementary phenomena of nonlinear optics and present the various methods used in experiment and theory. In essence, this book can be considered an expanded version of the new aspect of quantum mechanics and atomic physics that in time will be incorporated into textbooks on this subject.

By the middle of this century the interaction of light with atoms had become one of the most investigated branches of physics. However, in the mid-sixties the development of high-power lasers changed this situation completely. It is a well-known fact that lasers are essentially new sources of light with high intensity, sharp directivity, and practically ideal monochromaticity. Entirely new phenomena came up in the studies of the interaction of light with atoms.

In an intense light field, multiphoton transitions become important. The field disturbs the atomic levels, shifting, broadening, and mixing them. In an extremely strong field the atom ceases to be a bound system. These and similar phenomena on the atomic (microscopic) level determine the variations in the averaged, macroscopic properties of the medium, variations that cause nonlinear-optics phenomena, which radically change the fundamental classical laws of the interaction of light with matter.

The scope of this monograph is outlined in the Introduction and the Conclusion. We would like to note, however, that we had no intention of describing all the phenomena. Rather, we chose typical phenomena and studied these in greater detail. In discussing these phenomena we usually start with the theoretical aspect and then illustrate the conclusions with experimental observations. Such a presentation, we believe, reflects the essence of physics, which is a theoretical interpretation of experimental data that makes it possible to predict new phenomena and properties of the material world.

The present book is a result of many years of collaboration by an experimentalist (Delone) and a theoretician (Krainov). We have tried to interpret phenomena in such a way that the experimental physicist will understand all the formulae and to present the results of experiments in such a way that the theoretician will understand all the details. We hope that this style of presentation will make the book interesting for theoretical physicists, for experimentalists, for engineers working on applications of these ideas in technology, and for students as well.

This English edition does not represent a direct translation of our earlier book in Russian. The treatment has been rearranged, extended, and updated. In particular, new findings—both experimental and theoretical—have been incorporated for clarity and completeness.

We take pleasure in expressing our gratitude to Mr. E. Yankovsky, who undertook the translation and, being a physicist himself, has done it with a thorough understanding of the subject and terminology. The book would not have been realized, if it were not for Dr. H. Lotsch of Springer-Verlag who has worked on the manuscript at all stages of its preparation, and we express our deep gratitude to him for this engagement.

Moscow, December 1984   *N.B. Delone · V.P. Krainov*

Contents

1. Introduction 1
  1.1 The Strong Light Field 4
    1.1.1 The Classical Nature of the Field 4
    1.1.2 The Parameters of a Light Field 5
  1.2 The Atom 7
    1.2.1 One-Electron and Multi-Electron Approximations 7
    1.2.2 The Structure of Atomic Spectra 8
    1.2.3 Selection Rules 10
  1.3 Interaction of an Atom and a Light Field 11
    1.3.1 The Dipole Approximation 11
    1.3.2 The Hamiltonian of the Dipole Interaction 13
    1.3.3 An Atom in a Circularly Polarized Electromagnetic Field 16
    1.3.4 The Floquet Theorem 18
    1.3.5 Monochromatic Perturbation as a Stationary Problem ... 21
    1.3.6 Exact Solutions 23

2. Time-Dependent Perturbation Theory 24
  2.1 First-Order Perturbation Theory 25
    2.1.1 The System of Equations 25
    2.1.2 The Probability of One-Photon Transitions 25
    2.1.3 Monochromatic Perturbations 26
    2.1.4 Sudden Perturbations 28
    2.1.5 Large Perturbation Times 29
    2.1.6 The Role of the Shape of the Envelope of the Electro-
          magnetic Field 30
    2.1.7 Criteria of Applicability 30
  2.2 Second-Order Perturbation Theory 31
    2.2.1 Probability of a Two-Photon Transition 32
    2.2.2 Transitions Induced by Two Perturbations 35
    2.2.3 Large Perturbation Times 35
    2.2.4 Sudden Perturbations 37
    2.2.5 Criteria of Applicability 38
  2.3 The Diagrammatic Technique for Monochromatic Perturbations .. 38
    2.3.1 First-Order Perturbation Theory 39
    2.3.2 Second-Order Perturbation Theory 40
    2.3.3 The Rules for Constructing Diagrams 41
    2.3.4 Partial Summation of Diagrams 42
  2.4 Perturbation Theory of Arbitrary Order 43
    2.4.1 The Amplitudes of Different Processes 43
    2.4.2 Large Perturbation Times 43
    2.4.3 Criteria for Applying Perturbation Theory of Arbitrary
          Order 44
    2.4.4 Convergence of Expansion Series in Perturbation Theory 46

2.5 Monochromatic Perturbation and Degenerate States 46
    2.5.1 A Single Degenerate Level 47
    2.5.2 Mixing Degenerate Levels in a Field Due to the Presence
        of Other Levels 48
    2.5.3 Near-Degenerate Levels 51
    2.5.4 Time-Dependent Diagrams 52
    2.5.5 Two-Photon Mixing of Degenerate States in Low-Frequency
        Fields 53
    2.5.6 One-Photon Mixing of Degenerate States in Low-Frequency
        Fields 56
    2.5.7 Competition Between One- and Two-Photon Mixing 57
    2.5.8 Approximate Degeneracy 57
    2.5.9 Criteria of Applicability 57
2.6 The Green's Function in Time-Dependent Perturbation Theory .. 59
    2.6.1 The Green's Function 60
    2.6.2 Application of the Green's Function 61
    2.6.3 Realistic Atomic Potentials 62

3. The Resonance Approximation 64
3.1 A Two-Level System in a Resonance Field 65
    3.1.1 Wave Functions 66
    3.1.2 Criteria of Applicability 68
    3.1.3 Adiabatic Introduction of Perturbations 69
    3.1.4 Sudden Perturbations 70
    3.1.5 Adiabatic or Sudden Perturbation? 71
    3.1.6 Quasi-Energies in the Resonance Case 72
3.2 Multi-Photon Resonance 73
    3.2.1 Two-Photon Resonance 74
    3.2.2 Criteria of Applicability 76
    3.2.3 Multi-Photon Resonance 77
3.3 Degeneracy in a Resonance Field 79
    3.3.1 Equations 79
    3.3.2 Basis Solutions 80
    3.3.3 Adiabatic Introduction of a Perturbation 81
    3.3.4 Approximate Degeneracy 82
    3.3.5 Degeneracy and the Influence of Light on an Atom 83
3.4 A Two-Level System in a Circularly Polarized Electromagnetic
    Field 83
    3.4.1 Statement of the Problem 84
    3.4.2 Solution and Discussion 84
3.5 A Two-Level System in a Resonance Field: Time-Dependent
    Parameters 85
    3.5.1 The System of Equations 86
    3.5.2 An Exactly Solvable Problem 86
    3.5.3 The General Solution 87
3.6 A Three-Level System in Two Fields 89
    3.6.1 Mixing in a Three-Level System 89
    3.6.2 Zero Detuning 91
    3.6.3 Population Inversion in a Three-Level System 91

4. The Adiabatic Approximation 93
4.1 General Theory 94
    4.1.1 The Landau-Dykhne Adiabatic Approximation 94
    4.1.2 The Born-Fock Adiabatic Approximation 95

4.2 Bound-Bound Resonance Transitions 96
     4.2.1 Transition Probability per Unit Time 96
     4.2.2 Criterion of Applicability 98
     4.2.3 The Limiting Transition in Perturbation Theory 99
     4.2.4 Resonance Mixing in a Two-Level System 100
     4.2.5 Transitions in Multi-Level Systems 102
4.3 A Two-Level System in a Strong Field of Arbitrary Frequency . 105
     4.3.1 Adiabatic Wave Functions 105
     4.3.2 Average Populations 109
     4.3.3 Results, Limiting Cases, and a Comparison with
            Numerical Calculations 110
4.4 Transitions Between Degenerate States 112
     4.4.1 Reducing the Problem Concerning Degenerate States to
            One Involving a System that Has a Constant Dipole
            Moment 113
     4.4.2 Degeneracy and Transition Probability 114
4.5 Bound-Free Transitions 115
     4.5.1 Comparison of the Perturbation in Discrete and
            Continuous States 116
     4.5.2 Probability of a Transition in a Low-Frequency Field . 116
     4.5.3 Limiting Cases 118

5. Laser Radiation 120

    5.1 Intensity of Radiation 121
         5.1.1 Spatial-Temporal Distribution of the Intensity of Laser
                Radiation 121
         5.1.2 Dependence of the Intensity on the Type, the Design,
                and the Mode of Operation of a Laser 123
         5.1.3 Increasing the Intensity of Laser Radiation 125
         5.1.4 Varying the Intensity of Laser Radiation 126
         5.1.5 Measuring the Intensity of Radiation 127
    5.2 Frequency of the Radiation 128
         5.2.1 Changes in the Tunability and Spectral Narrowing 129
         5.2.2 The Single-Mode Laser 130
         5.2.3 The Measurement of Laser Frequency 130
    5.3 The Polarization of the Radiation 131
    5.4 The Monochromaticity of Laser Radiation 132

6. Experimental Aspects 135

    6.1 Atomic Target 136
    6.2 Competing Processes 138
         6.2.1 Collisions Between Identical Atoms 139
         6.2.2 Collisions with Electrons 140
    6.3 The Self-Induced Distortion of Intense Light in Atomic Targets 143
    6.4 Experimental Methods for Studying the Phenomena Produced by
         Strong Electromagnetic Fields Interacting with Atoms 146
         6.4.1 Detecting Ions 146
         6.4.2 Absorption of Light 148
         6.4.3 One-Photon Absorption 149
         6.4.4 Two-Photon Absorption 150
         6.4.5 Multi-Photon Ionization Spectroscopy 152
         6.4.6 Detecting the Light Produced in the Target 154

6.5   The Measurement of the Main Parameters that Characterize the
      Interaction Between Intense Electromagnetic Radiation and an
      Atom 156
      6.5.1 Degree of Nonlinearity 156
      6.5.2 Multi-Photon Cross Sections 157

7. Nonresonant Phenomena 161

7.1   Nonlinear Atomic Susceptibilities 163
      7.1.1 The Scattering of Light and the Linear Susceptibility 164
      7.1.2 The Nonlinear Scattering of Light and Nonlinear
             Susceptibility 166
      7.1.3 Linear and Nonlinear Susceptibilities and the Pertur-
             bation of Atomic Spectra 168
7.2   Perturbation of Isolated Atomic States 170
      7.2.1 Dynamic Polarizability of an Isolated Nondegenerate
             Atomic State 171
      7.2.2 Dynamic Hyper-Polarizability 174
      7.2.3 Criteria for Applying Perturbation Theory 175
      7.2.4 The Dynamic Polarizability as a Function of the
             Perturbing Field 176
      7.2.5 Experimental Data 177
7.3   Perturbation of Degenerate States 182
      7.3.1 Two Adjacent Levels 182
      7.3.2 The General Case 184
      7.3.3 A Perturbation in an Elliptically Polarized Field 185
      7.3.4 Perturbation of the Hydrogen Atom Spectrum 186
7.4   Nonlinear Scattering of Light 195
      7.4.1 The Interrelation of Nonlinear Scattering Processes
             Due to an External Field 195
      7.4.2 Spontaneous Two-Photon Scattering 197
      7.4.3 Stimulated Two-Photon Emission 200
      7.4.4 Hyper-Raman Scattering 201
      7.4.5 Generation of the Third Harmonic 202
      7.4.6 Other Processes Determined by $\chi^{(3)}$ 205
7.5   Nonresonant, Nonlinear Ionization 208
      7.5.1 The Mechanisms of Nonlinear Ionization 209
      7.5.2 Numerical Estimates 209
7.6   Ionization in a Short-Range Potential 211
      7.6.1 Tunneling Ionization 212
      7.6.2 Ionization by a Circularly Polarized Field 212
      7.6.3 Ionization by an Elliptically Polarized Field 218
      7.6.4 The Intermediate Case ($\gamma^2 \sim 1$) 219
      7.6.5 Nonlinear Detachment of an Electron from a Negatively
             Charged Ion 221
7.7   The Ionization of Atoms 221
      7.7.1 The Power-Law Dependence of the Ionization Probability
             on the Field Strength 222
      7.7.2 Dependence of the Multi-Photon Cross Section on the
             Frequency and the Degree of Ellipticity of the Light 223
      7.7.3 Criteria for Applying Perturbation Theory 225
      7.7.4 Calculation of Multi-Photon Cross Section and Their
             Comparison with Experimental Data 227
      7.7.5 The Angular Distribution of the Emitted Photoelectrons 229
      7.7.6 Tunneling Ionization of Atoms ($\gamma \ll 1$) 231
      7.7.7 The Intermediate Case ($\gamma^2 \sim 1$) 232

8. Resonance Phenomena 233

8.1 Spontaneous Emission of Light by an Atom in a Resonance Field   236
    8.1.1 The Natural Line Width 236
    8.1.2 The Lorentzian Line Shape 237
    8.1.3 Spontaneous Emission of Photons in a Weak Resonance
          Field 239
    8.1.4 Spontaneous Emission of Photons in a Strong Resonance
          Field 240
    8.1.5 Introduction of an External Field into the Resonance
          Excitation 242
8.2 Resonance Fluorescence 243
    8.2.1 A Weak External Field 243
    8.2.2 Criterion for Applying Perturbation Theory 246
    8.2.3 The Density-Matrix Method 247
    8.2.4 Elastic (Rayleigh) Scattering in Strong Fields 249
    8.2.5 Inelastic Scattering 251
    8.2.6 Comparison with Experiment 257
    8.2.7 The Bloch-Siegert Shift 259
    8.2.8 The Test Field 261
    8.2.9 Multi-Photon Resonance 262
8.3 Multi-Photon Excitation and Emission 263
    8.3.1 Selection Rules for Multi-Photon Transitions 264
    8.3.2 The Probability of Multi-Photon Bound-Bound Transitions 265
    8.3.3 Quadrupole Transitions 267
    8.3.4 Multi-Photon Mixing of Resonance States 269
    8.3.5 Competing Processes in the Multi-Photon Excitation of
          Atoms 270
    8.3.6 Stimulated Multi-Photon Emission 270
8.4 Spontaneous Raman Scattering 271
    8.4.1 A Weak Perturbation 273
    8.4.2 Raman Scattering Frequencies for a Strong Perturbation 274
    8.4.3 The Probability of Raman Scattering Under the Influence
          of a Strong Perturbation 275
    8.4.4 Multi-Photon Raman Scattering 277
8.5 A Three-Level System in Two Resonance Fields 278
    8.5.1 The System of Equations 278
    8.5.2 Perturbation Theory 279
    8.5.3 A Strong External Field 282
    8.5.4 Two Strong External Fields 284
8.6 Resonance Ionization of Atoms 285
    8.6.1 Resonance Ionization in a Weak Field 286
    8.6.2 Mechanisms of Resonance Ionization in a Strong Field   288
    8.6.3 Multi-Photon Ionization in a Strong Field 290
    8.6.4 Resonance Ionization in the Adiabatic Inversion Field   291
    8.6.5 The Polarization of Electrons and Nuclei in the Event
          of Resonance Ionization 293
    8.6.6 Angular Distribution of the Electrons in the Resonance
          Ionizaton Process 298

9. Conclusion 300

9.1 The Role of the Non-monochromaticity of Laser Radiation 300
    9.1.1 Direct Multi-Photon Ionization 302
    9.1.2 Tunneling Ionization in a Variable Field 304
    9.1.3 Multi-Photon Excitation of Atoms 305
    9.1.4 Perturbation of Atomic Levels 306

9.2 Many-Electron Approximation 309
     9.2.1 The Dynamic Polarization of the Atomic Core 309
     9.2.2 Two-Electron Multi-Photon Ionization 311
9.3 Ultrahigh Fields 312
9.4 Highly Excited Atomic States in a Strong Electromagnetic Field 313
     9.4.1 Radiative Transitions Between Highly Excited Atomic
            States 313
     9.4.2 The Dynamic Polarizability of Highly Excited Atomic
            States 315
     9.4.3 Multi-Photon Ionization of Highly Excited States 317
     9.4.4 Tunneling Ionization from Highly Excited States 318
     9.4.5 Stochastic Instability of the Classical Electron in a
            Variable Field and the Diffusion Ionization of Highly
            Excited Atoms 320

Notation Index 323

References 325

Subject Index 337

1. Introduction

The title "Atoms in Strong Light Fields" does not precisely define the scope
of the phenomena to be studied. For this reason we shall first define the
scope of our presentation.

We shall consider isolated atoms. This means that we shall not study the
collisions of atoms, the macroscopic properties of the atomic medium or the
effects that can be observed when an atom interacts with light created in
the atomic medium by the initial light field. The model of an isolated atom
is not an abstract one; it can be realized in many experiments and its main
merit lies in the possibility of observing elementary acts without additional
distortions.

The word "light" is to be understood to include both the visible region
and the adjacent ultraviolet and infrared regions, i.e., wavelengths between
0.1 and 1 μm, a region in which optical phenomena at the atomic level are
usually studied. We shall always assume that the light is completely monochromatic, the wave is plane and the distribution of the wave's amplitude
over the wavefront is uniform. All of these three assumptions are clearly
idealizations. Even in the optimal case of a laser operating in a single-
mode, the radiation is only weakly quasi-monochromatic, its divergence is
very small but finite and the distribution of the amplitude over the wavefront is Gaussian. It should be noted, however, that the divergence and the
amplitude's distribution over the wavefront need only to be taken into account when one has to interpret experimental data obtained for a collection
of atoms occupying a definite space. Both factors are irrelevant for an isolated atom, since its size is small compared to the wavelength of the light
and since it only moves a small distance during the observation time. However,
the fact that real radiation is always non-monochromatic plays an important
role in the interaction of light with an isolated atom. Even the most monochromatic laser radiation has a spectral linewidth of the same order of magnitude as the natural linewidth of atomic levels. In other cases the spectral
linewidth may be considerably greater. On the other hand, multi-mode laser

radiation and radiation produced by incoherent sources are characterized by very rapid fluctuations of the amplitude, which the atom is either able or unable to respond to, depending on the concrete experimental situation. Consequently the following question must be answered: Which light field acts on the atom—the instantaneous one or the one averaged over a definite time interval? In other words, must a specific physical quantity (probability, level shift) be averaged over time *after* the instantaneous value of the field has been substituted for this quantity, or must the field be averaged first *before* being substituted? When the dependence of the physical quantities on the field is nonlinear, a distinction must be made.

In the analysis of the interaction of light and atoms from the theoretical viewpoint, the light will be assumed to be monochromatic. In a great number of experiments this assumption is realized quite satisfactorily. When effects associated with the non-monochromaticity of the radiation are observed, they will be taken into consideration. The physical meaning of such effects is considered in Sect.9.1.

It should also be noted that both the duration of the interaction of the light field with the atom and the time needed to introduce the perturbation (i.e., the time it takes the intensity of the radiation to grow from zero to its maximum value) are important in theory and in practice. The reader must bear in mind that since the topic of discussion is also pulsed laser radiation, these time intervals may be of the same order of magnitude as the lifetime of an excited atomic state. We will not restrict ourselves to any specific conditions concerning the time of introduction of the field or its duration, but in a theoretical analysis conditions that are more realistic from an experimenter's point of view are usually stressed.

Least unambiguous is the question as to what field can be considered strong. The strength of a field is of course not an abstract concept; it originates when specific phenomena related to the action of the field on an atom are studied.

We shall say that a field can be called strong when, during the time of its action, processes involving more than one photon become significant. For instance, when a monochromatic field interacts in resonance with a two-level system, the successive multiple absorption and emission of photons of the external field takes place, in which emission follows absorption, absorption follows emission, and so on. In accordance with our definition we call such a field a strong one.

Another example of a strong field is a nonresonant multi-photon ionization. In this process a number of photons of the external field are successively absorbed until an electron is ejected into the continuum. If follows from our definition that a field is considered weak when it stimulates a one-photon transition of an electron from the ground state to a bound or free final state or a scattering process in which a scattered photon is created instead of an incident one. Thus, under the influence of a strong field, in contrast to a weak one, not only is an electron transferred from one state to another but the parameters of the electronic states are also changed.

To underline the specific features of phenomena occurring in strong fields, we will, in many cases, proceed from phenomena occurring in weak fields. Sometimes such phenomena will be the object of the investigation. However, we will not consider the properties of atomic spectra in the absence of external fields or make a detailed study of one-photon processes, i.e., the excitation of atomic states and their spontaneous decay and ionization. The reader can find these topics discussed in the appropriate literature; see, for example, [1.1,2].

It is well known that an atom must be described using the laws of quantum mechanics. This follows from the fundamental fact that the existence of bound electronic states cannot be explained within the framework of classical physics, according to which the electron's energy can continuously change. Quite naturally, then, a theoretical description of the phenomena which occur when light interacts with an atom will be given in the language of quantum mechanics. Classical mechanics is only adequate in describing extremely highly excited states (Chap.9). The answer to the question as to what language should be used to describe light is not as simple to answer. Should one use the classical approach, in which light is an electromagnetic field, or the quantum approach, which describes light as a collection of photons? Here we shall follow the classical approach. The reason for this will be examined in Sect.1.1. Nevertheless, the words "quantum" and "photon" will also be used in quite different contexts. Such a terminology is justified by tradition and convenience, not by necessity.

From a different point of view, one can say that this book is devoted to elementary, nonlinear optical phenomena. The reader must bear in mind that in nonlinear optics, which describes the interaction of light and matter in which the nature of the interaction depends on the intensity of the light, three successive stages can be distinguished. The first stage deals with elementary, nonlinear optical phenomena, i.e., the microscopic nonlinear characteristics of an isolated atom (e.g., its polarizability). The second stage has to do with calculating, by means of the microscopic characteristics of an isolated atom, the macroscopic nonlinear characteristics of matter

3

which depend on the intensity of the external field (e.g., the index of refraction). The third stage deals with the study of the wave propagation in a medium whose macroscopic characteristics depend on the intensity of the light. From this outline it is clear that the book only deals with the first stage: even this stage is not exhaustively described. Of the different microscopic objects —isolated atoms, molecules and electrons, atomic particles bound in a crystal lattice, free electrons in solids, etc. —only isolated atoms are considered. Naturally, such an approach narrows the scope of the research subject and, further, it removes it from practice. On the other hand, it is thus possible to study all of the phenomena with maximum rigor and clarity and yet keep the volume of the book to within reasonable limits.

1.1 The Strong Light Field

1.1.1 The Classical Nature of the Field

This book studies the interaction between light and an atom using a semiclassical approach, in which the light is described classically and the atom quantum-mechanically.

It is not always possible to use such an approach. For example, there is no sense in trying to use the classical method to describe spontaneous transitions or the Lamb shift, since, in the latter case, the electron vibrates about its equilibrium position in a Bohr orbit due to fluctuations in the vacuum. This is a direct consequence of the quantization of the field.

Another example is the following. According to the classical theory, the periodic motion of an electron in a Bohr orbit creates an electromagnetic wave of the same frequency (the emission of a quanta). According to quantum theory the emission is attributed not to the periodic motion of the electron but to its transition from one orbit to another.

In a strict quantum approach to light, the electric and magnetic fields of the electromagnetic wave are quantum operators, not just functions of time and space. It is well known that, when an atomic system absorbs a quantum of light, these operators introduce a factor $(n_{k\alpha})^{\frac{1}{2}}$ in the transition amplitude [1.3]. Upon emission of a quantum a factor $(n_{k\alpha}+1)^{\frac{1}{2}}$ is introduced. Here $n_{k\alpha}$ is the number of quanta absorbed by a given vibrational model, i.e., a vibration with a given wave vector $\mathbf{k}$ and a polarization α. In the classical limit one considers the matrix elements of the direct and inverse processes to be the same. Consequently, an electromagnetic field can be treated

4

classically if $n_{\mathbf{k}\alpha} \gg 1$. Let us estimate the strength of the electric field of the wave which would correspond to such a condition.

In calculating the total energy of the field one finds that the total energy of the field becomes infinite when one sums over very high values of energy. Indeed, in a real system, e.g., a laser cavity, photons with a definite energy $\hbar\omega$ and a definite energy interval $\hbar\Delta\omega$ predominate.

If one estimates the number of oscillators in the field, the result is $\Delta\mathbf{k}/(2\hbar)^3 \sim \omega^2\Delta\omega/c^3$. It was assumed that every oscillator emits the same number of quanta $n_{\mathbf{k}\alpha}$ and that the energy of each quantum is $\hbar\omega$. Thus, the total energy in a unit volume of the cavity is $n_{\mathbf{k}\alpha}\hbar\omega^3\Delta\omega/c^3$. On the other hand, from the classical point of view it is estimated to be $\mathbf{E}^2$. Hence, the condition $n_{\mathbf{k}\alpha} \gg 1$ is equivalent to the following condition imposed on the electric field strength of the wave:

$$E \gg (\omega^3\Delta\omega\hbar/c^3)^{\frac{1}{2}} \sim (\hbar c\Delta\lambda/\lambda^5)^{\frac{1}{2}} \ ,$$

where λ is the wavelength of the light, and $\Delta\lambda$ is the linewidth in the emission spectrum. For $\lambda = 5000$ Å and $\Delta\lambda \sim 0.1$ Å (typical values), $E \gg 1$ V/cm.

All in all one can say that the quantum characteristics of a laser as a light source do not manifest themselves if one is specifically not interested in the noise parameters of a laser or in a laser operating at the excitation threshold.

1.1.2 The Parameters of a Light Field

We have seen that the problems of interest to us can be discussed from the classical point of view. Let us now briefly discuss the parameters that charaterize the light field.

The first parameter is the frequency ω, which in the spectral region of interest to us is of the order of $10^{15}\mathrm{s}^{-1}$. This frequency corresponds to a quantum with an energy of approximately 1 eV.

The second parameter that is a vital characteristic of light is its degree of polarization. From now on it will be assumed that the light is completely polarized and that the degree of elliptical polarization can change from linear to circular, taking on all of the intermediate values. Such assumptions are fulfilled by present-day sources of light (Chap.5).

The third parameter, which must be fixed in all cases, is the field strength of the light. For the sake of clarity, we shall always use the term "field strength." The atomic field strength is equal to $E_{at} \simeq m^2 e^5/\hbar^4 \simeq 5 \times 10^9$ V/cm, and simple comparison with specific values for the field strength leads to an immediate conclusion about the character of the phenomena. The condition $E \gg E_{at}$ reaffirms that the field strength is a small parameter, whose use may constitute the basis for solving many problems. Finally, the principal difference between linear and circular polarized light should be noted. In the latter case the amplitude of the field does not vary with time. This makes it possible to separate the temporal and spatial variables by changing to a rotating frame of reference connected with the field vector. This possibility greatly simplifies the solution of a range of problems (for more details see Sect.1.3).

The qualitative criteria, i.e., which fields should be considered strong and which weak, cannot be reduced to one quantitative criterion. Evidently a numerical criterion depends on the specific problem. From a quantitative point of view the strength of a field varies within broad limits. The following are two examples. When resonance between the frequency of the external field and the frequency of the transition between two atomic levels occurs, a mixing of bound electronic states can occur for field strengths of about 10 V/cm (Chap.8). Hence, in this case, a field of so small an absolute strength must be considered strong. On the other hand, for fields of about 10^6V/cm $\sim 10^{-3}E_{at}$ the probability of a two-photon ionization of an atom is almost equal to the probability of a one-photon ionization by a field of double frequency (Chap.7). Thus, in this case, the critical field strength is 10^5 times greater than in the first example.

If we now turn to the practical aspects of creating a strong light field, we must first note that it was lasers as sources of coherent light which wrote a new chapter in physics, to which this book is devoted. The main advantages of a laser beam as a source of light over incoherent sources are well known: they are the extremely high degree of monochromaticity and the sharp directivity of the laser radiation, which produce the extremely high spectral brightness of lasers (the corresponding quantitative characteristics are given in Chap.5). Laser radiation can be effectively focussed, even down to one wavelength.

Finally, the fourth important parameter is the time δT needed for the amplitude of the field to reach its maximum value (switch-on time) and the time T during which the amplitude is a constant (the operation time of the perturbation). The effect of a light field on an atomic system depends to a significant extent on these two parameters. In present-day lasers it is not possible to vary δT and T continuously. Because of this, for purely practical reasons that will be discussed in Sect.5.1, we will deal with pulsed radiation, for which $\delta T \sim T \sim 10^{-8}$s, or with the radiation of continuous lasers, $\delta T \gg T$.

1.2 The Atom

An isolated atom is a sufficiently well-known quantum mechanical system.
Let us recall some basic characteristics of this system, which will often
be used in the study of various phenomena which occur in strong light fields.

1.2.1 One-Electron and Multi-Electron Approximations

When light interacts with an atom, the optical electrons, which populate
the outermost shell, change their states. Usually we will speak of one electron rather than many electrons, i.e., the one-electron approximation will
be used regardless of the specific structure of the atom. The legitimacy of
such an approximation for more complex atoms is not evident a priori, since
in some atoms (for example, the inert gases), the outermost shell consists
of several equivalent electrons. In all atoms other than hydrogen there
also exist a certain number of electrons in the inner shells (the atomic
core). A perturbation of the atomic core by an external field may in turn
perturb the optical electron.

Let us examine this effect, using as an example an alkali atom. When an
external periodic field acts on such an atom, the atomic core vibrates. Naturally, these vibrations are not of a resonant nature since the natural
frequencies are very high (of the order of the atomic number Z, in atomic
units)[1]. They are simply forced dipole oscillations with the frequency of
the applied perturbation. In the Thomas-Fermi approximation a variable dipole moment d(t) appears:

$$d(t) \sim (E/Z) \cos\omega t \; , \qquad\qquad\qquad (1.1)$$

where Z is the number of electrons in the atom. The dipole moment, via the
interaction with the valence electron, can also initiate transitions of
the latter. The corresponding potential energy in the charge-dipole system
is

$$(E/Z) \cos\omega t \; , \qquad\qquad\qquad (1.2)$$

since the distance between the valence electron and the core is of the
order of unity (in atomic units). Hence, it can be seen that the perturbation of the electron due to the dynamic polarization of the atomic core is
Z times smaller than the direct perturbation by the field $E \cos\omega t$, i.e.,
the effect is small.

---

1 As a rule, we shall use the atomic system of units, in which $e = \hbar = m_e = 1$

Let us turn our attention to an atom with several electrons in the outer-most shell. The effect of the field created by several optical electrons, which are perturbed by an external light field, on one of the optical elec-trons cannot be described using the Thomas-Fermi approximation. With many electrons, the most reasonable way to tackle this problem is to use the ran-dom-phase approximation [1.4,5]. The natural frequencies of the vibrations of these electrons are certainly greater than the frequency of the perturba-tion, especially in the multi-photon case. Therefore, the oscillations of the dipole moment and, hence, the polarization perturbation of the valence elec-tron are not resonant processes. Unfortunately no studies have been made in this field, which makes it impossible to judge to what extent the one-elec-tron approximation can be applied to atoms with several electrons in the valence shell.

In general, it is incorrect to use a statistical factor that depends on the degeneracy of a level to take into account the equivalent electrons in the outermost shell. The interelectronic interaction in the valence shell is the reason the one-electron approximation cannot be used. For example, it is known that the interelectronic interaction strongly influences the photoionization cross section [1.6]. When applied to calculations of the interelectronic interaction in noble gas atoms, the random-phase approxima-tion (with exchange) can be used to obtain correct values for the photo-ionization cross sections. The interelectronic interaction must also mani-fest itself in the multi-photon ionization of complex atoms. One should expect effects of the same order of magnitude as those in photoionization (the error is about 100 percent in cross section), since in both cases the correction is due to changes in the wave function of the ground state. A specific feature of the multi-electron case is the change in the selection rules, which must bring about changes in the multi-photon matrix elements.

1.2.2 The Structure of Atomic Spectra

The excited electronic states of an optical electron are quasi-stationary due to the existence of the field of the electromagnetic vacuum. Each state n is characterized by an energy $E_n^{(0)}$ which corresponds to the maximum of the probability distribution of an electron with a given energy and by the half-width Γ_n of this distribution, given by the Lorentz formula. We recall that $E_n^{(0)}$ is about $1 - 10$ eV, or $10^4 - 10^5 cm^{-1}$, and Γ_n is of the order of $10^{-3} cm^{-1}$. Consequently, the lifetime of excited states is of the order of $10^{-8} s$.

The separation between bound states and continuous states (i.e., the ionization potential) starts somewhere between 4 and 24 eV. It is well known that as the energy of the excited states increases the separation between energy levels decreases. Only in hydrogen is the energy of the bound states described by a simple law (the Rydberg series formula). As one gets closer to the continuum the density of the levels increases with n^3 (n is the principal quantum number).

In complex atoms there is no simple law for the distribution of energy levels. However, different groups of atoms have marked qualitative differences in the structure of the excited states. In alkali atoms, for example, the energies of the lowest excited states are about one electron volt; in noble gases, on the other hand, the value is about ten electron volts. Correspondingly, alkali atoms can be excited to the first excited state by one-photon processes, while the noble gases can only be excited by multi-photon processes with visible light.

The above numerical values show that it is possible to excite but impossible to ionize an atom in its ground state by a one-photon process. It is this fact that justifies the claim that atoms are transparent to incident light. Of course, this is only true within the framework of linear optics, i.e., only when one-photon absorption processes are taken into account. Multi-photon absorption processes are one of the main topics of this book, and the fact that they have a high probability (compared to one-photon processes) is an effect of a strong field (Sect.1.1). Multi-photon processes are not threshold processes with respect to any parameter that characterizes the interaction if, of course, the conservation of energy holds true for transitions between the initial and final states (but not for transitions between intermediate states (see Chaps.2 and 7). The probability of a multi-photon processes is nonzero for any external field. Only for weak fields does this probability become infinitely small, so that multi-photon processes can be neglected.

For weak fields the unperturbed atomic spectrum is the basis for the initial Hamiltonian. This is also possible for strong fields, but not in all cases. One must bear in mind that a strong field not only induces multi-photon transitions between bound states but also changes the energy of these states, the quasi-stationary width and the probability of finding an electron with a given energy. A considerable portion of this book is devoted to a study of the perturbation of an atomic spectrum by a strong field. A field is strong if, for one, the matrix element of the interaction of bound states with the field is larger than their spontaneous width (Chap.8). When solving problems involving a strong field in which the perturbation of the spectrum is great, one can use the spectrum of the atom-field system as an initial basis. This approach to the problem of the interaction of an atom and a strong field will be considered later.

1.2.3 Selection Rules

It is well known that the transitions of electrons are determined by selection rules, which reflect the conservation of different parameters. Since a strong field changes the atomic spectrum, it also changes the selection rules. First the selection rules in a weak field will be recalled, and then the selection rules in a strong field will be postulated without going into the nature of the phenomena that change the spectra. A considerable portion of this book is devoted to such phenomena.

Let us first briefly examine the selection rules in a weak field. It is assumed that the perturbation V is an electric dipole interaction (Sect.1.3). If the perturbation operator is written in the form $V = zE \cos\omega t$ (a linearly polarized field) and the matrix element of the transition initiated by this perturbation is calculated, it can be seen that allowed transitions obey the following selection rules: $\Delta J = J - J' = 0, \pm 1$; $\Delta M = M - M' = 0$. Here J and J' are the total angular momenta of the initial and final states, and M and M' are the magnetic quantum numbers of these states. In addition, the transition $J = 0 \rightarrow J' = 0$ is forbidden. It also follows that in a transition the parity π changes.

The above rules can be generalized to encompass multi-photon transitions brought about by linearly polarized light. Using the selection rules for every absorption of a photon, one gets for a transition involving K photons, the following rules: $\Delta J = K, K - 1, \ldots, -K$; $\Delta M = 0$. The parity of a state changes as $(-1)^K$. We will not dwell on the selection rules for magnetic dipole transitions, since, for multi-photon transitions, they can be found in an analogous manner as for electric dipole transitions.

We now turn to circularly polarized light. For one-photon processes the transition operator is $V = E(x \cos\omega t \pm y \sin\omega t)$. The matrix elements for this transition are not zero if the following selection rules hold: $\Delta J = 0, \pm 1$; $\Delta M = \pm 1$. The sign in the latter formula depends on the direction of the circular polarization. For transitions involving K quanta the selection rules are $\Delta J = K, K - 1, \ldots, -K$; $\Delta M = \pm K$. (The sign here is chosen in the same manner.) The parity of the states also changes as $(-1)^K$. For elliptically polarized light there are no selection rules for M.

Finally, let us turn to strong fields and consider the general case of several valence electrons. It is assumed that LS coupling is applicable.

In strong fields, with which the researcher has to deal when studying multi-photon processes, the interaction between the atom and the field may become greater than that part of the interaction among the electrons that produces the multiplet splitting of the unperturbed atomic terms, V_{LS}. A very strong field disrupts the LS coupling. The selection rules for the quantum numbers of the whole atom are reduced to selection rules for only the spatial part of the wave function. In other words, the electromagnetic field does not involve the spin part of the wave function. Thus, for one-photon transitions in a linearly polarized field the selection rules for LS coupling are [1.7]:

$$\text{for } V \ll V_{LS}: \begin{aligned} &\Delta J = 0, \pm 1 \\ &\Delta L = 0, \pm 1 \\ &\Delta S = 0 \\ &\Delta M = 0 \\ &J + J' \geq 1 \\ &L + L' \geq 1 \\ &\Delta \pi = 1 \end{aligned} \qquad \text{for } V \gg V_{LS}: \begin{aligned} &\Delta L = 0, \pm 1 \\ &\Delta S = 0 \\ &\Delta m_S = 0 \\ &\Delta M = \Delta m_L = 0 \\ &L + L' \geq 1 \\ &\Delta \pi = 1 \end{aligned} \qquad (1.3)$$

where m_L and m_S are the projections of the orbital and spin angular momenta on the z axis, S is the spin angular momentum, and L the orbital angular momentum of the system. One can examine the j-j and j-ℓ coupling schemes in the same manner. Appropriate tables were given in [1.7].

The same selection rules hold for circularly polarized light, with the exception of those that apply to magnetic quantum numbers. It is obvious that for elliptical and partial polarizations there are no selection rules for magnetic quantum numbers.

1.3 Interaction of an Atom and a Light Field

Since this monograph is entirely devoted to this question, here only topics of general interest will be mentioned.

1.3.1 The Dipole Approximation

The vector potential **A** that acts on the electron generally changes from point to point in space: $\mathbf{A} \sim \cos(\mathbf{k} \cdot \mathbf{r} - \omega t)$. But due to the small size of the atom, the spatial variation of the electromagnetic field in the optical frequency range within the atom is small, i.e., $\mathbf{k} \cdot \mathbf{r} \ll 1$.

If $\mathbf{A}$ is expanded in a power series in $\mathbf{k} \cdot \mathbf{r}$, the first term, which does not depend on $\mathbf{k} \cdot \mathbf{r}$, is responsible for the electric dipole interaction between the light and the electron. In the dipole approximation, the vector potential $\mathbf{A}$, and hence the electric field $\mathbf{E} = -c^{-1}(\partial \mathbf{A}/\partial t)$, do not depend on the radius vector of the electron, $\mathbf{r}$. They only depend on time.

In the next term in $\mathbf{k} \cdot \mathbf{r}$ the electric quadrupole and magnetic dipole interactions come into play. Numerically $\mathbf{k} \cdot \mathbf{r}$ is of the order of $1/c \sim 1/137$. It is with this parameter that the multipole expansion is performed. This estimate would be so if the energy of the quantum were of the order of the atomic unit of energy, i.e., one rydberg. In the optical region this energy is actually much less. $\mathbf{r}$ is of the order of atomic dimensions (about 1 Å), whereas the wavelength of the light, $1/k$, is about 10^3Å. Thus the smallness parameter $\mathbf{k} \cdot \mathbf{r}$ is about 10^{-3}. It can be seen that the electric dipole interaction is a good approximation even in the far ultraviolet region.

Exceptions are transitions which are forbidden by the selection rules (Sect.1.2). Let us consider, for example, a transition through an intermediate level whose parity coincides with that of the initial state. A dipole transition through such a level is forbidden. With a sufficiently good tuning to resonance, the smallness of the probability of a quadrupole transition is compensated by the resonance nature of the quadrupole interaction, so that such a transition is possible (for a thorough study of this question see Sect.8.5). In strong fields the relative roles of quadrupole interactions can change. For example, in the above-mentioned example of an intermediate resonance in a strong field, the competition between the saturation process with respect to the first transition and the ionization of the intermediate level is quite strong. This competition determines the width of the resonance. When both transitions are dipole transitions, everything is determined by the multiplicity of the first and second transitions because the dipole matrix elements are of the same order of magnitude (Chap.8). If the first transition is a quadrupole one, the differences between the matrix elements of the dipole and quadrupole transitions may exceed the differences due to the different multiplicities of the resonances, and hence can determine the resonance width.

What we have just said about expansion in terms of $\mathbf{k} \cdot \mathbf{r}$ can be reformulated in terms of electric and magnetic fields, which in an electromagnetic wave are mutually perpendicular. The interaction between the magnetic field H and the atom can be determined from the expansion of $\mathbf{A}$ in a power series in

$\mathbf{k} \cdot \mathbf{r}$, since $\mathbf{H} = \mathrm{curl}\mathbf{A}$. In particular, this expansion leads to the magnetic dipole interaction, which can be formulated in terms of the interaction of the magnetic moment of the atom with the vector $\mathbf{H}$. Such an interaction results in a small effect, which has the same order of magnitude (10^{-3}) as the electric quadrupole interaction.

1.3.2 The Hamiltonian of the Dipole Interaction

In the one-electron approximation, the Hamiltonian of an electron in the potential field $U(\mathbf{r})$ of an atom and interacting with an electromagnetic field with a vector potential $\mathbf{A}(\mathbf{r},t)$ is

$$H = \frac{1}{2} \left(\hat{\mathbf{p}} + \frac{1}{c} \mathbf{A} \right)^2 + U \quad , \tag{1.4}$$

where $\hat{\mathbf{p}} = -i\partial/\partial\mathbf{r}$. In the dipole approximation the vector potential $\mathbf{A}$ does not depend on $\mathbf{r}$, i.e., $\mathbf{A} = \mathbf{A}(t)$.

Equation (1.4) does not represent the only form of the Hamiltonian. Let us introduce the following transformation of the wave function:

$$\Psi(\mathbf{r},t) = \exp\left[-\frac{i}{c} \mathbf{A}(t) \cdot \mathbf{r} \right] \phi(\mathbf{r},t) \quad . \tag{1.5}$$

Such a transformation does not change the value of the physical observables (e.g., the probability). The matrix elements between different states Ψ do not change either because the phase factor in (1.5) is the same for all of the states and, hence, does not appear.

Substituting (1.5) into the Schrödinger equation $i(\partial\Psi/\partial t) = H\Psi$ with the Hamiltonian (1.4), one finds that ϕ satisfies the following equation:

$$i \frac{\partial\phi}{\partial t} = \left[\frac{1}{2} \hat{\mathbf{p}}^2 - \frac{1}{c} \mathbf{r} \cdot \dot{\mathbf{A}}(t) + U(\mathbf{r}) \right] \phi \quad . \tag{1.6}$$

Since $-c^{-1}\dot{\mathbf{A}}(t)$ is equal to $\mathbf{E}(t)$, the electric field vector of the wave is given by,

$$i \frac{\partial\phi}{\partial t} = \left[\frac{1}{2} \hat{\mathbf{p}}^2 + \mathbf{r} \cdot \mathbf{E}(t) + U(\mathbf{r}) \right] \phi \quad . \tag{1.7}$$

Obviously, if one were to take into account the dependence of $\mathbf{A}$ on $\mathbf{r}$, the above transformation would not be valid.

Thus, in the dipole approximation the electric dipole interactions $V(t)$ represented by

$$V(t) = \frac{1}{c} \hat{\mathbf{p}} \cdot \mathbf{A}(t) + \frac{1}{2c^2} \mathbf{A}^2(t) \qquad \text{or by} \tag{1.8}$$

$$V(t) = \mathbf{r} \cdot E(t) \tag{1.9}$$

are equivalent [1.8].

One must be careful in analyzing the transition probabilities in the framework of the adiabatic approximation—an approximation that uses adiabatic wave functions (see Sect.4.3). The matrix elements will be determined by saddle-point methods and will contain complex-valued moments of time t. The phase factor in (1.5) will have an absolute value that is generally not equal to unity. If one rewrites the dipolar interaction Hamiltonian in the form (1.9), one must not forget to transform ϕ back to ψ via (1.5) when determining transition probabilities.

Equations (1.8,9) are equivalent if the wave functions ψ and ϕ are exact solutions of the Schrödinger equation. If approximations are permitted, the equivalence will be destroyed. Let us illustrate this using as an example a two-level system.

The calculation of a three-photon transition matrix element for a two-level system will be considered (see Sect.3.2). It will be assumed that the perturbation $V(t) = V^{(1)}$ $\cos\omega t$, that the two levels n and m are related to each other through an electric dipole matrix element, that their unperturbed energies are $E_n^{(0)}$ and $E_m^{(0)}$, respectively, and that $\omega_{mn} = E_m^{(0)} - E_n^{(0)}$. Then, see (2.43),

$$V_{mn}^{(3)} = \frac{1}{16\omega} \frac{V_{nm}^{(1)} V_{mn}^{(1)} V_{nm}^{(1)}}{\omega_{mn} - \omega} \quad . \tag{1.10}$$

Using (1.8) for V and substituting the vector potential $\mathbf{A}(t) = \mathbf{A} \cos\omega t$ $= (cE/\omega) \cos\omega t$ into (1.10), the numerator of (1.10) takes the form

$$\left| \frac{1}{c} \mathbf{p}_{nm} \cdot \mathbf{A} \right|^3 = \left| \mathbf{p}_{nm} \cdot \frac{E}{\omega} \right|^3 = \left| \frac{\omega_{nm}}{\omega} \mathbf{r}_{nm} \cdot \mathbf{E} \right|^3 \quad . \tag{1.11}$$

Here we have used the well-known quantum mechanical formula $\mathbf{p}_{nm} = i\omega_{nm}\mathbf{r}_{nm}$. Due to parity considerations in (1.10) there are no matrix elements for the term with $\mathbf{A}^2(t)$.

If the interaction in the form (1.9) is chosen, the denominator will be

$$\left| \mathbf{r}_{nm} \cdot \mathbf{E} \right|^3 \quad . \tag{1.12}$$

It can be seen that (1.11,12) differ by the factor $(\omega_{nm}/\omega)^3$. Since $\omega_{nm} \simeq 3\omega$, such a difference is quite significant.

This discrepancy is due to the fact that the eigenfunctions $\psi_m^{(0)}$ and $\psi_n^{(0)}$ and the corresponding eigenvalues $E_m^{(0)}$ and $E_n^{(0)}$ belong to the atomic

Hamiltonian, not to the Hamiltonian of the model two-level system. This contradiction can be eliminated in the following way [1.9]. If one adds a third level $E_k^{(0)}$ to the two-level system for which $\omega_{kn} \gg \omega_{mn}$, the third level will contribute a small term to the three-photon matrix element.

If one chooses the perturbation in the form (1.9), the contribution of level k to $V_{mn}^{(3)}$ is

$$\delta V_{mn}^{(3)} = \frac{1}{16\omega} \frac{V_{mn}^{(1)} V_{nk}^{(1)} V_{kn}^{(1)}}{\omega_{kn} - \omega} \simeq \frac{r_{mn} r_{nk} r_{kn} E^3}{16\omega\omega_{kn}} \quad . \tag{1.13}$$

As $E_k^{(0)} \to \infty$, the contribution tends to zero, i.e., the use of the perturbation in the form (1.9) is correct. But if one chooses (1.8), then

$$\delta V_{mn}^{(3)} = \frac{\omega_{mn} r_{mn} \omega_{nk} r_{nk} \omega_{kn} r_{kn} (E/\omega)^3}{16\omega(\omega_{kn} - \omega)} \simeq \frac{\omega_{mn} \omega_{kn} r_{mn} r_{nk} r_{kn} E^3}{16\omega^4} \quad , \tag{1.14}$$

which, due to the dipole sum rule, does not vanish as $E_k^{(0)} \to \infty$ and, hence, is not correct.

Thus, it can be concluded that when studying model systems with a finite number of levels (e.g., two-level systems in which a three-photon transition takes place, or a two-photon transition through an intermediate state in which all of the states except the intermediate one are neglected in the two-photon matrix elements), one must choose the dipole interaction in the form (1.9), not in the form (1.8).

However, for high frequencies, i.e., $\omega \gg \omega_{mn}$, it is more convenient to use (1.8), since the absolute value of the vector potential is small $|A| \sim c/E\omega$. This is convenient for power expansions in ω_{mn}/ω and will be used in Chap.7 in determining the dynamic polarizability of atomic levels.

Equation (1.8) is also more convenient in determining the wave functions of the continuum in a variable field, since in this case they are obtained by a shift of momentum $\hat{p} \to \hat{p} + A/c$ from the unperturbed wave function.

The interaction in form (1.8) or (1.9) leads to transitions between the atomic states. In a sufficiently weak monochromatic field the transition probabilities are described by the well-known Fermi rule [1.10] — if the field is not effective over a very long period of time.

Quantum transitions may be spontaneous if they are caused by fluctuations in the vacuum. In contrast to stimulated transitions, in which the amplitude of the perturbation is given (i.e., the transition is described classically), spontaneous transitions are essentially of a quantum nature. In such transitions the number of quanta of a given type changes from zero to unity. In other words, the vector potential **A** in (1.8) is quantized and is a super-position of the creation and annihilation operators of a quantum. It goes without saying that stimulated transitions can also be interpreted to oc-cur under the influence of a large number of photons. This is sometimes done in describing the interaction between light and an atom. However, in prac-tice the strength-of-field terminology is more convenient than the number-of-photons terminology.

Apart from spontaneous transitions, it is sometimes still necessary to use the number-of-photons terminology, e.g., when describing the fluctuations of laser radiation [1.11]. The same terminology is used in solving the prob-lem of the transitions in a two-level system placed in a resonant field [1.12], so that the results can be generalized for an arbitrary number of photons.

1.3.3 An Atom in a Circularly Polarized Electromagnetic Field

The interaction of an atom with an electromagnetic wave is, in general, a nonstationary problem of quantum mechanics. If there are no small perturba-tion parameters, it is extremely difficult to solve, because in this case the temporal and spatial coordinates are not separable in the Schrödinger equation.

When a circularly polarized electromagnetic field acts on an atom, the electric field vector, E, of the wave does not change in magnitude but simply rotates in the plane perpendicular to the direction of propagation. This sug-gests that the problem of a circularly polarized wave interacting with an atom can be considerably simplified if one reduces it to a stationary prob-lem by shifting to a reference frame that rotates with E. In such a frame the field E acting on the atom will be constant. Naturally, an additional centrifugal term will appear in the Hamiltonian because the rotating frame is noninertial.

Let us formulate this method of reducing the problem to a stationary one in general form [1.13]. The Schrödinger equation that describes the state $\Psi(\mathbf{r},t)$ of the electron in a circularly polarized monochromatic field is

$$i \frac{\partial \Psi}{\partial t} = [H_0 - E(x \cos\omega t - y \sin\omega t)]\Psi \quad , \tag{1.15}$$

with $H_0 = -(1/2)\nabla^2 + U(\mathbf{r})$. Here $U(\mathbf{r})$ is the atomic potential, which is to be assumed spherically symmetric. Equation (1.15) states that the electromag-netic field is circularly polarized in the xy plane.

The new wave function, in the rotating frame, is

$$\Psi_{rot}(\mathbf{r},t) = \exp(i\omega t \hat{L}_z)\Psi(\mathbf{r},t) \quad , \tag{1.16}$$

where $\hat{L}_z = x(\partial/\partial y) - y(\partial/\partial x)$ is the operator of the projection of the angular momentum on the z axis (the direction of propagation). If (1.16) is then substituted into (1.15) and the easily verifiable relationship,

$$\exp(i\omega t \hat{L}_z) \ (x \ \cos\omega t - y \ \sin\omega t) \ \exp(-i\omega t \hat{L}_z) = x \quad ,$$

is used, the following equation for $\Psi_{rot}(\mathbf{r},t)$ is obtained:

$$i \ \frac{\partial \Psi_{rot}}{\partial t} = H_{rot}\Psi_{rot} \quad , \qquad \text{in which} \tag{1.17}$$

$$H_{rot} = H_0 - \omega \hat{L}_z - Ex \quad . \tag{1.18}$$

Thus the above problem is reduced to a stationary problem using a stationary potential $V_{rot} = -\omega \hat{L}_z - Ex$. Obviously, such a type of perturbation is due to the presence of a constant electric field E directed along the x axis and a constant magnetic field $H = \omega/\mu_0$, where μ_0 is the Bohr magneton directed along the z axis.

It is important to note that this result is only true for atoms with a potential $U(\mathbf{r})$ that is spherically symmetric. If $U(\mathbf{r})$ depends on the direction of $\mathbf{r}$, then when we shift from (1.15-17), new time-dependent terms appear in the effective Hamiltonian (1.18), because the commutator $\hat{L}_z U(\mathbf{r}) - U(\mathbf{r})\hat{L}_z$ no longer vanishes.

What is the difference between an atom in constant crossed electric and magnetic fields and an atom in a circularly polarized electromagnetic field? Since the magnetic quantum numbers are not conserved, the difference in the wave function Ψ_{rot} in the first problem and the wave function,

$$\Psi = \exp(-i\omega t \hat{L}_z) \ \Psi_{rot} \quad , \tag{1.19}$$

of the second cannot be simply reduced to a phase factor. Thus, these functions differ substantially, in particular, for the ionization probabilities. In both problems the electric field brings about a decay of the system. Introduction of a magnetic field stabilizes the particle by reducing its drift along the direction of the electric field. Hence, in the first problem, as the magnetic field grows, the ionization probability decreases. This is because the magnetic field tends to rotate the particle in the plane perpendicular to the axis of the field. In the second problem one still has to shift to the laboratory frame of reference, which differs from the one in the first case in that it rotates with a frequency proportional to the strength of the magnetic field. Under such a rotation an additional centrifugal force that tends to ionize the particle appears, i.e., increases the ionization probability. The final result is that the ionization probability grows with increasing frequency.

In Sects.3.4 and 7.3 the above results are used in the problems concerning the ionization of an electron in a short-range potential well by a circularly polarized field and the action of such a field on a two-level system.

The trick of the method lies in shifting to a rotating reference frame with an angular velocity such that the Hamiltonian is no longer dependent on time. This makes it possible to introduce stationary states which considerably simplify the problem. In the above case this transition was exact and contained no approximations. In other cases a similar approach is carried out. In other words, in a rotating frame the Hamiltonian contains rapidly oscillating time-dependent terms. If they are neglected because of the rapidity of the oscillations, one again obtains a stationary Hamiltonian. Such an approach is called the "rotating wave" approximation [1.2,14] and is widely used (Sect.3.1).

1.3.4 The Floquet Theorem

Let us consider some of the general features of the solution of the Schrödinger equation for a particle in a periodic electromagnetic field. It is assumed that at any given time t the system of functions $\Psi_s(\mathbf{r},t)$ constitutes a complete orthonormal basis. Any function $\Psi(\mathbf{r},t)$ can be expanded in terms of this basis:

$$\Psi(\mathbf{r},t) = \sum_s A_s \Psi_s(\mathbf{r},t) \quad . \tag{1.20}$$

Due to the periodicity of the Hamiltonian, the basis functions $\Psi_s(\mathbf{r},t+2\pi/\omega)$, i.e., taken over a period, are also solutions of the Schrödinger equation and constitute a new basis. They can also be expanded in terms of the basis functions given at time t:

$$\Psi_s(\mathbf{r},t + 2\pi/\omega) = \sum_{s'} a_{ss'} \Psi_{s'}(\mathbf{r},t) \quad . \tag{1.21}$$

Let us expand the function $\Psi(\mathbf{r},t+2\pi/\omega)$ in terms of the new basis functions $\Psi_s(\mathbf{r},t+2\pi/\omega)$. This is equivalent to writing (1.20) for the time $t+2\pi/\omega$:

$$\Psi(\mathbf{r},t + 2\pi/\omega) = \sum_s A_s \Psi_s(\mathbf{r},t + 2\pi/\omega) \quad . \tag{1.22}$$

Substituting (1.21) into the right-hand side of (1.22),

$$\Psi(\mathbf{r}, t + 2\pi/\omega) = \sum_{ss'} A_s a_{ss'} \Psi_{s'}(\mathbf{r},t) \quad . \tag{1.23}$$

Next, a function that is not as arbitrary as $\Psi(\mathbf{r},t)$ will be considered. It contains certain coefficients A_s, which will be chosen. It is assumed that in (1.20),

$$\sum_s A_s a_{ss'} = kA_{s'} \quad . \tag{1.24}$$

If (1.24) is substituted into the system of equations (1.23), and (1.20) is used, then

$$\Psi(\mathbf{r},t + 2\pi/\omega) = k\Psi(\mathbf{r},t) \quad . \tag{1.25}$$

The system of h homogeneous linear equations in h unknowns (1.24) has a non-trivial solution if, and only if,

$$\begin{vmatrix} a_{11} - k & & \cdots & a_{1h} \\ a_{21} & a_{22} - k & \cdots & a_{2h} \\ \cdots\cdots\cdots\cdots\cdots\cdots\cdots\cdots \\ a_{h1} & a_{h2} & & a_{hh} - k \end{vmatrix} = 0 \quad , \tag{1.26}$$

where h is the dimension of the basis Ψ_s. If one of the solutions of (1.26) is designated k_α, one can construct a function $\psi^{(\alpha)}(\mathbf{r},t)$ that is a solution of the Schrödinger equation and that satisfies the following condition:

$$\psi^{(\alpha)}(\mathbf{r},t + 2\pi/\omega) = k_\alpha \psi^{(\alpha)}(\mathbf{r},t) \quad . \tag{1.27}$$

The obvious condition that a wave function conserves its normalization yields

$$k_\alpha = \exp(-i2\pi E_\alpha/\omega) \quad , \tag{1.28}$$

where E_α is real. This can also be proven by using the hermiticity of the matrix $[a_{ss'}]$.

A function Ψ that satisfies (1.27) can also be written in the form,

$$\psi^{(\alpha)}(\mathbf{r},t) = \exp(-iE_\alpha t) \, \phi^{(\alpha)}(\mathbf{r},t) \quad , \tag{1.29}$$

where $\phi^{(\alpha)}(\mathbf{r},t)$ is periodic in time, i.e.,

$$\phi^{(\alpha)}(\mathbf{r},t + 2\pi/\omega) = \phi^{(\alpha)}(\mathbf{r},t) \quad . \tag{1.30}$$

Indeed, if (1.30) is applied to (1.29), then

$$\psi^{(\alpha)}(\mathbf{r},t + 2\pi/\omega) = \exp(-i2\pi E_\alpha/\omega) \, \psi^{(\alpha)}(\mathbf{r},t) \quad , \tag{1.31}$$

which coincides with (1.27) if (1.28) is taken into account.

Equations (1.29,30) are the essence of the Floquet theorem [1.15], i.e., the Schrödinger equation with a periodic perturbation has particular solutions (h in number) that satisfay (1.29,30) (of course, there are other solutions that do not satisfy these conditions). These solutions constitute an orthogonal set.

The state $\psi^{(\alpha)}(\mathbf{r},t)$ is described as *quasi-energetic*, while E_α is the *quasi-energy* of this state [1.16]. Since the periodic function $\phi^{(\alpha)}(\mathbf{r},t)$ can be expanded in a Fourier series in t,

$$\psi^{(\alpha)}(\mathbf{r},t) = \exp(-iE_\alpha t) \sum_{n=-\infty}^{\infty} C_n(\mathbf{r}) \exp(-in\omega t) \quad . \tag{1.32}$$

Hence, a quasi-energetic state can be expressed as a superposition of stationary states with energies equal to $(E_\alpha + n\omega)$.

The concept of quasi-energy will be widely used in this book. In Sects. 3.1,6 the quasi-energies of a two- and a three-level system, respectively, will be determined in a resonance approximation.

From the physical point of view, $(E_\alpha + n\omega)$ is the energy of the system "atom + field", i.e., quasi-energy E_α of the atom and the energy of the n quanta. There is no interaction between the two. The initial interaction between the field and the atom, V, was incorporated into the definition of the quasi-energy E_α. Hence, the concept of an atom "dressed" in a field is born [1.17], analogous to the concept of quasi-particles in the many-body problem. The crossing of the levels of a dressed atom are examined using degenerate perturbation theory by diagonalizing the Hamiltonian of the dressed atom, which is time-independent, in the neighborhood of the intersection point [1.18]. In the more conventional time-dependent perturbation theory this corresponds to actual transitions between atomic states.

It has already been seen that the transition to the stationary Hamiltonian is often achieved by using the "rotating wave" approximation. In other words, this approximation enables one to find the quasi-energies of a dressed atom. Hence, monochromatic perturbations can be studied in the framework of stationary theory. Exactly how this is done is shown in Sect.1.3.5.

Aside from the usual perturbation theory, the idea of quasi-energy is useful in the resonance approximation, in which only a small number of harmonics in (1.32) appear to be essential. As a result the computation of the amplitudes of these harmonics becomes straightforward (see Chap.3). From the point of view of stationary theory, the resonance approximation is similar to the problem of a doubly degenerate level in a constant field. It is also obvious that the resonance approximation is equivalent to the "rotating wave" approximation.

In the most general case of the Schrödinger equation for a periodic perturbation with an arbitrary amplitude, the introduction of quasi-energetic
states does not lead to any considerable simplifications. This is evident
a priori, since the procedure does not contain any small perturbation parameters that could be used to develop an approximate computational method.

1.3.5 Monochromatic Perturbation as a Stationary Problem

The quasi-energetic functions (1.32) can be taken as an orthonormal basis
for the Schrödinger equation. Hence, the Schrödinger equation can be re-
written in a form similar to the time-independent equation [1.19].

The functions $C_n^{\alpha}(\mathbf{r})$ of (1.32) can be expanded in a series involving the
unperturbed basis functions $\psi_k^{(0)}(\mathbf{r})$ of the Schrödinger equation without an
electromagnetic field:

$$C_n^{(\alpha)}(\mathbf{r}) = \sum_k b_n^{\alpha k} \psi_k^{(0)}(\mathbf{r}) \quad . \tag{1.33}$$

When the wave function,

$$\psi^{(\alpha)}(\mathbf{r},t) = \exp(-iE_{\alpha}t) \sum_{n=-\infty}^{\infty} \sum_k b_n^{\alpha k} \psi_k^{(0)}(\mathbf{r}) \exp(-in\omega t) \quad , \tag{1.34}$$

is substituted into the Schrödinger equation, $i(\partial\psi^{(\alpha)}/\partial t) = \hat{H}(\mathbf{r},t)\psi^{(\alpha)}$, the
following system of equations is obtained:

$$\sum_{n=-\infty}^{\infty} \sum_{k'} [<k|\hat{H}^{(m-n)}|k'> - (E_{\alpha} + m\omega)\delta_{mn}\delta_{kk'}]b_n^{\alpha k'} = 0 \quad , \tag{1.35}$$

where k varies from 1 to h. By definition,

$$\hat{H}^{(q)}(\mathbf{r}) = \frac{\omega}{2\pi} \int_0^{2\pi/\omega} \hat{H}(\mathbf{r},t) \exp(iq\omega t) \, dt \quad . \tag{1.36}$$

Although (1.36) formally contains an infinite sum over n, for a monochromatic perturbation this sum actually only contains three terms. This can
be seen from (1.36). Since H_0 is time-independent, only the matrix elements
in (1.35) with q = 0, or n = m, do not vanish, and since the perturbation
$V \sim \cos\omega t$, the matrix elements in (1.35) with q = ±1, or n = m ± 1, do not vanish.
Due to the coupling of the different n's in (1.35), there is need to solve
a system of equations with an infinite set of n's.

The system of equations (1.35) is similar to the system for a constant
perturbation [1.20]. The difference lies in the fact that the matrix elements
of the Hamiltonian are averaged over the perturbation period, that quasi-
energies are used instead of normal energies, and that functions such as

$$u_{nk}^{(0)}(\mathbf{r},t) = \psi_k^{(0)}(\mathbf{r}) \, \exp(-in\omega t) \qquad (1.37)$$

are used instead of the basis wave functions. The term "matrix element", which earlier meant integrating with respect to $\mathbf{r}$, e.g.,

$$<k|\hat{H}(\mathbf{r},t)|k'> = \int \psi_k^{(0)^*}(\mathbf{r})\hat{H}(\mathbf{r},t)\psi_{k'}^{(0)}(\mathbf{r}) \, d\mathbf{r} \quad , \qquad (1.38)$$

must now be generalized in accordance with (1.35) by allowing for an integration with respect to t in (1.36). It is obvious from (1.36) that such a generalization is nothing more than an averaging over the perturbation period $T = 2\pi/\omega$:

$$<k|\hat{H}^{(m-n)}|k'> = \ll u_{mk}^{(0)}(\mathbf{r},t)|\hat{H}(\mathbf{r},t)|u_{nk'}^{(0)}(\mathbf{r},t) \gg \quad , \qquad (1.39)$$

where the double angle brackets signify an integration with respect to $\mathbf{r}$ and an averaging over time.

This makes it possible to quickly determine the energy corrections brought about by perturbation theory (see Sect.2.4). One only needs to write down the known expression for the correction due to a constant perturbation, substitute the variable perturbation into this expression, average the matrix elements of the variable perturbation over the perturbation period [according to (1.36)] and, finally, substitute expressions of the type $(E_k + m\omega)$ for the quasi-energies in the energy denominators (for the sake of simplicity the differences in energies in the denominators of series which turn up in perturbation theory will be called *energy denominators*). When the variable field is switched off, the quasi-energies become the eigenvalues of the stationary Hamiltonian H_0, and the quasi-energetic states become the eigenfunctions which correspond to these eigenvalues. Hence, $E_\alpha^{(0)} + n\omega$ is substituted into the formulas of time-independent perturbation theory.

Perturbation theory which contains exact values of energy in the energy denominators is developed in a similar manner. To do this the exact quasi-energies, $E_\alpha + n\omega$, must be substituted in the formulas of time-independent perturbation theory rather than $E_\alpha^{(0)} + n\omega$. These exact quasi-energies are formally defined, in analogy with time-independent perturbation theory, as the mean values of the Hamiltonian $\hat{H}(\mathbf{r},t)$ operating over the exact quasi-energetic states $\psi^{(\alpha)}(\mathbf{r},t)$:

$$E_\alpha = \ll \psi^{(\alpha)}(\mathbf{r},t)|\hat{H}(\mathbf{r},t)|\psi^{(\alpha)}(\mathbf{r},t) \gg \quad . \qquad (1.40)$$

It is assumed that the $\psi^{(\alpha)}$ are normalized:

$$\ll \psi^{(\alpha)}(\mathbf{r},t)|\psi^{(\alpha)}(\mathbf{r},t) \gg \; = 1 \quad . \qquad (1.41)$$

Using this approach it is easy to see that the energies E_α and $E_\alpha + n\omega$ are equivalent, so that E_α can always be restricted to the interval $E_0 - \omega/2 < E_\alpha < E_0 + \omega/2$, where E_0 is an arbitrarily chosen quantity (the reference energy).

Time-independent perturbation theory can similarly be used to find the corrections to the wave functions. This technique is very convenient from the practical side, because the formulas up to the fourth order are well known [1.20].

1.3.6 Exact Solutions

When there are no small perturbation parameters, one must determine the effect of a monochromatic perturbation on a quantum mechanical system by using exact methods. However, only rarely is an exact solution possible. An exact solution is one that can be expressed either in an analytical form or as an integral. The best known examples of exact solutions pertinent to this text are: (a) a free electron in the field of a monochromatic wave [1.21], (b) a two-level system in a circularly polarized external field (see Sect. 3.4), and (c) a harmonic oscillator in a monochromatic field [1.22]. It is sometimes possible to find a differential or integral equation whose solutions, for definite values of the parameters, are obtained with the aid of computers. Examples are discussed in Chap.7. However, they cannot be called exact solutions because the solutions usually depend on a large number of parameters. Hence, it is impossible to plot the results for arbitrary values of these parameters.

Most of the problems are solved by using a variety of small parameters in the series expansions. Often, after a specific perturbation parameter has been used, there are physical quantities for some quantities of which the problem has an analytical solution. In such cases the problem can often be solved exactly. For example, the problem of the interaction of a two-level system and a monochromatic perturbation having a pulse envelope equal to $1/\cosh(t/T)$ can only be solved exactly in the resonance approximation (Chap.3).

2. Time-Dependent Perturbation Theory

The equations of motion can be solved exactly in the framework of quantum mechanics for only a very limited number of systems. This is why approximate methods play such an important role in practically all of the cases in which quantum theory is used. The best-known method is the perturbation method. The goal of perturbation theory is to determine how the states of a system are changed under the influence of a small perturbation. The idea behind this method is quite simple, namely, that a small perturbation imposed on the system does little "damage" to the states. However, this is not always true. Consider, for example, the ionization of an atom in a constant electric field. The field may be infinitely weak, but if one waits long enough the probability of ionizing the atom approaches unity. In other words, under the influence of a small perturbation the initial state is considerably changed. Another example is the excitation of a two-level system by a monochromatic field whose frequency is equal to the energy difference of the levels [2.1]. For an infinitely weak field but over a sufficiently long interval of time the initial state is noticeably changed. Hence, in both examples the state is hardly changed only when the perturbation acts for a short time. It stands to reason that there are many cases in which the states are hardly changed at all for any length of time.

Perturbation theory also assumes that the wave function of the initial state can be expanded in a Taylor series. This is not always the case either (e.g., the ionization of an atom in a constant weak electric field [2.2]).

It is assumed that the reader is familiar with time-independent perturbation theory [2.3]. The aim of this chapter is to set forth the theory of time-dependent perturbations and its limits. When applying this theory to an atom in a strong light field, the focus will be on the monochromatic perturbation. After that time-dependent perturbation theory will be discussed based on expansions in terms of the perturbed basis.

The existence of spontaneous atomic transitions caused by vacuum fluctuations makes the system in an external light field an open system, so that a correct description is to be found in terms of a density matrix rather than a wave function [2.4,5]. Such a description is given in Chap.8.

2.1 First-Order Perturbation Theory

For perturbations that explicitly depend on time, the notion of energy levels, as used in time-independent perturbation theroy, is no longer meaningful. The problem lies in the calculation of the wave functions knowing the stationary-state wave functions of the unperturbed system.

2.1.1 The System of Equations

We assume $\psi_k^{(0)} = \psi_k^{(0)} \exp(-iE_k^{(0)}t)$. An arbitrary solution Ψ of the perturbed Schrödinger equation, $i(\partial\Psi/\partial t) = [H_0 + V(t)\Psi]$, in which $V(t)$ is the perturbation, is required in the form, $\Psi = \sum_k a_k \psi_k^{(0)}$. It is readily seen that the coefficients of the expansion satisfy the following system of ordinary differential equations:

$$i\frac{da_m}{dt} = \sum_k V_{mk} \exp(i\omega_{mk}t)a_k \quad , \tag{2.1}$$

where

$$\omega_{mk} = E_m^{(0)} - E_k^{(0)} \quad , \quad V_{mk} = \int \psi_m^{(0)*} V\psi_k^{(0)}dq \quad , \tag{2.2}$$

and q is the product of all of the variables on which the wave function depends (except time).

2.1.2 The Probability of One-Photon Transitions

Let the unperturbed wave function be that of the nth stationary state. To zeroth order in V, $a_n^{(0)} = 1$ and $a_{k \neq n}^{(0)} = 0$. First of all we are interested in the probability of transitions from state n to other states k. In a first-order approximation one looks for the a_k in the form $a_k = a_k^{(0)} + a_k^{(1)}$, and then substitutes $a_k = a_k^{(0)}$ into the right-hand side of (2.1). In accordance with the general principles of perturbation theory, the perturbation must be small, so that $a_k^{(1)} \ll a_n^{(0)} = 1$. This yields $i(da_k^{(1)}/dt) = V_{kn} \exp(i\omega_{kn}t)$. After integrating this equation one obtains the first-order wave functions:

$$a_k^{(1)}(t) = -i \int^t V_{kn} \exp(i\omega_{kn}t') \, dt' \quad . \tag{2.3}$$

In all of the above formulas it was assumed that only a discrete spectrum of the unperturbed energy levels exists. However, one can immediately generalize the formulas to take into account a continuous spectrum. One can do this by adding the corresponding integrals over the continuous spectrum to the sum of the energy levels of the discrete spectrum.

Let us assume that the initial n^{th} state of the system belongs to the discrete spectrum. The perturbation usually acts over a finite time interval, i.e., $V(t)$ vanishes as $t \to \pm\infty$. Hence, the probability of a transition from the initial state n to the final stationary state k in the discrete spectrum is:

$$W_{kn} = |a_k^{(1)}(\infty)|^2 = \left| \int_{-\infty}^{\infty} V_{kn} \exp(i\omega_{kn}t) \, dt \right|^2 \, . \tag{2.4}$$

Equation (2.4) shows that the transition probabilities from state n to state k and from state k to state n are the same. The symmetry of the process with respect to the interchange of the initial and final states is a general property. It is valid for the discrete levels n and k in perturbation theory of any order (see also Sect.2.2).

One can easily generalize (2.4) for the case in which the state k belongs to the continuous spectrum by multiplying (2.4) by the number of states, dk, inside the interval [k, k +dk]. This yields:

$$dW_{kn} = \left| \int_{-\infty}^{\infty} V_{kn} \exp(i\omega_{kn}t) \, dt \right|^2 dk \, , \tag{2.5}$$

where dW_{kn} is the probability of finding the system in a state with quantum numbers between k and k +dk, and k contains all of the quantities needed for a complete description of the system.

2.1.3 Monochromatic Perturbations

Let us consider the important case of a perturbation monochromatic in time: $V = V^{(1)} \cos\omega t$, where $V^{(1)}$ is a function of the coordinates q. For example, in the dipole interaction of linearly polarized light with the electron of an atom, $V^{(1)} = zE$, where E is the amplitude of the electric field of the wave. The perturbation is substituted into (2.3). To evaluate the integral the initial conditions must be known. It is assumed that the monochromatic perturbation is switched on at $t \to -\infty$ and grows exponentially, i.e., $\exp(\lambda t)$, where λ is a small positive quantity (the adiabatic introduction of the perturbation). Integrating (2.3) with respect to time yields

$$a_k^{(1)}(t) = -\frac{V_{kn}^{(1)} \exp[i(\omega_{kn} - \omega)t + \lambda t]}{2(\omega_{kn} - \omega - i\lambda)} - \frac{V_{kn}^{(1)} \exp[i(\omega_{kn} + \omega)t + \lambda t]}{2(\omega_{kn} + \omega - i\lambda)} \, . \tag{2.6}$$

This formula can be used if none of the denominators is so small or none of the numerators so big that a_k approaches the order of unity. Hence, V_{kn} must be a small perturbation, i.e., the perturbation field must be weak, and no intermediate resonances must be present, i.e., $\omega_{kn} = \pm\omega$. When the state k belongs to the discrete spectrum, the condition $|a_k^{(1)}| \ll 1$ is fulfilled.

What will happen if the states in the continuous spectrum are perturbed? It is assumed that the continuous spectrum starts at zero, which is usually the case. If $-E_n^{(0)} > \omega$, not one of the $\omega_{kn} \pm \omega$ in (2.6) vanishes, so that according to (2.6) the criterion for applying perturbation theory, $|a_k^{(1)}| \ll 1$, is reduced to the condition that the perturbation $V^{(1)}$ is small. From the physical point of view this case describes a situation in which the one-photon ionization probability per unit time is zero.

A more complicated case occurs when $-E_n^{(0)} < \omega$, for which one-photon ionization is allowed. Equation (2.6) shows that the main role will be played by states in the continuous spectrum with energies $E_k^{(0)}$ close to the resonance energy, $E_n^{(0)} + \omega$. This is why only the first term on the right-hand side of (2.6) does not vanish. Hence,

$$|a_k^{(1)}|^2 = \frac{1}{4} |V_{kn}^{(1)}|^2 \frac{\exp(2\lambda t)}{(\omega_{kn} - \omega)^2 + \lambda^2} \quad . \tag{2.7}$$

Since the denominator is small, it seems at first that $|a_k^{(1)}|^2$ should not at all be small. How does this agree with the general principles of perturbation theory? When the final state k belongs to the continuous spectrum, not the probability of the transition to the fixed state k, but the probability

$$\sum_{k < k_i < k + dk} |a_{k_i}^{(1)}|^2$$

of finding the system in a state with quantum numbers between k and k + dk has a physical meaning. This probability will be low, because the term in which the exact equality, $\omega_{k_i n} = \omega$, holds is represented in the above sum with a zero statistical weight. Also, the further ω gets from $\omega_{k_i n}$ the more rapidly the denominator in (2.7) grows. The transition probability per unit time is determined by the derivative with respect to time of $|a_k^{(1)}|^2$:

$$w_{kn} = \frac{d}{dt} |a_k^{(1)}|^2 = 2\lambda |a_k^{(1)}|^2 \quad . \tag{2.8}$$

This must be multiplied by the number of states dk and integrated over the final states. As $\lambda \to 0$, this yields

$$w_{kn} = \frac{1}{2} |V_{kn}^{(1)}|^2 \int_{-\infty}^{\infty} \frac{\lambda}{(\omega_{kn} - \omega)^2 + \lambda^2} \frac{dE_k^{(0)}}{dE_k^{(0)}/d\mathbf{k}}$$

$$= 2\pi |\frac{1}{2} V_{kn}^{(1)}|^2 \rho_k |_{\omega_{kn}=\omega} \tag{2.9}$$

$$= \frac{\pi}{2} |z_{kn}|^2 E^2 \rho_k \quad ,$$

where $\rho_k = d\mathbf{k}/dE_k^{(0)}$, and ρ_k is the energy density of the unperturbed states. If the level is degenerate (which is usually the case for the states of a continuous spectrum), then (2.9) must also be summed over the different quantum numbers corresponding to a given energy. If the emitted electron is defined by its momentum $\mathbf{p}$, then $d\mathbf{k} = d\mathbf{p}/(2\pi)^3$ and $E_k^{(0)} = \mathbf{p}^2/2m$. Hence,

$$\rho_k = \frac{mp}{(2\pi)^3} d\Omega_{\mathbf{p}} \quad , \tag{2.10}$$

where $d\Omega_{\mathbf{p}}$ is the solid angle that the emitted electrons circumscribe.

The condition $E_k^{(0)} - E_n^{(0)} = \omega$ in (2.9) is called the energy conservation law for the absorption of a photon. If state k does not satisfy this law, there is a nonzero absolute probability for the transition from state n to state k. If state k does satisfy the law, there is a nonzero transition probability per unit time. It is thus natural to call ω_{kn} the one-photon ionization probability. Let us estimate its value by assuming that the dipole matrix element z_{kn} is equal to one atomic unit. Obviously,

$$w_{kn} \sim \left(\frac{E}{E_{at}}\right)^2 \frac{1}{\tau_{at}} \quad , \tag{2.11}$$

where $E_{at} = 5 \times 10^9 \text{V/cm}$, and τ_{at} is the characteristic atomic time, i.e., $\tau_{at} = \hbar^3/me^4 = 2.4 \times 10^{-17} \text{s}$. For example, the one-photon ionization probability of an excited state becomes comparable to the probability of its spontaneous relaxation, i.e., $w_{kn} \sim 1/\tau_n \sim 10^7 \text{s}^{-1}$ at a field strength of $E \sim 10^5 \text{V/cm}$.

2.1.4 Sudden Perturbations

Equation (2.9) is the result of a perturbation which was switched on adiabatically and slowly. The result is the same if the perturbation is suddenly switched on, i.e., assuming $a_n^{(0)} = 1$ and $a_k^{(0)} = 0$ at $t = 0$, then the amplitude is

$$a_k^{(1)} = -\frac{V_{kn}^{(1)}}{2} \frac{\exp[i(\omega_{kn} - \omega)t - 1]}{\omega_{kn} - \omega} \quad . \tag{2.12}$$

That the probabilities for the adiabatic and the sudden perturbations should coincide is not obvious a priori. It will be seen in Sect.2.2 that the results are different in second-order perturbation theory. At that point the reason for such a discrepancy will be discussed.

What is the actual perturbation due to switching in an experiment with atoms in the strong laser-light field? Even the switch-on time of a picosecond laser (10^{-12}s) is a thousand times longer than the length of the period of an electromagnetic wave. Consequently, the actual switch-on period is clearly adiabatic. This statement only holds for nonresonant transitions, though it will be seen (Chap.8) that in the event of resonance the switch-on is sudden.

2.1.5 Large Perturbation Times

From (2.9) it follows that the ionization probability is proportional to time, i.e., $W_{kn}(t) = w_{kn}t$. For longer times this formula is still valid, obviously as long as $|a_k^{(1)}| \ll 1$, i.e., $w_{kn}t \ll 1$. This means that the time T during which the field operates must not be too long, so that only a small fraction of the atoms become ionized. This may be mathematically stated as follows:

$$\rho_k |z_{kn}|^2 E^2 T \ll 1 \quad . \tag{2.13}$$

For a field of 10^6V/cm, $T \ll 10^{-9}$s, which is only satisfied by picosecond lasers. For nanosecond lasers we cannot speak of the ionization per unit time for the above case, but only of the absolute probability $W_{kn}(t)$, which is of the order of unity. The exact expression for $W_{kn}(t)$ has the following form [2.6]:

$$W_{kn}(t) = 1 - \exp(-w_{kn}t) \quad . \tag{2.14}$$

As $t \to \infty$, $W_{kn}(\infty) = 1$. In Sect.8.1 we will derive (2.14) for the case in which the perturbation is the electromagnetic field in the vacuum. The quantity $1/w_{kn}$ characterizes the time for the one-photon ionization of an atom. For perturbation theory to be applicable this time must be large compared to the atomic time, 10^{-17}s.

It must be noted that in the ionization processes, apart from the transitions to state k for which $\omega_{kn} = \omega$, there are also transitions to the state ℓ for which $\omega_{\ell n} \neq \omega$. These transitions also represent ionization processes if the state l belongs to the continuous spectrum. Unlike a transition to state k, the probability of a transition to ℓ is not directly proportional to time. To a first-order approximation it is given by $(V_{\ell n}^{(1)}/\omega_{\ell n})^2$. If we are in a regime with $W_{kn}(1) \sim 1$, then obviously transitions to states such as ℓ can be neglected, because with $E \ll E_{at}$, $(V_{\ell n}^{(1)}/\omega_{\ell n})^2 \ll 1$. If the operation time for the perturbation is very small, such that the transition probability is directly proportional to time, i.e., $W_{kn}(t) = w_{kn}t$, then the time of the perturbation must satisfy the following condition:

$$w_{kn}T \gg (V_{1n}^{(1)}/\omega_{1n})^2 \quad , \qquad \text{or,} \tag{2.15}$$

$$\rho(\omega_{1n})^2 T \gg 1 \quad . \tag{2.16}$$

This in fact means that the perturbation time T must be much larger than typical atomic times, $\tau_{at} \sim 10^{-17}$ s. This condition is always fulfilled in actual experiments with atoms in strong light fields. However, the situation is more complicated in processes involving two or more photons.

2.1.6 The Role of the Shape of the Envelope of the Electromagnetic Field

As shown in Sect.2.1.4, the transition probability (2.9) may be obtained by imagining that the atom is acted upon by a square pulse of a monochromatic electromagnetic field switched on at time zero and switched off at time T. Equation (2.9) then determines the transition probability per unit time. Actually the pulse is not square but has a definite envelope $E(t)$ that vanishes as $t \to \pm\infty$ and whose characteristic pulse time is large compared to the length of the period of the perturbation $1/\omega$. Then, (2.9) yields the total ionization probability of one laser pulse:

$$W_{kn}(\infty) = \frac{\pi}{2} |z_{kn}|^2 \rho_k|_{\omega_{kn}=\omega} \times \int_{-\infty}^{\infty} E^2(t) \, dt \quad . \tag{2.17}$$

This expression is valid only if the probability is much smaller than unity i.e., $W_{kn}(\infty) \ll 1$. If this is not so, then in accordance with the results of Sect.2.1.5 the absolute probability of a one-electron ionization with one laser pulse is

$$W_{kn}(\infty) = 1 - \exp\left[-\frac{\pi}{2} |z_{kn}|^2 \rho_k|_{\omega_{kn}=\omega} \times \int_{-\infty}^{\infty} E^2(t) \, dt\right] \quad . \tag{2.18}$$

The last two equations are equivalent if the expression in brackets is much less than unity.

2.1.7 Criteria of Applicability

Let us discuss the criteria for applying first-order perturbation theory. For all of the states, $|a_k^{(1)}| \ll 1$. Then, using (2.7) or (2.12),

$$\frac{z_{kn}E}{(\omega_{kn} \pm \omega)} \ll 1 \quad . \tag{2.19}$$

Applied to the discrete states k and n this means that the perturbation is of a nonresonant nature, i.e.,

$$|\omega_{kn} \pm \omega| \gg z_{kn}E \quad . \tag{2.20}$$

This, in turn, implies that for the perturbation to be small, either

$$\frac{z_{kn}E}{\omega_{kn}} \ll 1 \quad , \qquad \text{or} \qquad \frac{z_{kn}E}{\omega} \ll 1 \quad , \tag{2.21}$$

i.e., the matrix elements of the external perturbation must be small compared to the typical energy differences in the unperturbed system or to the frequency of the external field.

If for the first of the excited states and for radiation of the visible range (when $\omega_{kn} \sim \omega$) these two conditions are identical, only one condition is significant for the ionization from an excited state (when $\omega_{kn} \ll \omega$) or for transitions between highly excited states ($\omega_{kn} \gg \omega$). The latter case is important in the study of atomic multiplet levels for which the energy differences are small and the first condition is not met (Sect.2.5.1).

For an adiabatic perturbation, for which $\omega \ll E_n^{(0)}$, (2.21) is no longer a criterion (Sect.4.2).

The only necessary condition for the formulas of first-order perturbation theory to be valid is that, $a_k^{(1)} \ll a_n^{(0)} = 1$. Obviously, it is also necessary that the inequality, $a_k^{(1)} \gg a_k^{(2)}$, be fulfilled. This will be examined in Sect.2.2.5.

2.2 Second-Order Perturbation Theory

The second term in the time-dependent perturbation theory series shall be discussed in this section. This discussion is expedient for the development of the general diagrammatic technique, which describes the arbitrary orders of perturbation theory (Sect.2.3). It is extremely important to know the behavior of the second term in problems where for some reason the matrix element of the first order, $V_{kn}^{(1)}$, is equal to zero or is small (e.g., in a strictly or almost strictly forbidden transition, or for the case for which no one-photon exit channel into the final state exists; see Sect.2.1).

To keep the presentation simple, arbitrary perturbations V(t) will not be discussed. Only the monochromatic perturbation $V^{(1)}\cos\omega t$ will be dealt with keeping in mind future applications of the theory of the interaction of atoms and light.

Section 2.1 showed that in first-order perturbation theory the interaction of an atom with light can be expressed in terms of the absorption or emission of one quantum ω. Naturally then, second-order perturbation theory takes into account the possibility of two-photon absorption or emission and those processes not accompanied by the emission or absorption of light (elastic scattering).

2.2.1 Probability of a Two-Photon Transition

Let us assume that the perturbation is switched on adiabatically according to the exponential law $\exp(\lambda t)$ with $\lambda \to +0$ (Sect.2.1). Substituting into the right-hand side of (2.1) the expression for first-order perturbation theory (2.6) and integrating, it follows that

$$a_k^{(2)} = V_{kn}^{(2)}(\omega)\left(\frac{\exp(i\omega_{kn}t + \lambda t)}{\omega_{kn} - i\lambda} + \frac{\exp[i(\omega_{kn} - 2\omega)t + \lambda t]}{\omega_{kn} - 2\omega - i\lambda}\right)$$

$$+ V_{kn}^{(2)}(-\omega)\left(\frac{\exp[i(\omega_{kn} + 2\omega)t + \lambda t]}{\omega_{kn} + 2\omega - i\lambda} + \frac{\exp(i\omega_{kn}t + \lambda t)}{\omega_{kn} - i\lambda}\right) . \qquad (2.22)$$

Here the following notation has been introduced:

$$V_{kn}^{(2)}(\omega) = \sum_m \frac{V_{km}^{(1)}V_{mn}^{(1)}}{4(\omega_{mn} - \omega)}$$

$$= \frac{1}{4} E^2 \sum_m \frac{z_{km}z_{mn}}{\omega_{mn} - \omega} \qquad (2.23)$$

$$= z_{kn}^{(2)}(\omega)E^2 .$$

The term $V_{kn}^{(2)}$ is called the two-photon matrix element. It contains a sum over the intermediate states m. Unlike the energies of the initial and final states, which are connected by energy conservation, the energies of the states m may be arbitrary. Transitions $n \to m$ and $m \to k$ are called virtual transitions. The selection rules for virtual transitions are obviously the same as those for one-photon transitions, which were studied in Sect.2.1. They determine the corresponding selection rules for the initial (n) and final (k) states. These rules are such that the parities of states n and k coincide and the orbital angular momenta differ by 0, ±1, or ±2. The selection rules for the magnetic quantum numbers depend on the degree of elliptical polarization of the perturbing field: for linear polarization $\Delta M = 0$, for circular polarization $\Delta M = \pm 2$, and for the general case of elliptical polarization $\Delta M = 0, \pm 1$, or ± 2.

According to the time-energy uncertainty relation, the time that an electron spends in a virtual state m is of the order $\Delta t \sim (\omega_{mn} - \omega)^{-1}$. Since $\omega_{mn} \gg \omega$, this time is about 10^{-17} s, i.e., small compared to the time required for a transition from the initial to the final state. It is for this reason that the states m are called virtual states.

The first term in (2.22) describes a process involving the absorption and subsequent emission of one quantum ω, the second term the absorption of two quanta, the third term the emission of two quanta, and the fourth term the emission and subsequent absorption of one quantum ω.

Even for such a simple case involving a monochromatic perturbation, an explicit expression for the wave function in second-order perturbation theory is very complicated. To write the numerous higher-order terms in a unified and simple way it is therefore expedient to develop the diagrammatic technique. This is done in Sect.2.3.

In some cases (2.22) can be simplified. For example, let us examine the case in which there is a channel open for two-photon transitions into the continuum. This can be described as follows: $2\omega > -E_n^{(0)} > \omega$. For some states k in the continuum the difference $\omega_{kn} - 2\omega$ vanishes. Hence, the second term in (2.22) is much larger than the other terms. Only this term is taken into consideration in evaluating the probability of a two-photon absorption per unit time. Repeating the calculations of Sect.2.1, the following formula for the transition probability per unit time is obtained:

$$
\begin{aligned}
w_{kn}^{(2)} &= 2\pi \left| V_{kn}^{(2)}(\omega) \right|^2 \rho_k \Big|_{\omega_{kn}=2\omega} \\
&= 2\pi \left| z_{kn}^{(2)}(\omega) \right|^2 E^4 \rho_k \Big|_{\omega_{kn}=2\omega} \quad .
\end{aligned}
\tag{2.24}
$$

The equation for the probability of a two-photon transition from a discrete state m to a discrete state k, which can occur when ω_{kn} is almost equal to 2ω, has a similar structure. The difference lies in the expression for the density of the final state k which depends on the specific decay process for state k. This decay process will be studied in Chap.8.

Equation (2.22) enables one not only to describe the absorption or emission of two photons, but also to describe the two-photon transition process in which a monochromatic wave is scattered by an atom (Rayleigh scattering; see Sect.8.2). Since the state of the system does not change in such a process, $\omega_{kn} \to 0$ and the first and fourth term in (2.22) must be considered. Unlike the previous case of two-photon absorption (or emission), the frequency ω of the incident light is not related to the atomic frequencies by any law of conservation and thus may be quite arbitrary. As a result the probability of scattering is

$$w_{nn}^{(2)} = 2\pi |z_{nn}^{(2)}(\omega) + z_{nn}^{(2)}(-\omega)|^2 E^4 \rho \quad , \tag{2.25}$$

where ρ is the density of the final states. Naturally, this result is valid only if the first-order diagonal matrix element $V_{nn}^{(1)}$ is equal to zero (the usual case because of selection rules) and the denominators in (2.23) are not small. The case for which the denominators are small (resonance fluorescence) is studied in Sect.8.2.

One can determine the dependence of the two-photon ionization probability on the angle of emission of the electrons by performing the following substitution:

$$\rho_k \to \frac{mp}{(2\pi)^3} d\Omega_p \quad . \tag{2.26}$$

Since $E \ll E_{at}$, the two-photon ionization probability is approximately $(E/E_{at})^2$ times smaller than the one-photon ionization probability, unless the latter process is forbidden by energy conservation. This is why two-photon ionization is not observed if one-photon ionization takes place. If one-photon ionization is forbidden, much stronger fields than those used in ordinary linear optics are required for the observation of two-photon ionization. For example, to attain the same probability, $w_{kn} \sim 10^7 s^{-1}$, as in Sect.2.1, a field of approximately 10^7V/cm is needed. In the one-photon case 10^5V/cm are needed. Only lasers can produce such fields.

How is the matrix element $V_{kn}^{(2)}(\omega)$, which describes a two-photon absorption from the initial state n to a final state k related to the inverse process involving the stimulated emission of two photons? The two-photon matrix element for the inverse process will be

$$V_{nk}^{(2)} = \sum_m \frac{V_{nm}^{(1)} V_{mk}^{(1)}}{4(\omega_{mk} + \omega)} \quad . \tag{2.27}$$

This matrix element coincides with (2.23) [more exactly, with the complex conjugate of (2.23)] because $\omega_{kn} = 2\omega$, see (2.24).

2.2.2 Transitions Induced by Two Perturbations

Equation (2.23) must be generalized so that it includes the case in which a two-photon transition is induced by two perturbations: $V^{(1)}\cos\omega t$ and $V^{(1)'}\cos\omega' t$. Then, in addition to the absorption of two photons of the first perturbation, two-photon transitions associated with the absorption of two photons of the second perturbation and the absorption of photons from each beam may be observed. Hence, there are two such matrix elements: one describes the initial absorption of a photon from one beam and the subsequent absorption of a photon from the second beam, while the other element describes the reverse process. If (2.22) is generalized to encompass this case, then instead of (2.23) one obtains

$$V^{(2)}_{kn}(\omega) \equiv \sum_m \frac{V^{(1)'}_{km} V^{(1)}_{mn}}{4(\omega_{mn} - \omega)} \quad , \qquad \text{and} \tag{2.28}$$

$$V^{(2)'}_{kn}(\omega') \equiv \sum_m \frac{V^{(1)}_{km} V^{(1)'}_{mn}}{4(\omega_{mn} - \omega')} \quad . \tag{2.29}$$

The two-photon absorption probability is

$$w^{(2)}_{kn} = 2\pi \left| V^{(2)}_{kn}(\omega) + V^{(2)'}_{kn}(\omega') \right|^2 \rho_k \Big|_{\omega_{kn}=\omega+\omega'} \quad . \tag{2.30}$$

These formulas can also describe processes involving the absorption of photons from one beam and the emission of photons of the other beam. Then ω and ω' have opposite signs. They serve as a basis for describing two-photon transitions in a non-monochromatic field.

With two perturbations, one described by first-order perturbation theory, the other by second-order theory, (2.25) is needed to describe the Raman scattering (Sect)3.4). Here the field of the electromagnetic vacuum plays the role of the second perturbation. Hence,

$$w^{(2)}_{kn} = 2\pi \left| V^{(2)}_{kn}(\omega) + V^{(2)'}_{kn}(-\omega') \right|^2 \rho_k \Big|_{\omega_{kn}=\omega-\omega'} \quad . \tag{2.31}$$

2.2.3 Large Perturbation Times

Equation (2.24) predicts that the absolute probability of the two-photon ionization of an atom is proportional to time, i.e., $W_{kn}(t) = w^{(2)}_{kn} t$. Obviously, this is only true if the probability is much less than unity. In accordance with what was said in the brief summary at the beginning of Chap.2, for large values of t the above formula must be changed to

$$W_{kn}(t) = 1 - \exp(-w_{kn}^{(2)}t) \quad , \tag{2.32}$$

which is analogous to the result for one-photon absorption (Sect.2.1).

What has just been said is only applicable to ionization by a rectangular pulse of duration t. If the pulse has an envelope, $E(t)$, of arbitrary form,

$$W_{kn}(\infty) = 1 - \exp\left[-\frac{\pi}{2} \int_{-\infty}^{\infty} E^4(t) \, dt \left| \sum_m \frac{z_{km}z_{mn}}{\omega_{mn} - \omega} \right|^2 \right] \quad . \tag{2.33}$$

As the strength of the electric field grows, the probability approaches unity much faster than in the one-photon case. In other words, the range of field strengths for which the probability changes from zero to unity is much narrower than in the one-photon case. This range, however, lies considerably higher than the similar range for one-photon ionization.

The time required for two-photon ionization is of the order of $(w_{kn}^{(2)})^{-1}$, which is much longer than for one-photon ionization. The difference in times is about $(E_{at}/E)^2$.

Suppose a two-photon ionization is to be studied in a situation where energy conservation forbids one-photon ionization. Then, if one wants to use (2.24), which is the analog of (2.9) for two-photon ionization, one must make sure that the absolute probability of ionization and a simultaneous transfer to a state k which satisfies the energy conservation law is high, compared to the absolute probabilities of transitions to states that do not satisfy the energy conservation law. A similar requirement for one-photon ionization yielded (2.16) (Sect.2.1.5). In the present case, of the nonresonant transitions the most probable are one-photon transitions between states n and k that do not satisfy the energy conservation law. If one insists that $w_{kn}^{(2)}T \gg |V_{kn}^{(1)}/\omega_{kn}|^2$, the criterion for the perturbation time is

$$T \gg \tau_{at}\left(\frac{E_{at}}{E}\right)^2 \quad . \tag{2.34}$$

As for (2.16), this criterion only has meaning if the probability is defined per unit time. For large perturbation times it is satisfied automatically.

What is the meaning of (2.34)? For a specific field, $E = 10^5$V/cm, $T \gg 10^{-8}$s. Even for nanosecond lasers this is not so. It follows that nonresonant one-photon ionization instead of two-photon ionization takes place in such a field.

Thus, (2.34) is quite a demanding criterion for the perturbation time.

2.2.4 Sudden Perturbations

How are the results of second-order perturbation theory modified when the perturbation is switched on suddenly? In the nonresonant case it is usually the adiabatic regime (switch-on mode) that is realized. It is nevertheless advantageous to dwell on this point because the sudden regime is used in many theoretical works.

It is assumed that a periodic perturbation is applied at $t = 0$, at which time the system is in a state with energy $E_n^{(0)}$. Instead of (2.6),

$$a_k^{(1)}(t) = -\frac{V_{kn}^{(1)}}{2}\left(\frac{\exp[i(\omega_{kn} - \omega)t] - 1}{\omega_{kn} - \omega} + \frac{\exp[i(\omega_{kn} + \omega)t] - 1}{\omega_{kn} + \omega}\right) . \quad (2.35)$$

Substitution into the right-hand side of (2.1) and a subsequent integration with respect to time yields $a_k^{(2)}$ with four terms of the same type as in (2.22) and other new terms, which appear because of the 1's in the numerators of (2.35):

$$a_k^{(2)} = \sum_m \frac{V_{km}^{(1)}V_{mn}^{(1)'}}{4}\left(\frac{\exp[i(\omega_{km} + \omega)t] - 1}{\omega_{km} + \omega}\right.$$
$$\left. + \frac{\exp[i(\omega_{km} - \omega)t] - 1}{\omega_{km} - \omega}\right)\frac{2\omega_{mn}}{\omega^2 - \omega_{mn}^2} . \quad (2.36)$$

The singularities are understood in the principal-value sense.

The terms in (2.22) describe effects which were considered in the adiabatic switch-on mode, whereas the terms in (2.36) are a reflection of the sudden switch-on mode. One can always choose states m and k in the continuum for which $\omega_{mk} \pm \omega \rightarrow 0$. For this reason the terms in (2.36) also contribute to the transition probability per unit time regardless of whether energy is conserved, i.e., $E_k^{(0)} - E_n^{(0)} = 2\omega$. Thus, the terms in (2.35) describe transitions that are not induced by a monochromatic perturbation but by the higher Fourier components, which emerge due to the suddenness of the perturbation.

From the qualitative point of view the terms in (2.36) are a result of the nonresonant population of level m in the framework of first-order perturbation theory and the subsequent resonant transition to state k. In this case the nonresonant population is also a result of the suddenness of the perturbation.

Thus, when the perturbation is switched on suddenly, in second-order perturbation theory (in contrast to first-order theory) there is a strong overlap of effects associated with the monochromatic field itself and the switching-on of the perturbation.

2.2.5 Criteria of Applicability

The criteria for applying second-order perturbation theory shall now be discussed, see (2.24). Suppose that for all of the states, $|a_k^{(2)}| \ll |a_k^{(1)}|$. Then (2.22) yields the following criteria:

$$\frac{z_{kn}E}{\omega_{kn} \pm 2\omega} \ll 1 \quad , \qquad \frac{z_{kn}E}{\omega_{kn}} \ll 1 \quad . \tag{2.37}$$

If these conditions are applied to the discrete levels k and n, one sees that, as in the case of first-order perturbation theory, the perturbation is of a nonresonant nature:

$$|\omega_{kn} \pm 2\omega| \gg z_{kn}E \quad . \tag{2.38}$$

If this criterion is met, the phrase "smallness of perturbation" obviously describes a situation in which,

$$\frac{z_{kn}E}{|\omega_{kn}|} \ll 1 \quad \text{or} \quad \frac{z_{kn}E}{\omega} \ll 1 \quad , \tag{2.39}$$

which coincides with (2.21).

Equation (2.24) cannot be used when the matrix element $V_{kn}^{(2)}$ becomes anomalously large due to the small difference in ω_{mn} and ω. This can happen when the system absorbs one quantum ω and in the process shifts from the initial state n to another discrete state m (intermediate resonance; for more details see Chap.8).

2.3 The Diagrammatic Technique for Monochromatic Perturbations

In Sect.2.2 it was shown that already in first-order perturbation theory the amplitude $a_k^{(2)}$ of the state $\psi_k^{(0)}$ consists of four terms. As the order of perturbation theory grows, the number of such terms increases drastically. It is therefore desirable to have a universal rule to describe the amplitude in any order of perturbation theory. The rule must contain instructions for the construction of all the graphs (Feynman diagrams [2.7,8]) for a given order of the perturbation series and write the mathematical expressions which describe the different parts of a diagram.

For an arbitrary time-dependent perturbation the diagrammatic technique is rather cumbersome to use because an interaction representation is used for the perturbing Hamiltonian and because chronologically ordered products emerge [2.8,9]. For a given order it is also hard to decide whether all of the diagrams in the perturbation series have been drawn, i.e., whether or not some diagrams have been left out. This difficulty is due to the varying topology of the diagrams. A harmonic perturbation is simpler to describe because all of the integrals with respect to time can be evaluated explicitly. An analysis of first- and second-order perturbation theory has shown (Sects.2.1,2) that it is convenient to use the concept of the perturbation quanta i.e., of the emitted and absorbed photons. It is natural then to introduce these concepts into the diagrammatic technique by classifying the diagrams according to the total number of photons and by distinguishing between emitted and absorbed photons. Moreover, it is also natural to develop a diagrammatic technique in which the absorption and emission of photons of a harmonic perturbation is introduced adiabatically. This is done so as not to mix the effects associated with the harmonic perturbation and those associated with the switch-on mode (for example, when the perturbation is suddenly switched on). These effects must be taken into account just as they were taken into account for a non-monochromatic perturbation.

For the sake of convenience a diagrammatic technique will be developed for calculating the matrix elements of processes involving several photons.

2.3.1 First-Order Perturbation Theory

The probability of a transition, given by (2.9), is determined by the square modulus of the matrix element represented by the following diagram

where n and k represent the initial and final states, respectively. The dot, i.e., the vertex, represents the matrix element $V_{kn}^{(1)}/2$. The dotted line to the left represents the absorption of one photon of frequency ω. Similarly, the probability of a one-photon emission process is determined by the square modulus of the matrix element represented by the diagram,

The dotted line to the right of the line signifies the emission of a photon. In the diagrams it is understood that time increases from left to right, which ensures that absorption and emission are indeed pictorial.

In the first of the above diagrams the initial and final states have energies such that

$$E_k^{(0)} = E_n^{(0)} + \omega \; , \tag{2.40}$$

while in the second diagram,

$$E_k^{(0)} = E_n^{(0)} - \omega \; . \tag{2.41}$$

2.3.2 Second-Order Perturbation Theory

The matrix element, $V_{kn}^{(2)}$, corresponding to the absorption of two photons, can be represented by the diagram,

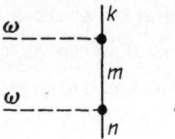

The solid vertical line labeled m symbolizes the free-particle propagator (the one-particle Green function) [2.9,10], which has a shape determined by $(\Delta_m - i\lambda)^{-1}$, where Δ_m is the energy of the mth state relative to the energy of the initial state n minus the energy of the absorbed photon. Hence, $\Delta_m = \omega_{mn} - \omega$.

The total second-order amplitude, $a_k^{(2)}$, is a sum of four two-photon matrix elements, represented by the following diagrams,

The second has just been discussed. The first and fourth diagrams describe (with k = n) elastic two-photon scattering (2.25), while the third symbolizes two-photon emission.

In accordance with (2.24), the conservation of energy for the absorption of two photons requires that the energy of the final state, $E_k^{(0)}$, differ from the energy of the initial state, $E_n^{(0)}$, by the energy of two absorbed photons, 2ω. Similarly, for elastic scattering the final state must coincide with the initial state (Sect.2.2.1).

In addition to the rules given for first-order diagrams, one must introduce a summation over the states of the inner lines, i.e., over m, which is usually done for Feynman diagrams.

One can easily generalize the above diagrams for two-photon matrix elements to include transitions stimulated by photons from different beams. For example, the matrix element that describes the transition, whose probability is given by (2.31), can be symbolized by the sum of two diagrams:

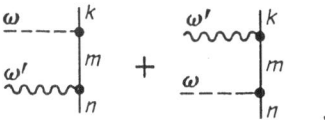

The conservation of energy requires that,

$$E_k^{(0)} = E_n^{(0)} + \omega + \omega' \quad . \tag{2.42}$$

In this technique it should be noted that the intermediate states m are virtual states which do not necessarily obey the laws of energy conservation.

2.3.3 The Rules for Constructing Diagrams

Now we can formulate the rules for constructing the diagrams for any order of monochromatic perturbation theory. Each diagram that symbolizes a multiphoton matrix element consists of a vertical line from which dotted lines radiate outwards to the left and right in an arbitrary order. Such a diagram resembles a tree. The matrix elements in K^{th}-order perturbation theory are represented by various diagrams, each including K dotted lines (branches). A branch entering the trunk from the left symbolizes the absorption of a quantum of the monochromatic perturbation, while a branch entering from the right symbolizes the emission of such a quantum.

Each segment of the vertical line (trunk) between neighboring branches has a corresponding factor, i.e., the Green function, $(\Delta_m - i\lambda)^{-1}$, where Δ_m is the energy of the mth state of the system, relative to the energy of the initial state n less the energy of the absorbed quanta plus the energy of the emitted quanta $(\lambda \to + 0)$. Of course, one can determine the energy of the state m relative to the final state k instead of the initial state. From the point of view of the final state the absorbed photons must be treated as emitted photons and vice versa. The equivalence of both approaches was discussed in Sect.2.2.1.

Dots in the diagrams symbolize a factor, $V_{ms}^{(1)}/2$, which characterizes the interaction of the system with the external field. The states m and s correspond to the upper and lower sections, respectively, of a given segment of the trunk.

The energies of the initial and final states differ by a quantity equal to the difference in the energy of all the absorbed and emitted quanta. There are no restrictions on the energies of the intermediate states. The generalization of the diagrammatic technique to include the absorption or emission of quanta from different beams is obvious. Such quanta are represented by different lines (e.g., wavy and dotted lines). Each diagram starts with the initial state n at the bottom and ends with the final state k at the top.

It is easy to see that in K^{th}-order perturbation theory there are 2^K diagrams in all.

2.3.4 Partial Summation of Diagrams

When two electromagnetic fields simultaneously interact with an atomic electron, one field is often strong while the other one is weak. The reason a field is strong is not necessarily because its amplitude is large (Chap.1). It may be that the detuning of a resonance with this field is small. Under such circumstances one must not limit oneself to diagrams of the lowest order. One must sum over all of the orders of the perturbation expansion in this field.

The method consists of replacing the free-particle propagator by the propagator that takes into account diagrams of all orders (of course only when possible):

$$\left|\begin{smallmatrix}n\\ \\n\end{smallmatrix}\right. + --\bullet\!\!\left|\begin{smallmatrix}n\\ \\n\end{smallmatrix}\right. + \left|\begin{smallmatrix}n\\ \\ \end{smallmatrix}\right.\!\!\bullet-- + \left\{\begin{smallmatrix}n\\ \\m\\ \\n\end{smallmatrix}\right. + \left|\begin{smallmatrix}n\\ \\m\\ \\n\end{smallmatrix}\right\} + \ldots \equiv \left|\begin{smallmatrix}n\\ \\ \\n\end{smallmatrix}\right. .$$

The result is a perturbed, or "thick", propagator, which differs from the free-particle propagator by a new energy spectrum. A weak field is taken into account in the same way as was done in perturbation theory, only that "thick" propagators are used instead of free-particle propagators.

Note that the summation of diagrams for the perturbation expansion in the strong field is selective. Only diagrams whose contribution is large are considered. Those that contribute very little are neglected. Such a selection is determined by the circumstances of the specific problem.

It is clear from the above summation that, when such an approach is used, diagrams with long "tails" are the result corresponding to free-particle propagators. Their analytical form is found according to the rules discussed in Sect.2.3.3.

2.4 Perturbation Theory of Arbitrary Order

2.4.1 The Amplitudes of Different Processes

The amplitudes of different processes with K photons can clearly be illustrated in the diagrammatic approach. For example, the matrix element for the multi-photon excitation of an atom which absorbs K photons is represented by a "lopsided" tree,

$$V_{kn}^{(K)} = K$$

or analytically,

$$V_{kn}^{(K)} = \frac{1}{2^K} \sum_{ms\ldots p} \frac{V_{km}^{(1)} V_{ms}^{(1)} \ldots V_{pn}^{(1)}}{[\omega_{mn} - (K-1)\omega]\ldots[\omega_{pn} - \omega]}$$

$$\equiv z_{kn}^{(K)} E^K , \qquad \text{where} \qquad (2.43)$$

$$z_{kn}^{(K)} = \sum_{ms\ldots p} \frac{z_{km} z_{ms} \ldots z_{pn}}{[\omega_{mn} - (K-1)\omega]\ldots[\omega_{pn} - \omega]} . \qquad (2.44)$$

Thus, the transition rate for a K-photon process will be,

$$w_{kn}^{(K)} = 2\pi |V_{kn}^{(K)}|^2 \rho_k \Big|_{\omega_{kn} = K\omega}$$

$$= 2\pi |z_{kn}^{(K)}|^2 E^{2K} \rho_k . \qquad (2.45)$$

2.4.2 Large Perturbation Times

Large perturbation times have already been discussed in connection with one- and two-photon ionization processes. Here the results will be generalized to include multi-photon transitions of any order K.

Equation (2.45) yields the multi-photon ionization probability per unit time only if the absolute probability $W_{kn} = w_{kn}^{(K)} t \ll 1$. For large values of t, the probability of ionization is

$$W_{kn}(t) = 1 - \exp(-w_{kn}^{(K)} t) . \qquad (2.46)$$

The above equation gives the ionization probability for a rectangular pulse of duration t. In reality, if it is assumed that the pulse has a sufficiently smooth envelope, $E(t)$, whose variation with time is small compared to ω^{-1}, then the ionization probability per pulse is:

$$W_{kn}(\infty) = 1 - \exp\left[-2\pi|z_{kn}^{(K)}|^2\rho_k \int_{-\infty}^{\infty} E^{2K}(t) \, dt\right] \quad . \tag{2.47}$$

Obviously, this equation is only valid if it can be assumed that the stochastic properties of multi-mode laser radiation do not affect the process (for more details see Chap.9), i.e., is only valid for a single-frequency laser.

The above equation shows that since the curve $E^{2K}(t)$ gets narrower as K grows (Chap.6), the effective ionization time decreases. It turns out that for processes with $K \leq 20$ total ionization is realized in fields with $E \ll E_{at}$ and $T \sim 10^{-8}$s (a typical time for pulsed lasers).

Just as in two-photon ionization processes, the perturbation time must be sufficiently large for multi-photon ionization. One can neglect one-photon ionization, which, of course, proceeds without the conservation of energy provided that $\omega < |E_n^{(0)}|$. All of this makes the absolute transition probability proportional to the square of the field strength. Thus,

$$w_{kn}^{(K)}T \gg (V_{1n}^{(1)}/\omega_{1n})^2 \quad , \qquad \text{i.e.,} \qquad T \gg (E/E_{at})^{2(1-K)}\tau_{at} \quad . \tag{2.48}$$

For large K's and small T's these conditions become quite demanding. For example, with $K = 5$ and $E = 5 \times 10^7$V/cm, $T \gg 10^{-1}$s. Naturally, under these conditions (2.45) is not valid, because for real lasers T is much less than 10^{-1}s. Thus, for weak fields, nonresonant one-photon ionization occurs. The probability of such a process is small because it does not increase with time. For a certain field strength E, determined by setting the exponent in (2.47) equal to unity, the probability jumps almost instantaneously from zero to unity. Hence, especially for large values of K, (2.47) should be used not as an equation for determining the ionization probability but as an equation for determining the critical value of the field strength at which the ionization probability increases drastically.

2.4.3 Criteria for Applying Perturbation Theory of Arbitrary Order

In accordance with what has been said about the criteria for applying first-order (Sect.2.1.7) and second-order (Sect.2.2.5) perturbation theory, one can draw a general conclusion for perturbation theory of arbitrary order.

First of all, the following condition must be fulfilled:

$$a_k^{(K+1)} \ll a_k^{(K)} \quad , \tag{2.49}$$

which in fact says that the correction due to the $(K+1)^{th}$ approximation is small compared to the correction introduced by the Kth approximation.

Secondly, there must be no resonances, including multi-photon resonances, of multiplicity K or less between the initial, final, and intermediate states:

$$|\omega_{kn} \pm K\omega| \gg z_{kn}E \quad . \tag{2.50}$$

Also, the matrix elements of the external perturbation must be small compared to differences of the energies in the unperturbed system or compared to the energy of the photons:

$$\frac{z_{kn}E}{\omega_{kn}} \ll 1 \qquad \text{or} \qquad \frac{z_{kn}E}{\omega} \ll 1 \quad . \tag{2.51}$$

These criteria are not valid for (a) an adiabatic perturbation ($\omega \ll E_n^{(0)}$), for which criteria are given in Chap.4, (b) the tunneling of an electron through a potential barrier into the continuum (Chap.4), and (c) for degenerate electronic states (Sect.2.5.8).

Special consideration must be given to the above criteria when the initial and final states have large principal quantum numbers (n), i.e., for states of a quasi-classical nature. It appears that the criteria for applying perturbation theory largely depend on whether the problem is a one-dimensional or three-dimensional one. In the one-dimensional case, which is the simplest to consider, for perturbation theory to be applied, a matrix element of the perturbation must be small compared to the distance between close levels, i.e., between levels whose quantum numbers differ very little: $z_{n,n-1}E \ll E_n^{(0)} - E_{n-1}^{(0)}$. Thus the criterion virtually remains the same. The situation for the three-dimensional case is different. Even if a matrix element of the perturbation is of the same order of magnitude as the distance between adjacent levels, whose principal quantum numbers are almost identical but whose other quantum numbers differ greatly, the overlap of the unperturbed wave functions due to the application of the perturbation is vanishingly small. In other words, perturbation theory can be applied in this case. This is not so if a matrix element of the perturbation is of the order of the energy of the perturbed levels. Thus, the criterion for applying perturbation theory in the three-dimensional case is: $z_{n,n-1}E \ll E_n^{(0)}$. All that has been said in this paragraph also applies to a perturbation in the form of a constant field.

2.4.4 Convergence of Expansion Series in Perturbation Theory

It is known that when a constant electric field acts on an atomic system the
expansion series of perturbation theory approaches an asymptote. This means
that, no matter how small the amplitude of the perturbation, all terms following a certain term become very large due to the growth of the numerical
coefficients of successive terms. Such a behavior is due to the continuum
spectrum of the system.

The existence of a continuum makes a tunneling through the potential barrier possible under the influence of a constant electric field. The dependence of the ionization probability on the field strength, E, exhibits an
essential singularity at $E = 0$, i.e., in the neighborhood of this field
strength the ionization probability cannot be expanded in a power series.
For a system with a discrete spectrum the series converges for not very large
field strengths.

In a variable monochromatic field the situation changes. Although, as
will be shown in Sect.4.5, a tunneling through the potential barrier is
also possible here, tunneling stops playing an important role for $\omega \gg E$ (a
fixed field strength). The ionization probability is then determined by the
multi-photon transition of the electron to the continuum, i.e., by (2.45).
This means that for $\omega \neq 0$ the expansion series converge, i.e., do not approach
an asymptote. For small frequencies the radius of convergence is very small,
at least much smaller than ω.

2.5 Monochromatic Perturbation and Degenerate States

Up to this point time-dependent perturbation theory has been applied to
nondegenerate states. However, as a rule, atomic states are completely degenerate (e.g., in the magnetic quantum number) or partially degenerate
(e.g., the hydrogen atom, where levels with the same principal quantum number but different orbital quantum numbers are split because of a small spin-
orbit coupling).

If the perturbation is time independent, the solution is well known
[2.3]. Generally speaking, such a perturbation lifts the degeneracy (totally
or partially). The initial states are strongly mixed by the perturbation. As
a result the zeroth-order wave functions differ drastically from the unperturbed functions.

2.5.1 A Single Degenerate Level

Let us see how a time-dependent perturbation acts on completely or partially degenerate states. Only perturbations that are monochromatic with time will be considered. The simplest case is a system with one degenerate level. To the first order in the perturbation, the amplitude $a_k^{(1)}$, determined by (2.30), remains finite as $\omega_{kn} \to 0$. This is not the case for time-independent perturbations, where $a_k^{(1)}$ approaches infinity. This can also be easily seen from (2.7), if, as $\omega_{kn} \to 0$, ω is set equal to zero.

In second-order perturbation theory there are two terms in (2.22) that increase indefinitely as $\omega_{kn} \to 0$, namely, the first and fourth terms. But because $\omega_{mn} = 0$, their sum vanishes. It is easy to see that for higher orders the sum of the large terms also vanishes.

Let us consider the special case of a two-fold degeneracy. It is easy to solve (2.1) exactly:

$$a_n(t) = \cos\left(\frac{z_{nk}E}{\omega} \sin\omega t + \phi_0\right) \quad , \tag{2.52}$$

$$a_k(t) = -i \sin\left(\frac{z_{nk}E}{\omega} \sin\omega t + \phi_0\right) \quad , \tag{2.53}$$

where the phase, ϕ_0, is a constant determined by the initial conditions.

A similar exact solution can be obtained for a level with an arbitrary degree of degeneracy. It is assumed that the dipole matrix elements between the different states of the degenerate level are nonzero. Such levels possess constant dipole moments. For example, for a two-fold degenerate level z_{nk} plays the role of such a constant dipole moment.

To solve the system of differential equations for the amplitude a_n of the probability of finding the electron in state n,

$$i\dot{a}_n = E \cos\omega t \sum_{n'} z_{nn'} a_{n'} \quad , \tag{2.54}$$

a change in variables is undertaken:

$$a_n(t) = A_n \exp\left(-is \frac{E}{\omega} \sin\omega t\right) \quad . \tag{2.55}$$

For A_n and s one then obtains a system of algebraic equations with constant coefficients:

$$sA_n = \sum_{n'} z_{nn'} A_{n'} \quad . \tag{2.56}$$

(Note that neither s nor A_n depends on the strength or the frequency of the field.) One can solve for s by nullifying the determinant of the system:

$$\det[s\delta_{nn'} - z_{nn'}] = 0 \quad . \tag{2.57}$$

As a result one obtains p values for s, where p is the degree of degeneracy of the level. After this one can determine the A_n, which correspond to a set of p solutions. The general solution is a superposition of these solutions. The coefficients of each solution are found using the initial conditions. Before the field is switched on all of the degenerate levels are likely to be equally populated. An alternative is to average over all of the possible initial conditions. With the above example, one can easily verify that, on the average, all of the degenerate levels remain equally populated. This result is valid for any degree of degeneracy.

The eigenstates, obtained by diagonalizing the perturbation on the basis states of the degenerate levels, are superpositions of the latter. Thus, the "old" quantum numbers that characterized the unperturbed states are no longer "good" quantum numbers. For the new, "good", quantum numbers one can use the different labels s.

The solutions are periodic, which means that in this case the quasi-energy is equal to zero. Also, the solutions clearly show that perturbation theory can be used in determining amplitudes only if $E \ll \omega$.

In conclusion, one can say that the solutions represent oscillations in the populations of the degenerate levels. These solutions are valid for fields which act for any length of time. A degeneracy makes any formulation of the problem using an adiabatic perturbation incorrect.

2.5.2 Mixing Degenerate Levels in a Field Due to the Presence of Other Levels

Let us discuss the more general case in which levels other than a given degenerate level are also present. As usual, no problems are encountered using first-order perturbation theory (2.7). However, in second-order perturbation theory (2.22), the first and fourth terms do not cancel out because of the intermediate states m of the other levels, for which $\omega_{mn} \neq 0$. Since $a_k^{(2)} \to \infty$ as $\omega_{kn} \to 0$, one cannot use perturbation theory in its simplest form. In other words, the degenerate levels mix strongly, i.e., these states enter the wave function with comparable weights because of the presence of the other levels.

Let us determine the degree to which the degenerate levels mix. It is assumed that there is no mixing of the type discussed in Sect.2.5.1. In the initial system of (2.1) the terms associated with the degenerate levels are isolated from the terms associated with the other levels. Then, (2.1) can be written in the form

$$i\dot{a}_n = \sum_m V_{nm} \exp(i\omega_{nm}t)a_m \quad , \tag{2.58}$$

$$i\dot{a}_m = \sum_n V_{mn} \exp(i\omega_{mn}t)a_n + \sum_{m'} V_{mm'} \exp(i\omega_{mm'}t)a_{m'} \quad , \tag{2.59}$$

where the a_n are the amplitudes of the degenerate levels, and the a_m and $a_{m'}$ are the amplitudes of the other levels. It is evident that if the system found itself in the degenerate level $E_n^{(0)}$ before the perturbation was switched on, the a_n will become comparable with one another after the switch-on and of the order of unity. In addition the a_m will become small. Needless to say, it is assumed that the criterion given in (2.21) is met in the process, i.e., the perturbation is small compared to the difference in energy of the levels n and m.

With this in mind it is evident that the second sum on the right-hand side of (2.59) is small compared to the other sum. Neglecting it, integrating, and substituting the result into (2.58), one gets

$$i\dot{a}_{n'} = -i \sum_{nm} V_{n'm}(t) \int_{-\infty}^{t} V_{mn}(t')a_n(t') \exp[i\omega_{mn}(t' - t)] \, dt' \quad . \tag{2.60}$$

As will be evident later, the $a_n(t)$ vary slowly with time, i.e., their variation is determined by the perturbation itself, not by the frequencies of the unperturbed system. Because of this, $a_n(t')$ can be taken outside of the integral in (2.59), since it varies slowly with time compared to the exponential factor. This procedure is correct if the criterion given by (2.21) is met.

After this the second sum on the right-hand side of (2.59) is integrated (it is assumed that the perturbation is adiabatic),

$$i\dot{a}_{n'} = -\frac{1}{2} \sum_{nm} V_{n'm}^{(1)} \cos\omega t \times V_{mn}^{(1)} \left[\frac{\exp(i\omega t + \lambda t)}{\omega_{mn} + \omega - i\lambda} \right.$$

$$\left. + \frac{\exp(-i\omega t + \lambda t)}{\omega_{mn} - \omega - i\lambda} \right] a_n \quad . \tag{2.61}$$

In (2.61) one can substitute for cos t,

$$\cos\omega t = \frac{1}{2} [\exp(i\omega t) + \exp(-i\omega t)] \quad . \tag{2.62}$$

Of the four terms that appear after multiplication of the exponentials the two that are time independent can be neglected. The oscillating terms of the type $\exp(\pm 2i\omega t)$ lead to small variations in the a_n. Finally, from (2.61) the following equation is obtained [2.11]:

$$i\dot{a}_{n'} = -E^2 \sum_n [z_{n'n}^{(2)}(\omega) + z_{n'n}^{(2)}(-\omega)]a_n \quad . \tag{2.63}$$

It is easy to see that the same equations are obtained for sudden perturbations. The two-photon matrix element, $z_{n'n}^{(2)}$, is defined in (2.23). Due to the smallness of E, the solutions to (2.63) are slowly varying functions of time. From (2.63) it follows that a strong mixing of the degenerate levels takes place if

$$T \gtrsim (V_{n'n}^{(2)})^{-1} \sim E^{-2} \quad . \tag{2.64}$$

For nanosecond lasers ($T \sim 10^{-8}$s) this criterion is met for fields with $E > 10^5$V/cm.

Let us consider the general form of the solution of (2.63). For the sake of brevity the column of amplitudes a_n will be denoted by $\mathbf{a}$. An arbitrary solution of a system of first-order differential equations with constant coefficients can be written in terms of an expansion in the eigenfunctions (eigenvectors) of the system:

$$\mathbf{a}(t) = \sum_N C_N \mathbf{a}^{(N)}(t) \quad . \tag{2.65}$$

The coefficients C_N are determined from the initial conditions and N is used to label the eigenvalues of (2.63), i.e., the new quantum numbers.

It is a well-known fact that the eigenfunctions, $\mathbf{a}^{(N)}(t)$, of the system of differential equations with constant coefficients must have the following form:

$$\mathbf{a}^{(N)}(t) = \mathbf{g}^{(N)} \exp(-i\varepsilon_N t) \quad . \tag{2.66}$$

Substitution into (2.63) yields

$$-E^2 \sum_n [z_{n'n}^{(2)}(\omega) + z_{n'n}^{(2)}(-\omega)]g_n^{(N)} = \varepsilon_N g_{n'}^{(N)} \quad . \tag{2.67}$$

This equation is used to find the eigenvalues, ε_N, and the eigenvectors, $\mathbf{g}^{(N)}$. For an adiabatic perturbation the final solution consists of eigenfunctions because they describe states with defined energies ε_N. For other cases combinations of these eigenfunctions must be used.

50

One could also obtain the eigenvalue equation (2.67) by considering the stationary problem of the "atom-field" system (Sect.1.3). Equation (2.67) follows from the secular equation of time-independent perturbation theory if second-order terms and degeneracy are taken into account.

Note that the ε_N are small due to the smallness of E since $\varepsilon_N \sim E^2$. This confirms the earlier hypothesis concerning the slowness of the variation of $a_n(t)$ with time. Moreover, the dependence of $a^{(N)}$ on time, (2.66), justifies taking $a_n(t')$ outside the integral in (2.60). If this had not been done, one would have had to change the energy denominators in $z_{n'n}^{(2)}$ from $\omega_{mn} \pm \omega$ to $\omega_{mn} \pm \omega + \varepsilon_N$. Since in this section the possibility of small energy denominators due to resonances was excluded, ε_N can be neglected provided that the criterion given in (2.21) is met .

Hence, in the zeroth approximation, the wave function of a particle with fixed initial conditions is given by

$$\psi^{(0)}(t) = \sum_{Nn} C_N g_n^{(N)} \psi_n^{(0)} [\exp -i(E_n^{(0)} + \varepsilon_N)t] \quad . \tag{2.68}$$

If the initial conditions prior to the introduction of the perturbation consist of a homogeneous distribution of the particles over the degenerate levels, this distribution is maintained after the perturbation is switched on. This can easily be verified by explicitly solving (2.67) for a two-fold degenerate level. Obviously, the mixing of the degenerate levels may be neglected, although mixing does take place. The same conclusion was made in Sect.2.5.1.

Now the case for which the second-order expansion of the operator in (2.67) is insufficient shall be considered. It is easy to see that by applying the procedure used to derive (2.63) to the next order in the perturbation, a similar equation is obtained for $a_n(t)$, in which terms of the type $V_{n'n}^{(4)} \sim E^4$ were added to the operator in (2.67).

The perturbation, $V_{n'n}^{(4)}$, will strongly mix degenerate states if $T \gtrsim E^{-4}$, — a criterion similar to (2.64). In Sect.2.4.2 it was shown that such mixing is significant when analyzing an ionization involving three, four, or more photons.

Similar criteria can be formulated for higher-order time-dependent perturbation theory.

2.5.3 Near-Degenerate Levels

A level that is approximately degenerate shall now be considered. Obviously, the above results can be applied to nearly degenerate levels if

$$z^{(2)}_{n'n}E^2 \gg \omega_{n'n} \quad , \tag{2.69}$$

where n and n' denote states with almost identical energies. Before the perturbation is switched on the particles are fixed in definite states (nonequilibrium population). For the case of an equilibrium population (all of the nearly degenerate levels are populated with equal probabilities) the transition probability averaged over the degenerate levels does not depend on the diagonalization of (2.67) or the use of traditional perturbation theory without degeneracy.

If $z^{(2)}_{n'n}E^2 \ll \omega_{n'n}$, which is the opposite of (2.69), no mixing occurs and one is dealing with the nondegenerate case just considered. In this case the perturbation $z_{nm}E$ may be greater than the distance between the closely spaced levels, $\omega_{n'n}$, i.e., $z_{nm}E \ll \omega_{n'n}$, and still mixing occurs.

Let us now briefly examine the intermediate case in which $z^{(2)}_{n'n}E^2 \sim \omega_{n'n}$. In (2.58) there will appear a factor equal to $\exp(i\omega_{n'n}t)$. The same factor will appear in (2.60). This factor will also appear in (2.63), which takes on the following form:

$$i\dot{a}_{n'} = -E^2 \sum_n [z^{(2)}_{n'n}(\omega) + z^{(2)}_{n'n}(-\omega)] \exp(i\omega_{n'n}t)a_n \quad . \tag{2.70}$$

Since $\omega_{n'n} \sim z^{(2)}_{n'n}E^2$, the $a_n(t)$ still behave smoothly. Let the new amplitudes be $A_n = a_n \exp(-iE^{(0)}_n t)$. From (2.70) one can derive a system of differential equations with constant coefficients,

$$i\dot{A}_{n'} - E^{(0)}_{n'}A_{n'} = -E^2 \sum_n [z^{(2)}_{n'n}(\omega) + z^{(2)}_{n'n}(-\omega)]A_n \quad . \tag{2.71}$$

The above equation is a generalization of (2.63). In (2.71) it is advisable to measure $E^{(0)}_{n'}$ relative to some energy value in the neighborhood of the nearly degenerate level (e.g., the center of gravity of the multiplet). The solution to (2.71) should have a form similar to the one given in (2.66).

As a result an equation similar to (2.67) is obtained. However, the ε_N have been changed to $\varepsilon_N - E^{(0)}_{n'}$. The criterion for the correctness of the above theory with exact degeneracy is $\varepsilon_N \gg E^{(0)}_{n'}$, from which (2.69) follows.

2.5.4 Time-Dependent Diagrams

Let us now illustrate the basic Eq. (2.63) of Sect.2.5.3 using the diagrammatic language. It can be seen that the mixing of degenerate states is determined by the sum of two two-photon matrix elements:

$$\omega \dashv \begin{matrix} n' \\ m \\ n \end{matrix} \quad + \quad \omega \vdash \begin{matrix} n' \\ m \\ n \end{matrix} \omega \tag{2.72}$$

It is obvious that these matrix elements do not depend on time. In a similar manner, one-photon mixing (Sect.2.5.1) may be represented by the sum of two one-photon matrix elements:

$$\omega \dashv \begin{matrix} n' \\ n \end{matrix} \quad + \quad \vdash \begin{matrix} n' \\ n \end{matrix} \omega \tag{2.73}$$

For these diagrams to be synonymous with the analytical language, one must associate a factor $\exp(i\omega t)$ with each dotted line that represents the absorption of a photon and a factor $\exp(-i\omega t)$ with each dotted line that represents the emission of a photon. If one also bears in mind that for each vertex there exists a factor $V^{(1)}_{n'n}/2$ then one can see that the sum shown in (2.73) can be represented analytically by the term $V^{(1)}_{nn'}\cos\omega t$.

The introduction of the concept of time dependence into the diagrammatic language does not change the results of Sect.2.3 because all of the phase factors, when multiplied by their complex conjugates, are equal to unity. But the diagrammatic technique makes it possible to discuss transitions which proceed without the conservation of energy [e.g., the process shown in (2.73)].

Processes that take place without the conservation of energy are significant at low frequencies (in Sect.2.5.1 it was required that $\omega < V^{(1)}_{nn'}$). This is also clear from general considerations. Usually time-dependent diagrams can be neglected because their dependence is determined by rapidly oscillating factors of the type $\exp(\pm i\omega t)$. At low frequencies, i.e., when these factors become comparable to other quantities that characterize the time dependence of the amplitudes of the states, they cease oscillating rapidly. Then, one can no longer neglect the time-dependent diagrams.

2.5.5 Two-Photon Mixing of Degenerate States in Low-Frequency Fields

The results of Sect.2.5.4 will be illustrated using as an example the two-photon mixing of degenerate states (Sect.2.5.3). This time, however, the field is a low-frequency field, which means that in (2.60) one cannot neglect those two-matrix elements that depend on time. In other words, in addition to the diagram in (2.72) one must include the sum of the second-order diagrams,

$$
\begin{array}{ccc}
\omega & \overset{n'}{\underset{}{}} & \overset{n'}{\underset{}{}} \; \omega \\
\text{---} & \bullet & \bullet \text{---} \\
\omega & m & + \quad m \;\; \omega \\
\text{---} & \bullet & \bullet \text{---} \\
& n & n
\end{array}
\qquad ,
$$

(2.74)

which was neglected in Sect.2.5.3. In the process it will be assumed that first-order diagrams of the type shown in (2.73) simply vanish because of certain selection rules. For example, we must take into account the diagrams (2.74) in the consideration of the degeneracy with respect to the magnetic quantum numbers of the energy levels of complex atoms in an elliptically polarized field.

Bearing in mind that in a low frequency field $\omega \ll \omega_{mn}$, (2,60) yields

$$
i\dot{a}_n = 4E^2 \cos^2\omega t \sum_{n'} z_{nn'}^{(2)} a_{n'} , \tag{2.75}
$$

where the two-photon matrix element is

$$
z_{nn'}^{(2)} = \sum_m \frac{z_{nm} z_{mn'}}{4\omega_{mn}} . \tag{2.76}
$$

Equation (2.75) describes the mixing of degenerate levels in a low-frequency field. The structure of its solution may be easily understood if a two-fold degenerate level is considered. One must then consider two equations with the initial conditions $a_n(0) = 1$ and $a_{n'}(0) = 0$.

For the sake of simplicity it is assumed that the diagonal matrix elements $V_{nn}^{(2)}$ and $V_{n'n'}^{(2)}$ are identical. Thus,

$$
a_n = \cos\left[2|V_{nn'}^{(2)}|\left(t + \frac{\sin 2\omega t}{2\omega}\right)\right] \times \exp\left[-2iV_{nn}^{(2)}\left(t + \frac{\sin 2\omega t}{2\omega}\right)\right] , \tag{2.77}
$$

and

$$
a_{n'} = -i\,\frac{V_{n'n}^{(2)}}{|V_{n'n}^{(2)}|}\,\sin\left[2|V_{nn'}^{(2)}|\left(t + \frac{\sin 2\omega t}{2\omega}\right)\right]
$$

$$
\times \exp\left[-2iV_{nn}^{(2)}\left(t + \frac{\sin 2\omega t}{2\omega}\right)\right] . \tag{2.78}
$$

The second basis state, determined by the initial conditions $a_n(0) = 0$ and $a_{n'}(0) = 1$, is described by equations similar to (2.77,78). The average population densities of the levels n and n', which can be found from (2.41), are different. If at time zero they are the same, then they remain so, on the average, even after the field has been switched on.

If the amplitudes a_n and $a_{n'}$ are expanded in a Fourier series, then according to the Floquet theorem a large number of harmonics are obtained (Sect.1.3). The population densities of these harmonics are obtained (Sect. 1.3). The population densities of these harmonics are determined by the coefficients of the Fourier expansion. They can be directly measured by applying a low-intensity field and by varying the frequency of the light so that a resonance is achieved between a definite quasi-energetic state and some other (finite) state. If the probability of the transition to the finite state is plotted against the frequency, one observes resonance peaks that are spaced ωs^{-1} apart. The area under each peak is proportional to the square of the corresponding Fourier harmonic.

As expected, (2.77,78) lead to the results of Sect.2.5.2, provided that

$$\frac{z_{nn'}^{(2)} E^2}{\omega} \ll 1 \quad , \tag{2.79}$$

i.e., for sufficiently high frequencies. Mixing only occurs because of the time-dependent two-photon matrix elements. Obviously this criterion is of a general nature and is not dependent upon any specific assumptions concerning the degree of degeneracy.

If the criterion in (2.79) is not met, then instead of (2.63) one must use (2.75), in which $\cos^2 \omega t$ can no longer be replaced by $1/2$. In other words, one can say that when $z_{nn'}^{(2)} E^2 / \omega > 1$ one must add to the usual mixing, which does not depend on ω, the mixing that strongly depends on ω.

For the case of a two-fold degenerate level, the two types of mixing are taken into account in a_n and $a_{n'}$, i.e., in the arguments of the cosine, the sine, and the exponential in (2.77,78).

If the diagonal matrix elements $V_{nn}^{(2)}$ and $V_{n'n'}^{(2)}$ are not equal or if the degree of degeneracy is greater than two, (2.39) may only be solved using numerical methods. A difference in $V_{nn}^{(2)}$ and $V_{n'n'}^{(2)}$ does not, in principle, change the validity of the above results, because these matrix elements are chiefly responsible for the phases of the states n and n', not for the mixing of these states.

The reader must also bear in mind that the mixing of degenerate states, i.e., the development of a new basis with new quantum numbers, requires relatively long perturbation times T. As can be seen from (2.77,78), the mixing condition is

$$z_{nn'}^{(2)} E^2 T > 1 \quad , \tag{2.80}$$

which coincides with the condition given in (2.64).

2.5.6 One-Photon Mixing of Degenerate States in Low-Frequency Fields

A more complicated situation arises when the degenerate levels are similar to those of hydrogen, i.e., when the first-order matrix elements, $z_{nn'}$, that connect the states n and n', do not vanish. For example, in hydrogen this occurs because of the degeneracy with respect to the orbital angular momentum.

In this case, in generalizing (2.75) one obtains a system of equations for determining the mixing of degenerate states [2.12]:

$$i\dot{a}_n = 4E^2 \cos^2\omega t \sum_{n'} z^{(2)}_{nn'} a_{n'} + E \cos\omega t \sum_{n'} z_{nn'} a_{n'} \quad . \tag{2.81}$$

At high frequencies there is no contribution from $z_{nn'}$ to the mixing due to the rapid oscillation of $\cos\omega t$ (Sect.2.5.2). However, when $\omega \ll \omega_{mn}$, the oscillations are slow and $z_{nn'}$ may contribute considerably to the mixing. This contribution is represented by the diagram in (2.73).

It is very easy to obtain a solution for a two-fold degenerate level. For the sake of simplicity, it is assumed that $z^{(2)}_{nn} = z^{(2)}_{n'n'} = 0$. These matrix elements mainly influence the phase of the state amplitudes of n and n', not the mixing of these states (2.77,78). For this reason, their being equal to zero will change nothing from the qualitative point of view, although, of course, this is important in calculations. It will also be assumed that all of the matrix elements of the perturbations V are real. Bearing all of this in mind, one finds that (2.81) takes on the following form:

$$i\dot{a}_n = 4E^2 \cos^2\omega t \times z^{(2)}_{nn'} a_{n'} + E \cos\omega t \times z_{nn'} a_{n'} \quad , \tag{2.82}$$

$$i\dot{a}_{n'} = 4E^2 \cos^2\omega t \times z^{(2)}_{nn'} a_n + E \cos\omega t \times z_{nn'} a_n \quad . \tag{2.83}$$

With the initial conditions $a_n(0) = 1$ and $a_{n'}(0) = 0$, the following basis state is obtained:

$$a_n = \cos\left[2E^2 z^{(2)}_{nn'}\left(t + \frac{\sin 2\omega t}{2\omega}\right) + Ez_{nn'} \frac{\sin\omega t}{\omega}\right] \quad , \tag{2.84}$$

$$a_{n'} = -i \sin\left[2E^2 z^{(2)}_{nn'}\left(t + \frac{\sin 2\omega t}{2\omega}\right) + Ez_{nn'} \frac{\sin\omega t}{\omega}\right] \quad . \tag{2.85}$$

This is in agreement with (2.77,78) when $z_{nn'} = 0$. One can find the second basis state in a similar way.

2.5.7 Competition Between One- and Two-Photon Mixing

As illustrated in (2.84,85), one-photon mixing is insignificant at high frequencies (Sect.2.5.1), i.e.,

$$\frac{z_{nn'}E}{\omega} \ll 1 \quad . \tag{2.86}$$

When $\omega \sim \omega_{mn}$ the above condition is equivalent to a small perturbation relative to characteristic atomic fields. This is why, in this case, one can neglect $z_{nn'}$.

If $z_{nn'}E/\omega > 1$, a mixing of degenerate states occurs due to the one-photon matrix elements. Equation (2.81) remains valid even as the strength of the field grows, as long as $V_{nn'}^{(3)}/\omega \ll 1$. When this condition is no longer met, a three-photon mixing of degenerate states comes into play.

The condition given in (2.86) imposes a more stringent restriction on the field strength than the one given in (2.79). This is why, as the perturbation gradually increases, only the mixing due to the two-photon matrix elements that are time independent appears at first. When $z_{nn'}E > \omega$ and $z_{nn'}^{(2)}E^2 \ll \omega$, the frequency-dependent mixing due to the one-photon matrix elements also comes into play. Furthermore when $z_{nn'}^{(2)}E^2 > \omega$, the mixing due to the two-photon matrix elements which depend on time can take place. As the field is further increased, the mixing due to the three-photon matrix elements is observed; and so on.

2.5.8 Approximate Degeneracy

Let us briefly examine the case in which the degeneracy in a low-frequency field is approximate rather than exact (Sect.2.5.3). Then in (2.81) $a_{n'} \exp(i\omega_{nn'}t)$ is substituted for $a_{n'}$. Thus,

$$i\dot{a}_n = 4E^2 \cos^2\omega t \sum_{n'} z_{nn'}^{(2)} a_{n'} \exp(i\omega_{nn'}t)$$

$$+ E \cos\omega t \sum_{n'} z_{nn'} a_{n'} \exp(i\omega_{nn'}t) \quad . \tag{2.87}$$

It is seen that an additional oscillation of frequency $\omega_{nn'}$ appears. Everything that was said concerning exact degeneracies remains valid provided that $\omega_{nn'} \ll z_{nn'}^{(2)}E^2$ (the more so if $\omega_{nn'} \ll z_{nn'}E$).

2.5.9 Criteria of Applicability

For singly degenerate levels (Sect.2.5.1), time-dependent perturbation theory can be used if

$$\frac{z_{nn'}E}{\omega} \ll 1 \quad . \tag{2.88}$$

In the optical region this criterion virtually coincides with the criterion for applying time-dependent perturbation theory without degeneracy (2.21). It should be noted that (2.88) has no analog in time-independent perturbation theory. If the criterion given in (2.88) is met, the mixing of the degenerate states with the given initial state is small. In this case all of the results of Sects.2.1-3 can be used.

This criterion can easily be verified by using as an example a two-fold degeneracy, see (2.47). One can expand (2.47) in a perturbation series if this criterion is met.

Let us turn our attention to the criterion for applying (2.63), which describes the mixing of states due to the presence of two levels (Sect. 2.5.2). In deriving this equation it was assumed that no direct mixing of the degenerate levels occurs, i.e., no mixing of the type considered in Sect.2.5.1 takes place. In other words, it was assumed that the criterion given in (2.88) is fulfilled. In addition, in transferring from (2.61) to (2.63) the rapidly oscillating terms of the type exp($\pm i\omega t$) were neglected, which is justified if

$$\frac{z_{nn'}^{(2)}(\omega)E^2}{\omega} \ll 1 \quad . \tag{2.89}$$

This is certainly the case if (2.88) is true, because whenever time-dependent perturbation theory is used it is assumed that the perturbing fields are weak compared to the atomic fields.

The theory presented in Sect.2.5.2 is applicable when the channel for a one-photon transition to the continuum is closed ($\omega < |E_n^{(0)}|$). Otherwise, the two-photon matrix element, $V_{nn'}^{(2)}$, has an imaginary part and the operator in (2.67) is not hermitian. As a result the basis, $\mathbf{g}^{(N)}$, is no longer orthogonal and the solution is incorrect.

In the case of two-photon ionization and for sufficiently long times, T, which satisfy (2.64), there is a strong mixing of degenerate levels due to the presence of other levels in the system, because the absolute probability of two-photon ionization, $\sim E^4 T$, is low.

Finally, the theory presented in Sect.2.5.2 can only be used when the denominators of the two-photon matrix elements, $V_{n'n}^{(2)}$, are not small. Such small denominators are found if any intermediate resonances between a degenerate level, n, and some other level, m, take place.

The additional condition which emerges for approximate degeneracies (Sect. 2.5.3) arises from the condition that there be no resonances of the field's frequency ω and the separation $\omega_{nn'}$ between the multiplet components. Hence it must be assumed that $\omega \gg \omega_{nn'}$. The same is also true for approximate degeneracies in the low-frequency case.

In studying the low-frequency case in Sect.2.5.5 the condition given in (2.89) was dropped. The system of equations (2.75) that emerges is not valid for any arbitrarily large perturbation V, but only for such perturbations for which

$$\frac{V_{nn'}^{(4)}}{\omega} \ll 1 \quad , \tag{2.90}$$

where $V_{nn'}^{(4)}$ is the corresponding four-photon matrix element. If this is not so, then mixing due to time-dependent four-photon processes such as shown in (2.91),

$$\tag{2.91}$$

start to play a significant role. Fields for which (2.90) is no longer true are almost atomic fields.

2.6 The Green's Function in Time-Dependent Perturbation Theory

It has been shown that the expressions for the multi-photon matrix elements contain sums over intermediate virtual states. These sums incorporate an integration over the continuum as well. In practice, such a summation is difficult to perform because the number of terms is infinite. At the same time, to take only a finite number of terms is not, in general, a justifiable operation. Hence, the following trick is used. The atomic potential of the valence electron is given by a potential for which the Schrödinger equation can be solved exactly and the eigenfunctions, $\phi_n(r)$, and the eigenvalues, $E_n^{(0)}$, can be determined. It appears then that the summation can also be performed exactly. It will be explained how this is done in practice by using as an example the two-photon matrix element.

2.6.1 The Green's Function

The two-photon matrix element (2.23) can be written in the following form

$$V_{kn}^{(2)}(\omega) = \frac{1}{4} \iint \phi_k^*(\mathbf{r'})V^{(1)}(\mathbf{r'})G_\omega(\mathbf{r'},\mathbf{r})V^{(1)}(\mathbf{r})\phi_n(\mathbf{r}) \; d^3r \; d^3r' \quad , \qquad (2.92)$$

where

$$G_\omega(\mathbf{r'},\mathbf{r}) = \sum_m \frac{\phi_m^*(\mathbf{r'})\phi_m(\mathbf{r})}{\omega_{mn} - \omega - i\lambda} \quad , \qquad \lambda \to +0 \quad . \qquad (2.93)$$

The function $G_\omega(\mathbf{r'},\mathbf{r})$ is known as the Green's function (propagator) [2.10]. The Green's function can be represented by a thin line,

$$\left|\begin{matrix} r' \\ \\ r \end{matrix} \right. \quad . \qquad (2.94)$$

In the two-photon matrix element, it is illustrated as follows,

$$\begin{matrix} \underline{\boldsymbol{\omega}} \, {-}{-}{-}\!\bullet & \begin{matrix}k \\ r' \end{matrix} \\ \\ \underline{\boldsymbol{\omega}} \, {-}{-}{-}\!\bullet & \begin{matrix}r \\ n \end{matrix} \end{matrix} \quad . \qquad (2.95)$$

The above diagram describes a two-photon absorption. Hence one is simply dealing with Feynman diagrams in the coordinate representation instead of the energy representation. The vertices correspond to $V(\mathbf{r})/2 = d(\mathbf{r})E/2$ and $V(\mathbf{r'})/2$, the thin line represents the Green's function, and the upper and lower "tails" of the diagram represent the functions $\phi_n(\mathbf{r})$ and $\phi_k(\mathbf{r'})$. The integration is performed over the inner variables $\mathbf{r}$ and $\mathbf{r'}$ (integration replaces summation over the quantum numbers m for Feynman diagrams in the energy representation). The coordinate Green's function (2.93) corresponds to the thin line in (2.94) while the Green's function,

$$G_\omega(m) = \frac{1}{\omega_{mn} - \omega - i\lambda} \quad , \qquad (2.96)$$

corresponds (in the energy representation; Sect.2.2.3) to the diagram

$$\left|\begin{matrix} m \end{matrix}\right. \quad .$$

2.6.2 Application of the Green's Function

For the problems dealt with in this context the Schrödinger equation can be solved exactly for three cases: (a) for a free electron, (b) for an electron in a zero-range potential, and (c) for an electron in a Coulomb field.

In the first case (a free electron with an energy $p^2/2$) the Green's function has the simple form

$$G_\omega(\mathbf{r},\mathbf{r}') = \frac{1}{2\pi} \frac{\exp(ip|\mathbf{r} - \mathbf{r}'|)}{|\mathbf{r} - \mathbf{r}'|} \quad . \tag{2.97}$$

For an electron in a Coulomb field, the Green's function has a much more complicated structure and is expressed in terms of the product of two confluent hypergeometric functions [2.13].

The three-photon matrix element can easily be found using (2.43) and (2.93):

$$V_{kn}^{(3)}(\omega) = -\frac{1}{8} \iiint \phi_k^*(\mathbf{r}')V(\mathbf{r}')G_\omega(\mathbf{r}',\mathbf{r}'')V(\mathbf{r}'')G_\omega(\mathbf{r}'',\mathbf{r})$$

$$\times V(\mathbf{r})\phi_n(\mathbf{r})d\mathbf{r}\ d\mathbf{r}'\ d\mathbf{r}'' \quad . \tag{2.98}$$

The corresponding diagramm describing a three-photon ionization is

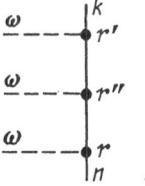

On the basis of (2.98) it is easy to see how the K-photon matrix element can be found: it is the product of K-1 Green's functions.

Naturally, even when Green's functions can be expressed in an analytical form, multiple integrals such as those found in (2.92) or (2.98) are calculated with the help of computers. The practical advantage or disadvantage of (2.92) over (2.23) is determined by the speed with which the integrals converge with respect to $\mathbf{r}$ in (2.92) or the sums converge with respect to m in (2.23). No definite conclusions can be drawn a priori, but practice has shown the obvious advantages of the method using the Green's function [2.14].

2.6.3 Realistic Atomic Potentials

Multi-photon transitions are usually calculated for cases in which the Green's function can be expressed analytically. Practically speaking, this means that one is dealing with Green's function of the Coulomb type.

First of all alkali atoms will be considered, in which one valence electron is surrounded by the field of the Coulomb atomic core. At small distances the potential differs from the Coulomb potential because of the influence of the atomic core. In the zeroth approximation one considers the interaction of the valence electron with the core as a whole, the interaction being of the $-1/r$ type. In the first-order approximation one must take into account the polarization of the core by the valence electron. The result is an interaction, proportional to $1/r^2$, between the valence electron and the dipole moment. The proportionality factor is dependent upon the orbital angular momentum of the electron, l, because l determines the degree of oblateness of the electron cloud and thus the strength of the interaction between the electron cloud and the dipole moment. This factor is labelled α_l. Thus one arrives at a model potential for which the Green function is reduced to the Coulomb-Green's function because the effective potential for a Coulomb interaction is $-1/r + l(l+1)/2r^2$ (the second term is the centrifugal potential). The model-potential approach yields a simple method for making valence-electron calculations [2.15]. However, in reality this approach does not correctly describe the potential near the atomic core. The reason is that, when the valence electron is near the atomic core, the next term in the multi-pole expansion of the electromagnetic interaction contains the dipole-dipole interaction. This interaction is proportional to $1/r^3$ and, hence, not small compared to the above-mentioned types of interaction. If one treats the proportionality factor, α_l, as a phenomenological constant, one can consider the model-potential approach to be a phenomenological method. The method is justified for highly excited states. In such states the valence electron spends only a small amount of time near the atomic core. Hence, the part played by the atomic core is small. For this reason α_l is also small and obviously the next term in the expansion, which describes the dipole-dipole interaction, will be considerably smaller.

Another approach would be to assume that the potential is a Coulomb potential and to assume that the energies $E_n^{(0)}$ were not determined from the Schrödinger equation but were determined phenomenologically, i.e., from the experimentally determined positions of the levels. This is equivalent to assuming that in the Coulomb formula for the energy levels the values of the principal quantum number n are not integral but are shifted by a quantity called the quantum defect. For this reason the method is known as the quantum defect method [2.16]. The quantum defect is also dependent upon the orbital angular momentum l (it becomes smaller as l increases). The reason is that for large values of l the electron cannot, due to the centrifugal barrier, "reach" into the inner regions of the atom and "feel" the potential of the atomic core.

Thus, one can conclude that both approaches, the model potential and the quantum defect, increase in effectiveness as the quantum numbers of the studied atomic states increase. With respect to precision, preference cannot be given to either approach. Both have found wide use in numerous calculations in higher-order time-dependent perturbation theory using the Green's function approach [2.14].

3. The Resonance Approximation

As has already been noted in Chap.2, time-dependent perturbation theory does not work when the denominators in the expressions for the matrix elements become small, i.e., $\omega_{kn} \approx K\omega$. In this case the resonance approximation is applicable. However, it must be ensured that

$$\Delta_K \equiv \omega_{kn} - K\omega \ll \omega \quad , \tag{3.1}$$

which implies that the detuning from resonance is much smaller than the frequency of the perturbing field. If the above condition is met, it becomes considerably simpler to describe the interaction between the light and an atom than if perturbation theory is used. Indeed, of the infinite sums given in (2.23), one can select only one term for which $m = k$ and neglect all of the others.

It is important to note that the above-mentioned condition is much less stringent than the condition for realizing resonance: $\Delta_K = \omega_{kn} - K\omega < \Gamma_{kn}$, where Γ_{kn} is the reduced width of the resonance transition, determined by the widths (natural or stimulated) of the resonance states k and n. For this reason there can also appear a detuning from resonance, Δ_K, which together with its width, Γ_{kn}, is one of the main characteristics of the resonance process.

In accordance with the assumptions of the resonance approximation, all of the calculations can be considerably simplified if only the term $(1/2)V^{(1)} \exp(-i\omega t)$ is taken into account in the perturbation $V^{(1)} \cos\omega t$ (this term leads to a small denominator) while the term $(1/2)V^{(1)} \exp(i\omega t)$ is neglected (this term does not lead to a small denominator). For example, when $\omega_{kn} \approx \omega$ the first large term in (2.7) must be considered and the second term neglected. By doing so the slowly oscillating terms of the amplitude are isolated while the rapidly oscillating terms are neglected.

However, one cannot restrict oneself exclusively to first-order perturbation theory in the formulas for the amplitudes, because in higher orders there are also terms with very small denominators. Nevertheless, the summation of figures for all orders of perturbation theory is considerably simpler than in the general case because only a fraction of the figures has to be accounted for. The actual quantum mechanical solution to resonance is simpler if one does not sum up an infinite number of "large" figures (i.e., those which correspond to large terms) but solves exactly the system of equations for the amplitudes in which the exponential in the sinusoidal perturbation can be neglected. Such a solution can be obtained in different ways, depending on the system under consideration.

In this chapter a variety of systems will be discussed. First of all in Sect.3.1 a two-level system in a resonance field will be considered. Then multi-photon resonance will be discussed in Sect.3.2. In Sect.3.3 the case in which the levels of such a system are degenerate will be examined. Section 3.4 contains the exact solution for a two-level system in a circularly polarized electromagnetic field. In Sect.3.5 the dependence of the amplitude and frequency of the field and the separation of the levels on time is taken into account. Finally, in Sect.3.6 the behavior of a three-level system in two resonance fields, each of which is in resonance with two of the three levels, is studied. In this way all of the typical cases that can be experimentally realized are covered. Each section also contains a discussion concerning the limits within which a certain approximation can be used for solving specific problems.

Whether or not the resonance approximation can be used to describe a specific experimental situation depends a lot on whether two (or three) atomic levels can be isolated in the spectrum while all of the other levels are neglected. When an atom interacts with visible light, it is not possible, as a rule, to unambiguously isolate several levels in a spectrum. The essence of the difficulties is as follows. As a rule atomic levels are many-fold degenerate. When an external field is applied and goes into resonance with a pair of levels, the degeneracy is lifted, the levels are split, and their energies change in such a way that the states each exhibit a different detuning with respect to the frequency of the perturbing field. In each case in which the resonance approximation is used (Chap.8) it will be discussed how the resonance levels in the spectrum can be isolated.

3.1 A Two-Level System in a Resonance Field

To start off, with the case in which a perturbation of frequency ω acts on system with two nondegenerate levels and the frequency of the transition between the levels, ω_{kn}, is close to ω, i.e., the detuning $\Delta = \omega_{kn} - \omega \ll \omega$, will be discussed.

A two-level system is essentially a quantum mechanical system because in classical physics energy is a continuous variable. Often, however, one can use the classical point of view if the two-level system is approximated by a classical electric dipole whose moment, according to the correspondence principle, is equal to the matrix element of the system, d_{kn}, and whose frequency is equal to ω_{kn}. The external monochromatic field induces an oscillation of the dipole with the frequency of the field ω. As a result the dipole moment becomes dependent upon time and, according to classical field theory, the dipole starts to emit electromagnetic waves at the frequency ω. If the frequency of the forced oscillations is almost equal to the frequency of the natural oscillations, one has classical resonance. Classical physics predicts that in the neighborhood of a resonance the amplitude of the oscillations is infinitely great (disregarding damping effects). Quantum physics, on the other hand, dealing as it does with probabilities, leads to a correct solution to the problem, i.e., makes it possible to avoid infinite amplitudes. From the physical point of view this fact is related to the quantum mechanical uncertainty relationship between the detuning and the observation time. Hence, the problem becomes an essentially quantum mechanical problem in the neighborhood of a resonance.

3.1.1 Wave Functions

The simplest case is one-photon resonance, i.e., when $K = 1$ and $\omega_{kn} \approx \omega$ [3.1]. For two levels, (2.1) assumes the following form:

$$i\dot{a}_n = V_{nk} a_k \exp(i\omega_{nk} t) \quad ,$$

$$i\dot{a}_k = V_{kn} a_n \exp(i\omega_{kn} t) \quad . \tag{3.2}$$

As in the previous chapter, here $V(\mathbf{r},t) = V^{(1)}(\mathbf{r}) \cos\omega t$ and $V^{(1)} = \mathbf{r} \cdot E = zE$. It is also assumed that $\omega > 0$.

The system of equations (3.2) may be reduced to Whittaker's equation [3.2]. This equation is a particular case of Hill's equation [3.3] in which the zeroth, first and second harmonics are incorporated in the coefficient. Mathieu's equation [3.3] only contains the zeroth and first harmonics. Thus Whittaker's equation is much more complicated than Mathieu's and for this reason has been studied much less.

Solving (3.2) when the resonance approximation is applied is much simpler. According to the rule discussed at the beginning of this chapter, $\cos\omega t$ is replaced by (1/2) exp(iωt) in the first equation of (3.2) and by (1/2) exp(-iωt) in the second. The coefficients in (3.2) will then only have slowly oscillating exponentials, which yield small denominators in the perturbation expansions. By introducing the detuning from resonance, $\Delta = \omega_{kn} - \omega$, into (3.2) one gets

$$i\dot{a}_n = \frac{1}{2} V_{nk}^{(1)} a_k \exp(-i\Delta t) \quad ,$$

$$i\dot{a}_k = \frac{1}{2} V_{kn}^{(1)} a_n \exp(i\Delta t) \quad . \tag{3.3}$$

Using the diagrammatic language (Sect.2.3), one can find the solution to (3.3) by summing up an infinite number of diagrams for different orders of perturbation theory. These diagrams consist of "elementary" segments of the type,

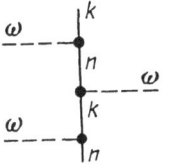

that are linked together. A typical such sum is:

$$\begin{array}{c}\end{array}$$

However, it has already been said that it is simpler to solve (3.3) exactly in analytical form. One can write the general solution to the Schrödinger equation in the form $\Psi = C_n \psi^{(n)} + C_k \psi^{(k)}$, where C_n and C_k are arbitrary constants determined from the initial conditions, and $\psi^{(n)}$ and $\psi^{(k)}$ are the orthonormal basis states of the system:

$$\psi^{(n)} = \left[\frac{1}{2}\left(1 + \frac{\Delta}{\Omega}\right)\right]^{\frac{1}{2}} \exp\left[i\left(\Omega - \frac{\Delta}{2}\right)t\right]\psi_n^{(0)}(t)$$

$$- S\left[\frac{1}{2}\left(1 - \frac{\Delta}{\Omega}\right)\right]^{\frac{1}{2}} \exp\left[i\left(\Omega + \frac{\Delta}{2}\right)t\right]\psi_k^{(0)}(t) \quad , \tag{3.4}$$

$$\psi^{(k)} = S^*\left[\frac{1}{2}\left(1 - \frac{\Delta}{\Omega}\right)\right]^{\frac{1}{2}} \exp\left[-i\left(\Omega + \frac{\Delta}{2}\right)t\right]\psi_n^{(0)}(t)$$

$$+ \left[\frac{1}{2}\left(1 + \frac{\Delta}{\Omega}\right)\right]^{\frac{1}{2}} \exp\left[i\left(-\Omega + \frac{\Delta}{2}\right)t\right]\psi_k^{(0)}(t) \quad ,$$

where

$$\Omega = \frac{1}{2}\left[\Delta^2 + |V_{nk}^{(1)}|^2\right]^{\frac{1}{2}} = \frac{1}{2}\left[\Delta^2 + z_{nk}^2 E^2\right]^{\frac{1}{2}} \quad , \tag{3.5}$$

is the Rabi frequency [3.4], $S = V_{nk}^{(1)}/|V_{nk}^{(1)}|$, and $\psi_n^{(0)}(t)$ and $\psi_k^{(0)}(t)$ are the unperturbed wave functions corresponding to the levels n and k. From (3.5) one can see that in exact resonance, i.e., when $\Delta = 0$, the Rabi frequency on resonance,

$$\Omega_{res} = \frac{1}{2} |V_{nk}^{(1)}| = \frac{1}{2} |z_{nk}|E \quad , \tag{3.6}$$

is determined by the perturbation of the states n and k in the field. It also gives the width of the resonance. From (3.4) it follows that each of the states $\psi^{(n)}$ and $\psi^{(k)}$ can be given in terms of the two initial states, $\psi_n^{(0)}(t)$ and $\psi_k^{(0)}(t)$, (without the field) using different weights. To find the constants C_n and C_k one defines the initial conditions. These, in turn, are dependent upon the relation between the time for switch-on and the detuning from resonance.

3.1.2 Criteria of Applicability

When can the resonance approximation be used? Obviously, the criteria of applicability are equivalent to the conditions under which the amplitudes slowly oscillate with time. As (3.4) shows, the solutions oscillate with the frequencies Δ and ω. Slow oscillations are oscillations whose frequencies are small compared to atomic frequencies ω_{kn}. If one substitutes (3.5) for Ω, one finds that the amplitudes $a_n(t)$ and $a_k(t)$ are slowly varying functions of time provided that the following conditions are fulfilled:

$$\Delta \ll \omega_{kn} \quad , \qquad |V_{kn}^{(1)}| = |z_{nk}|E \ll \omega_{kn} \quad , \tag{3.7}$$

i.e., both the detuning from resonance and the perturbation should be small compared to the difference between the resonating levels, while the ratio of the detuning to the width may take on arbitrary values.

If one expresses the second condition in terms of atomic quantities, then

$$E \ll \frac{\omega_{kn}}{\omega_{at}} E_{at} \quad , \tag{3.8}$$

where $E_{at} = 5 \times 10^9$ V/cm and $\omega_{at} = 2$ Rydbergs. Actually, in the frequency range of visible light, $\omega_{kn} \ll \omega_{at}$ and the ratio given in (3.8) is about ten or even more. Hence, the second condition in (3.7) implies that the electric field of the light wave must be very small compared to the atomic field. This is also the case in perturbation theory (Chap.2). If one is dealing with highly excited atomic states, the dipole matrix elements, z_{kn}, become very large, because $z_{kn} \propto n^2$, where n is the principal quantum number of a given level. The

second criterion in (3.7) is thus replaced by a criterion that is considerably more stringent:

$$E \ll \frac{\omega_{kn}}{\omega_{at}} \frac{1}{n^2} E_{at} \quad . \tag{3.9}$$

For the resonance approximation to work it must also be assumed that the
admixture of other levels to the states $\psi^{(n)}$ and $\psi^{(k)}$ (due to the influence
of the external field) is small. If, for example, there exists a third level
p near the level k, then, in first-order perturbation theory, its mixing with
the states $\psi^{(n)}$ and $\psi^{(k)}$ is proportional to the ratio $V_{pn}^{(1)}/(\omega_{pn} - \omega)$. This
implies that, from the viewpoint of the third level, a two-level system can
only be realized if

$$V_{pn}^{(1)} = z_{pn} E \ll |\omega_{pn} - \omega| = \Delta_{pn} \quad . \tag{3.10}$$

In the above analysis it has been assumed that the diagonal matrix ele-
ments of the perturbation vanish, i.e., $V_{nn}^{(1)} = V_{kk}^{(1)} = 0$. If they are nonzero,
then one must add $V_{nn}^{(1)} a_n \cos\omega t$ to the right-hand side of the first equation
in (3.2), and add $V_{kk}^{(1)} a_k \cos\omega t$ to the right-hand side of the second equation.
When the resonance approximation is used (3.2), these additional terms vanish because they do not contain slowly oscillating terms. It is for this
reason that in the one-photon case there are no first-order corrections in
V due to the detuning from resonance, Δ. In other words, the diagonal matrix
elements have no influence on the functions $\psi^{(n)}$ and $\psi^{(k)}$. In the multi-
photon case the situation is quite different (Sect.3.2).

3.1.3 Adiabatic Introduction of Perturbations

To find C_n and C_k, the initial conditions are fixed. They are determined by
the manner in which the perturbation is introduced. First of all the case in
which the perturbation is introduced adiabatically will be considered. As
$t \to -\infty$, $V \to 0$, and (3.4) yields

$$\psi^{(n)} \to \psi_n^{(0)} \quad , \quad \psi^{(k)} \to \psi_k^{(0)} \quad \text{if} \quad \Delta > 0 \quad ; \tag{3.11}$$

$$\psi^{(n)} \to -S\psi_k^{(0)} \quad , \quad \psi^{(k)} \to S^*\psi_n^{(0)} \quad \text{if} \quad \Delta < 0 \quad . \tag{3.12}$$

If the detuning from resonance, Δ, is positive and the system was in the
state n before the perturbation, then $C_n = 1$, $C_k = 0$, and the wave function of
the system is given by $\Psi = \psi^{(n)}$. If, however, as $t \to -\infty$ the system was in the
state k, then $C_n = 0$, $C_k = 1$, and $\Psi = \psi^{(k)}$. The case for which $\Delta < 0$ can be studied

in a similar manner. Combining the two cases, the probability of finding the system in the states n or k is

$$W_n = \frac{1}{2}\left(1 + \frac{|\Delta|}{2\Omega}\right) \quad , \quad W_k = \frac{1}{2}\left(1 - \frac{|\Delta|}{2\Omega}\right) . \tag{3.13}$$

When the perturbation is switched on adiabatically, the probability of finding the system in a particular state is a constant and does not depend on time.

Let us consider the extreme cases for the above formulas. If $|V_{nk}^{(1)}| \ll \Delta$, then according to (3.4) the amplitude of the transition to the state k is

$$a_k(t) = -\frac{V_{kn}^{(1)}}{2(\omega_{kn} - \omega)} \exp[i(\omega_{kn} - \omega)t] \quad , \tag{3.14}$$

which, obviously, coincides with that predicted by first-order perturbation theory, see (2.7). On the other hand, if $|V_{nk}^{(1)}| \gg \Delta$, the probability of finding the system in the states n or k is the same and equal to 1/2, i.e., the resonating states mix strongly (sometimes called saturation). For the given initial condition, $\Psi \xrightarrow[t \to -\infty]{} \Psi_n^{(0)}$, it is always true that $W_n \geq W_k$.

3.1.4 Sudden Perturbations

The opposite case of a sudden perturbation shall now be considered. At time zero it shall be assumed that the system is in the state n, i.e., $\Psi(0) = \Psi_n^{(0)}$. Substituting the functions $\Psi^{(n)}$ and $\Psi^{(k)}$ at $t = 0$ into $\Psi = C_n\Psi^{(n)} + C_k\Psi^{(k)}$ yields

$$C_n = \left[\frac{1}{2}\left(1 + \frac{\Delta}{2\Omega}\right)\right]^{\frac{1}{2}} \quad \text{and} \quad C_k = S\left[\frac{1}{2}\left(1 - \frac{\Delta}{2\Omega}\right)\right]^{\frac{1}{2}} . \tag{3.15}$$

The wave function of the system has the following form [3.1],

$$\Psi = \left(\cos\Omega t + \frac{i\Delta}{2\Omega} \sin\Omega t\right)\left[\exp\left(-\frac{i\Delta}{2} t\right)\right]\Psi_n^{(0)}(t)$$

$$- i \frac{V_{nk}^{(1)}}{2\Omega} \sin\Omega t\left[\exp\left(\frac{i\Delta}{2} t\right)\right]\Psi_k^{(0)}(t) \quad . \tag{3.16}$$

The probability of finding the system in the state n or k is

$$W_n = \cos^2\Omega t + \frac{\Delta^2}{4\Omega^2} \sin^2\Omega t \quad , \quad W_k = \frac{|V_{nk}^{(1)}|^2}{4\Omega^2} \sin^2\Omega t . \tag{3.17}$$

Both probabilities oscillate with time which was not the case for the adiabatic switch-on. These oscillations are not due to the periodicity of the perturbation after the switch-on but are due to the manner of the switch-on. They appear because of the presence of Fourier harmonics in the expansion of a perturbation that was zero before $t = 0$ and sinusoidal after $t = 0$.

From (3.16) it follows that, when $|V_{nk}^{(1)}| \gg \Delta$, the average probability of finding an electron in either level approaches the value 1/2. It is easy to see that, when $|V_{nk}^{(1)}| \ll \Delta$, we obtain (2.12) of first-order perturbation theory from (3.6). The limiting expression (2.12) can be derived for all values of $|V_{nk}^{(1)}|$ and Δ, but only if the perturbation time T is sufficiently small, i.e., $\Omega T \ll 1$.

Hence, the formulas which were derived using the resonance approximation correctly describe a two-level system for any perturbation time, $\Omega T > 1$ or < 1. One can only speak of resonance when $\Omega T \gg 1$, because when $\Omega T \ll 1$ the resonance denominator Ω vanishes, as can be seen in (3.17). For small perturbation times, $T \ll \Omega^{-1}$, no resonance occurs. In experiments with picosecond lasers this condition for one-photon resonance is only met by very weak fields, i.e., $E \ll 10^3 \text{V/cm}$.

3.1.5 Adiabatic or Sudden Perturbation?

An analytical criterion that will make it possible to determine whether the field is switched on suddenly or adiabatically will be developed. To this end it should be noted that the system "atom plus field" is quasi-degenerate with a level separation equal to Δ. In the adiabatic switch-on mode the initial state of the unperturbed system changes dynamically into one of the states of the system "atom plus field", i.e., into the state $\psi^{(n)}$ or $\psi^{(k)}$. However, when the perturbation is sudden, there appears a superposition of the two states $\psi^{(n)}$ and $\psi^{(k)}$. According to the uncertainty principle, the characteristic time required for transitions between these levels is of the order $1/\Delta$. The time needed for the perturbation to reach its maximum value (switch-on time) is denoted by δT. If $1/\Delta$ is large compared to δT, then no transitions between the levels of the system have time to take place during the switch-on time. Hence, the two-level system behaves as if the external field had been switched on at a definite time, e.g., at $t = 0$. Thus, if

$$\Delta \delta T \ll 1 \quad , \tag{3.18}$$

the field was switched on suddenly. The opposite condition implies that the field was switched on adiabatically. If $\Delta \delta T \sim 1$, we have an intermediate situation. To prove the above results rigorously, one must postulate a definite time dependence for E. An analytical solution [3.5] has been obtained for

$$E(t) \propto 1 + \tanh \frac{t}{\delta T} \quad . \qquad (3.19)$$

A change from $\delta T \rightarrow 0$ to $\delta T \rightarrow \infty$ corresponds to a transition from a sudden to an adiabatic switch-on. The analytical solution confirms the validity of (3.18).

In conclusion one can say that if $\Delta \gg |V_{nk}^{(1)}|$, then the adiabatic regime can always be realized with any feasible pulsed laser mode. In the case of resonance, i.e., when $\Delta \lesssim V_{nk}^{(1)}$, the situation is different. For typical detuning, $\Delta \sim V_{nk}^{(1)} \sim \gamma \sim 10^8 s^{-1}$, the sudden switch-on approximation can be realized because a typical switch-on time for pulsed lasers is $\delta T < T \lesssim 10^{-8} s$ (Chap.5).

3.1.6 Quasi-Energies in the Resonance Case

It will be remembered that the wave functions $\psi^{(n)}$ and $\psi^{(k)}$ are superpositions of two new stationary states. The energies of these two states are slightly shifted with respect to the unperturbed energies,

$$E_n = E_n^{(0)} - \Omega + \frac{1}{2} \Delta \quad , \qquad E_k = E_k^{(0)} - \Omega - \frac{1}{2} \Delta \quad , \qquad (3.20)$$

where E_n and E_k are called quasi-energies [3.6] (details in Chap.1). Their separation, $E_k - E_n$, differs from the separation, $E_k^{(0)} - E_n^{(0)}$, and is equal to $\omega_{kn} - \Delta = \omega$. This is a corollary of the Floquet theorem for quasi-energies in a periodic field [3.3] (see Chap.1).

Let us again return to the general solution, $\Psi = C_n \psi^{(n)} + C_k \psi^{(k)}$. The function Ψ, as shown in (3.4), is a linear combination of four stationary states. The quasi-energies of these states are respectively (Fig.3.1):

$$E_n^- = E_n^{(0)} + \frac{1}{2} \Delta - \Omega \quad , \qquad E_n^+ = E_n^{(0)} + \frac{1}{2} \Delta + \Omega \quad ,$$

$$E_k^- = E_k^{(0)} - \frac{1}{2} \Delta - \Omega \quad , \qquad E_k^+ = E_k^{(0)} - \frac{1}{2} \Delta + \Omega \quad . \qquad (3.21)$$

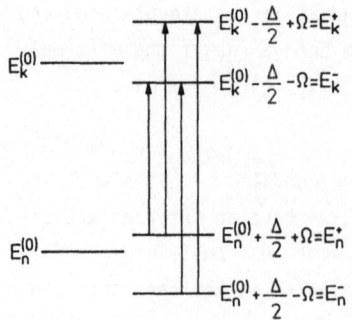

$E_k^{(0)} - \frac{\Delta}{2} + \Omega = E_k^+$

$E_k^{(0)}$

$E_k^{(0)} - \frac{\Delta}{2} - \Omega = E_k^-$

$E_n^{(0)} + \frac{\Delta}{2} + \Omega = E_n^+$

$E_n^{(0)}$

$E_n^{(0)} + \frac{\Delta}{2} - \Omega = E_n^-$

Fig.3.1. Quasi-energies of the four states whose linear combination yields the wave function for a two-level system in a resonance field

One can qualitatively explain the two-fold splitting of the levels by assuming that $\Delta = 0$. This splitting is due to the interaction of the two-level system with an external field. In the zeroth approximation the energy of the upper level is equal to the energy of the lower level plus the energy of one excitation quantum. The interaction between the field and the system removes this degeneracy. When $\Delta = 0$ the frequencies of the transitions between the states $E_n^{\pm}$ and $E_k^{\pm}$ are:

$$\omega_1 = \omega_{kn} - V_{kn}^{(1)} \quad , \quad \omega_2 = \omega_{kn} \quad , \quad \omega_3 = \omega_{kn} + V_{kn}^{(1)} \quad . \tag{3.22}$$

If the perturbation is introduced suddenly, then according to (3.16) the probability of finding the system in either of the levels E_k^+ and E_k^- (depicted in Fig.3.1) is the same. Such a splitting does not take place in the case of an adiabatic perturbation because ψ is equal to $\psi^{(n)}$ or $\psi^{(k)}$, i.e., is a superposition of two stationary states. Only when the perturbation is sudden does splitting occur. As shown in (3.21), the double Rabi frequency is equal to the splitting of each of the levels k and n. In a weak field $(z_{kn}E \ll \Delta)$ the splitting is proportional to the square of E. On the other hand, in a strong field $(z_{kn}E \gg \Delta)$, it is directly proportional to E:

$$\Omega = \Omega_{res} = \frac{1}{2} z_{kn} E \quad . \tag{3.23}$$

This effect of level splitting, which was first experimentally observed in the perturbation of molecular spectra by UHF radiation, has also been noted in the perturbation of atomic spectra by a field of light. The different manifestations of resonance splitting are examined in Chap.8.

3.2 Multi-Photon Resonance

The results of Sect.3.1, which describe the dynamics of a two-level system in an external field, will now be generalized to incorporate multi-photon resonance, i.e., the case in which the separation between levels is close to an integral multiple of the frequency of the external field. Naturally, such resonances occur in fields whose intensities are greater than those required to produce a one-photon resonance. Hence, the reason for studying multi-photon resonances is connected with the appearance of sources of intense light.

3.2.1 Two-Photon Resonance

For two-photon resonance to occur, the levels must have the same parity. To construct a two-photon matrix element that represents a transition, it is not sufficient to only consider the two given levels. It must be assumed that the atomic system has other levels, p, for which the transitions from the initial state to these levels and from these levels to the final state are allowed. Such transitions shall be considered to be virtual. It follows from Sect.2.2 that, when $2\omega \approx \omega_{kn}$, (3.3) can be modified in the following way:

$$i\dot{a}_n = V_{nk}^{(2)} a_k \exp(-i\Delta_2 t) + V_{nn}^{(2)} a_n \quad ,$$

$$i\dot{a}_k = V_{kn}^{(2)} a_n \exp(i\Delta_2 t) + V_{kk}^{(2)} a_k \quad ,$$

(3.24)

where $\Delta_2 \equiv \omega_{kn} - 2\omega$ is the detuning from the two-photon resonance, and the two-photon matrix element, $V_{nk}^{(2)}$, was defined in (2.23).

Diagonal terms appear in (3.24) because, in contrast to the one-photon case, they are slowly oscillating functions of time. Indeed, the diagonal one-photon matrix element of the perturbation, $V_{nn}^{(1)} \cos\omega t$, oscillates rapidly, whereas the diagonal two-photon matrix element is roughly proportional to $\cos^2\omega t$ and hence contains a nonoscillating part. More precisely, the non-oscillating diagonal matrix elements, $V_{nn}^{(2)}$ and $V_{kk}^{(2)}$, originate from the first and fourth terms in (2.22) when $k = n$, e.g.,

$$V_{nn}^{(2)} = -\sum_p \frac{1}{4} |z_{np}|^2 \left(\frac{1}{\omega_{pn} - \omega} + \frac{1}{\omega_{pn} + \omega}\right) E^2 \quad .$$

(3.25)

Equation (3.24) can be described exactly by an infinite sum of diagrams that consist of two kinds of segments (two-photon matrix elements) (Sect.2.3):

which represent the resonant part of the transition between n and k, while

represent the self-energy parts.

One can find a solution to (3.24) in the same way as was done for (3.3) if one first of all changes variables, i.e., $a_i \to a_i \exp(-iV_{ii}^{(2)}t)$, and then eliminates the terms $V_{ii}^{(2)}$ ($i = n,k$) by introducing certain additional terms to the energy eigenvalues. As a result one obtains the functions $\psi^{(n)}$ and $\psi^{(k)}$ which are the same functions as those in (3.4) except for the substitutions

$$\Delta \to \Delta_2' = \omega_{kn} - 2\omega + V_{kk}^{(2)} - V_{nn}^{(2)} \tag{3.26}$$

and

$$\Omega \to \Omega_2 = \frac{1}{2} [(\Delta_2')^2 + 4|z_{kn}^{(2)}|^2 E^4]^{\frac{1}{2}} \quad , \tag{3.27}$$

where Ω_2 is the two-photon Rabi frequency [3.7].

Comparing these parameters with the analogous ones for the one-photon case, one sees that the energy levels have shifted: $E_n^{(0)} \to E_n^{(0)} + V_{nn}^{(2)}$, $E_k^{(0)} \to E_k^{(0)} + V_{kk}^{(2)}$. Hence, $V_{nn}^{(2)}$ and $V_{kk}^{(2)}$ are the dynamic polarizabilities (Chap.7) of states n and k respectively.

A similar shift is observed in one-photon resonance. It is known as the Bloch-Siegert frequency shift (Chap.8). This effect is taken into account in the second- and higher-order terms in the expansion of the field strength, whereas the main effect manifests itself in the first-order term. Hence, within the framework of the resonance approximation the Bloch-Siegert shift is totally absent. In the exact solution it appears as a small correction to the main effect.

One must not mistake the shift of the unperturbed levels in an external field, i.e., the dynamic polarizability, to be the difference between energy and quasi-energy levels. A shift in levels can be observed in any constant (the Stark effect) or variable (dynamic polarizability) field, whereas the concept of a quasi-energy level only has meaning for a periodic perturbation. This concept is useful in describing a perturbed system in a resonance field in the common language of quasi-stationary states.

The width of the resonance in the two-photon case is $|z_{nk}^{(2)}E^2|$, which is much less than the width in the one-photon case, $|z_{nk}E|$.

The distance between the quasi-energy levels shall be defined in the same way as was done in the one-photon case:

$$E_k^+ - E_n^+ = \left(E_k^{(0)} + \Omega_2 - \frac{1}{2}\Delta_2'\right) - \left(E_n^{(0)} + \Omega_2 + \frac{1}{2}\Delta_2'\right) = 2\omega \quad . \tag{3.28}$$

It is equal to twice the frequency of the external field, which, as in one-photon resonance, is a direct consequence of the Floquet theorem (Chap.1).

The probability of finding the system in either states n or k when the perturbation is introduced suddenly or adiabatically slowly is given by the formulas of Sect.3.1 in which Δ_2 and Ω_2 have been substituted for Δ and Ω, respectively. All of the conditions concerning the mode of introducing the perturbation for the one-photon case can be applied, provided that Δ_2' is substituted for Δ (Sect.3.1.5). Since in resonance $\Delta_2' \sim E^2$, and in one-photon resonance $\Delta \sim E$, the typical values for a two-photon detuning from resonance, Δ_2', are small compared to those for a one-photon detuning, Δ. In other words, one can say that the range of two-photon resonance is considerably narrower than that of one-photon resonance. As a result, the range of the parameters that characterize a sudden external perturbation is broader. When $|V_{nk}^{(2)}| \ll |\Delta_2'|$, one can use perturbation theory, and for the probability of certain transitions one can find formulas similar to (2.24). However, if $|V_{nk}^{(2)}| \gg |\Delta_2'|$, the probability of finding the system in either state is the same, just as it was for one-photon resonance (regardless of how the perturbation is introduced).

3.2.2 Criteria of Applicability

What are the criteria for the resonance approximation to be valid in the two-photon case? The second condition in (3.9) is modified in the following way to incorporate this case:

$$|z_{nk}^{(2)}| E^2 \ll \omega_{kn} \quad . \tag{3.29}$$

In essence, this condition says that if the perturbation is of the order of the distance between the levels under consideration, the resonance approximation is no longer valid. The condition in (3.29) is less demanding than the corresponding condition in (3.7) for the one-photon case.

In a similar manner, the first condition in (3.7) is modified so that

$$\Delta_2' \ll \omega_{kn} \quad . \tag{3.30}$$

Finally, if there exists a third level, p, close to the two other levels, the third level must not contribute much to the wave function. A most dangerous situation arises when the level p lies between the levels k and n. The two-photon matrix element, $z_{kn}^{(2)}$, then increases sharply in resonance. When

$$z_{pn} E \ll |\omega_{pn} - \omega| \quad , \tag{3.31}$$

the above approach is correct. This is not so for a resonant increase of the diagonal matrix elements, $z_{nn}^{(2)}$ and $z_{kk}^{(2)}$, which enter (3.24). This can occur if the level p lies between levels n and k, as in the above situation, or if p lies above k or below n by an amount close to ω. Under such circumstances the problem becomes much more complex because the two-level approximation is no longer valid. The inclusion of the third level in the discussed resonance situations shall be considered in Sect.3.6.

Let us stress, in conclusion, that in the framework of the criteria in (3.29,30) the ratio of the detuning from two-photon resonance, Δ_2', to the width of the resonance, $z_{nk}^{(2)}E^2$, may be more or less than unity.

3.2.3 Multi-Photon Resonance

Let us now examine a resonance in a two-level system that is initiated by an arbitrary number of photons K of a monochromatic perturbation. In Sect.2.4 it was shown that when $\omega_{kn} \approx K\omega$ all of the above-mentioned results for the two-photon case remain valid if $V_{nk}^{(K)}$ is substituted for $V_{nk}^{(2)}$, see (2.45), and $\Delta_K' = \omega_{kn} - K\omega + \delta E_{kn}^{(K)}$ is substituted for Δ_2', where $\delta E_{kn}^{(K)}$ is the difference between the shifts of the levels k and n in the external field. Since the width of the resonance is of the Kth order in the perturbation, the detuning from resonance must also be known to within K terms. The calculation of the energy shifts for different orders of perturbation theory will be illustrated in Sect.7.2. Here, it is sufficient to know that, if K is even, the value of $\delta E_{kn}^{(K)}$ has to be known to within K terms, while, if K is odd, it has to be known to within the (K-1) terms.

One of the criteria of multi-photon resonance generalizes the second criterion in (3.7) and the criterion in (3.29):

$$z_{kn}^{(K)}E^K \ll \omega_{kn} \quad . \tag{3.32}$$

For resonances involving even harmonics (K is even), as well as for two-photon resonances, to occur, other nonresonant levels must be present to ensure that the multi-photon matrix element does not vanish. In addition the levels n and k must have the same parity. For resonances involving odd harmonics (K is odd) to occur there is no need for other levels to be present since the levels n and k must have different parities. A multi-photon matrix element can thus be constructed using only the two levels n and k.

In order to formulate a criterion similar to the first one in (3.7) and (3.30), let us examine a three-photon resonance. The criterion that $\Delta_K' \ll \omega_{kn}$ is undoubtedly necessary but not sufficient. In the framework of third-order perturbation theory the states n and k can mix resonantly. Moreover, nonresonant transitions between the two levels can occur in the framework of first-order perturbation theory. For the three-photon resonance mixing to be dominant, it must be true that

$$\frac{z_{nk}^{(3)}E^3}{\Delta_3^!} \gg \frac{z_{nk}E}{\omega_{kn}} \quad , \tag{3.33}$$

or expressed differently,

$$\Delta_3^! \ll \frac{E^2}{E_{at}^2} \omega_{kn} \quad . \tag{3.34}$$

This condition is much more demanding than the one in (3.7) or(3.30). But for all practical purposes it is usually realized, especially in strong fields, because the condition $\Delta_K^! \ll \Omega_{kn}$ is easily fulfilled. There are also similar criteria for resonance processes involving the absorption of an odd number of photons. However, a similar criterion does not exist for resonance processes involving the absorption of an even number of photons. Since the states n and k have the same parity, $z_{nk} = 0$.

Finally, in contrast to a one- or two-photon resonance, an additional criterion involving the perturbation time T has to be met in order to apply the resonance approximation to a resonance involving three (or any odd number greater than three) photons. For sufficiently short perturbation times (namely, when $\Omega_3 T \ll 1$), the results of this section are simplified to the usual formulas obtaining using third-order perturbation theory. In this case the transition amplitude is $z_{nk}^{(3)}E^3 T$. For this limiting case to be applicable the amplitude must be large compared to the amplitude of a nonresonant transition in first-order perturbation theory:

$$z_{nk}^{(3)}E^3 T \gg \frac{z_{nk}E}{\omega_{nk}} \quad . \tag{3.35}$$

Hence,

$$E \gg E_{at}(\tau_{at}/T)^{\frac{1}{2}} \quad . \tag{3.36}$$

Similar criteria for resonant processes involving an even numbers of photons do not exist because $z_{nk} = 0$. If the above criterion is applied to a nanosecond laser ($T = 10^{-9}$ s), the lower limit of the field strength that still allows a three-photon mixing of levels can be calculated: $E \gg 10^5$ V/cm.

The criterion in (3.18), which determines the switch-on mode, can be generalized to include multi-photon resonance by simply substituting $\Delta_K^!$ for Δ. Since the region in which multi-photon resonance can occur is much narrower than the region in which one-photon resonances can occur, the switch-on mode is always sudden for multi-photon resonances.

3.3 Degeneracy in a Resonance Field

In this section the changes that occur when two states (of which one or both are degenerate) mix resonantly will be discussed (Sect.3.1). Also, the trivial but nevertheless often encountered case in which two two-level systems interact independently with an external resonance field will be discussed briefly.

3.3.1 Equations

The effect of a monochromatic perturbation on degenerate states was studied in Sect.2.5 in the framework of time-dependent perturbation theory. The set of equations (2.63), which determines the amplitudes a_n of the degenerate states in an external field, contains the two-photon matrix element (2.23). It is easy to see that these equations cannot be used to analyze the resonance transitions between two degenerate states when the frequency of the external field is almost equal to the separation of the levels. Indeed, in this case the two-photon matrix element (2.23) becomes infinitely large. This means that in the presence of a resonance the mixing of degenerate levels is much more intense than in the absence of a resonance. The mixing in the former case is described by first-order perturbation theory, whereas in the latter case, it is described by second-order theory.

Mathematically, resonance with degeneracy is analogous to resonance without degeneracy (Sect.3.1). A generalization of (3.3) to include the case with degeneracy is as follows [3.8]:

$$
\begin{aligned}
i\dot{a}_{ni} &= \frac{1}{2} E \exp(-i\Delta t) \sum_j <\psi^*_{ni}|z|\psi_{ki}>a_{kj} \quad , \\
i\dot{a}_{kj} &= \frac{1}{2} E \exp(i\Delta t) \sum_i <\psi^*_{kj}|z|\psi_{ni}>a_{ni} \quad .
\end{aligned}
\tag{3.37}
$$

Here, the summation with respect to i and j is over the degenerate states of the levels n and k, respectively. Only slowly oscillating terms have been retained in (3.37). Note that in (3.37) there are no terms of the type $<\psi^*_{ni}|z|\psi_{ni'}>$ or $<\psi^*_{kj}|z|\psi_{kj'}>$ which directly relate the different states of a degenerate level. This is explained by the fact that such terms oscillate rapidly.

3.3.2 Basis Solutions

A general solution to (3.37) is sought which is a superposition of the basis orthonormalized solutions for the given system (just as in the nondegenerate case). To find these orthonormalized solutions the amplitudes a_{ni} and a_{kj} are written in the following form:

$$a_{ni}(t) = A_{ni} \exp\left[i\left(\Omega - \frac{1}{2}\Delta\right)t \right] \; ,$$

$$a_{kj}(t) = A_{kj} \exp\left[i\left(\Omega + \frac{1}{2}\Delta\right)t \right] \; ,$$

(3.38)

where Ω is the Rabi frequency of a two-level system with degeneracy. The value Ω can be found by substituting (3.38) into (3.37) and by setting the determinant of the resulting set of algebraic equations equal to zero. The number of Rabi frequencies is equal to the sum of the degrees of degeneracy of the upper and lower levels. The same set of algebraic equations determines the A_{ni} and A_{kj}.

Instead of studying the general case involving a degeneracy of arbitrary degree it is expedient to start with a particular case. Let us assume that the upper level k is two-fold degenerate while the lower level n is nondegenerate. If (3.38) is substituted into (3.37) one obtains a system of equations for A_{ni} and A_{kj}. Their solution yields a set of three orthonormal basis states, which replace the two basis states given in (3.4):

$$\Psi^{(n)} = \left[\frac{1}{2}\left(1 + \frac{\Delta}{2\Omega}\right)\right]^{\frac{1}{2}} \exp\left[i\left(\Omega - \frac{\Delta}{2}\right)t \right]\Psi_n^{(0)}(t)$$

$$- \frac{\Omega}{|\Omega|}\left[\frac{1}{2}\left(1 - \frac{\Delta}{2\Omega}\right)\right]^{\frac{1}{2}} \exp\left[i\left(\Omega + \frac{\Delta}{2}\right)t \right]$$

$$\times \left[\frac{z_{nk}}{(|z_{nk}|^2 + |z_{nk'}|^2)^{\frac{1}{2}}} \Psi_k^{(0)}(t) + \frac{z_{nk'}}{(|z_{nk}|^2 + |z_{nk'}|^2)^{\frac{1}{2}}} \Psi_{k'}^{(0)}(t) \right] \; ,$$

$$\Psi^{(k)}(\Omega) = \Psi^{(n)}(-\Omega) \; ,$$

(3.39)

$$\Psi^{(k')} = \frac{z_{nk'}}{(|z_{nk}|^2 + |z_{nk'}|^2)^{\frac{1}{2}}} \Psi_k^{(0)}(t) - \frac{z_{nk}}{(|z_{nk}|^2 + |z_{nk'}|^2)^{\frac{1}{2}}} \Psi_{k'}^{(0)}(t) \; .$$

The labels k and k' refer to the two states of the upper (degenerate) level, while the Rabi frequency is given by,

$$\Omega = (1/2)[\Delta^2 + (|z_{kn}|^2 + |z_{k'n}|^2)E^2]^{\frac{1}{2}} \; .$$

(3.40)

3.3.3 Adiabatic Introduction of a Perturbation

As in Sect.3.1.3, let us fix the mode of introducing the perturbation by assuming that $\Psi \to \Psi_n^{(0)}$ as $t \to -\infty$ (the particle populates the lower level prior to the perturbation). From (3.39) one can calculate the probabilities of finding the system in a certain state. Since $\Psi = \Psi^{(n)}$ when $\Delta > 0$ and $\Psi = \Psi^{(k)}$ when $\Delta < 0$,

$$W_n = \frac{1}{2}\left(1 + \frac{|\Delta|}{2\Omega}\right) \quad ,$$

$$W_k = \frac{1}{2}\left(1 - \frac{|\Delta|}{2\Omega}\right) \frac{|V_{kn}^{(1)}|^2}{|V_{kn}^{(1)}|^2 + |V_{k'n}^{(1)}|^2} \quad , \tag{3.41}$$

$$W_{k'} = \frac{1}{2}\left(1 - \frac{|\Delta|}{2\Omega}\right) \frac{|V_{k'n}^{(1)}|^2}{|V_{kn}^{(1)}|^2 + |V_{k'n}^{(1)}|^2} \quad .$$

As in the case without degeneracy, the probabilities are constants which do not depend on time.

If (3.41) is compared to the appropriate expressions for the case without degeneracy (Sect.3.1.3), it is evident that, because the upper level is degenerate, the particles are distributed over the states in accordance with the matrix elements for the transitions into these states. The absolute probability of finding a particle in a degenerate or nondegenerate state is the same. Also, (3.39) show that degeneracy does not change the number of quasi-energies present in a variable field.

The above reasoning is also valid for the case in which the upper level exhibits an arbitrary degree of degeneracy with only minor modifications. The result is

$$\Omega = \frac{1}{2}\left(\Delta^2 + E^2 \sum_j |z_{nk_j}|^2\right)^{\frac{1}{2}} \quad , \tag{3.42}$$

$$W_{k_j} = \frac{1}{2}\left(1 - \frac{|\Delta|}{2\Omega}\right) \frac{|z_{nk_j}|^2}{\sum_j |z_{nk_j}|^2} \quad . \tag{3.43}$$

If the perturbation is introduced suddenly, Ψ is a linear combination of the three states in (3.39) and the populations of the levels depend on time. Since the expressions are cumbersome, they will not be given here.

Let it be assumed that the lower level is also degenerate. This requires a knowledge of additional initial conditions. One must state explicitly in which state (of the lower degenerate level) a particle was prior to the perturbation. Another alternative is that the particle was in any one of these states with equal probability. The resulting probabilities will obviously depend on which of these two possibilities was chosen. However, the absolute probabilities of finding a particle in the upper or lower states are always determined using the formulas of Sect.3.1.

The label s used in connection with the basis states $\psi^{(s)}$, which can be found by the above-mentioned method, see (3.39), can be considered to be a new quantum number since it is conserved when the perturbation is switched on. For example, if for the above two-fold degenerate upper level it is assumed that $\Delta > 0$, then $\psi = \psi^{(n)}$. On the other hand, if it is assumed that $\Delta < 0$, then $\psi = \psi^{(k)}$. This new quantum number replaces the old quantum number m, which labeled the initial states $\psi_m^{(0)}$ and was not conserved after the perturbation was introduced. Naturally, the number of new and old quantum numbers is the same.

3.3.4 Approximate Degeneracy

In this chapter so far, exact degeneracies have been discussed. What are the consequences if the degeneracy is approximate? This is the case in the hyperfine splitting of levels. Instead of studying all of the fine details of the problem, the case in which one of the levels, e.g., the upper level k, is a doublet and the lower level n is nondegenerate will be analyzed. All of the features of the general case can be seen in this example. The states of the doublet will be denoted k and k'. The system of three equations is

$$i\dot{a}_k = \frac{1}{2} Ez_{kn} \exp(i\Delta t) a_n \quad ,$$

$$i\dot{a}_{k'} = \frac{1}{2} Ez_{k'n} \exp(i\Delta't) a_n \quad , \tag{3.44}$$

$$i\dot{a}_n = \frac{1}{2} Ez_{nk} \exp(-i\Delta t) a_k + \frac{1}{2} Ez_{nk'} \exp(-i\Delta't) a_{k'} \quad ,$$

where

$$\Delta = \omega_{kn} - \omega \quad , \qquad \Delta' = \omega_{k'n} - \omega \quad , \tag{3.45}$$

are the detunings from resonance. Since the states k and k' have the same parity, they are not directly related (because $z_{kk'} = 0$). They are only related through the level n.

The solutions to (3.44) should have the following form:

$$a_n(t) = A_n e^{i\Omega t} \quad , \qquad a_k(t) = A_k e^{i(\Omega+\Delta)t} \quad ,$$

$$a_{k'}(t) = A_{k'} e^{i(\Omega+\Delta')t} \quad . \tag{3.46}$$

82

Thus, one obtains the following cubic equation for the Rabi frequency:

$$\Omega(\Omega + \Delta)(\Omega + \Delta') - \frac{1}{4} E^2 |z_{nk}|^2 (\Omega + \Delta') - \frac{1}{4} E^2 |z_{nk'}|^2 (\Omega + \Delta) = 0 \quad . \quad (3.47)$$

This equation has three real roots which correspond to the three quasi-energy levels in the neighborhood of n and the three quasi-energy levels in the neighborhood of the doublet k, k'.

If $\Delta = \Delta'$, one is dealing with exact degeneracies, and the eigenstates are determined by (3.39). Two of the three roots coincide. In the other limiting case, i.e., when the detuning Δ' is so great that the state k' is no longer in resonance, one arrives at a situation which was considered in Sect.3.1: two roots are in the neighborhood of state k, while the third corresponds to the unperturbed state k'.

3.3.5 Degeneracy and the Influence of Light on an Atom

What role does degeneracy play when an atom is illuminated? For linearly or circularly polarized light, only one nonzero matrix element appears in the z_{nkj} due to the selection rules with respect to the magnetic quantum number ($\Delta m = 0$ or ± 1). For this reason, the effects due to degeneracy only manifest themselves with elliptically polarized light in many-electron atoms. The hydrogen atom is an exception. Because of a degeneracy with respect to the orbital angular momentum ℓ, transitions can occur in which ℓ changes but not m. Hence, with the hydrogen atom, the degeneracy manifests itself for any type of polarized electromagnetic field. It must be stressed once more that the effects associated with a degeneracy do not affect the absolute probabilities of populating the resonating levels.

3.4 A Two-Level System in a Circularly Polarized Electromagnetic Field

It is always simpler to study the behavior of an atom in a circularly polarized field than in a linearly polarized field because the magnitude of the field strength does not change with time. In a circularly polarized field, the problem may be reduced to a stationary one by using a rotating coordinate system in which one can separate the spatial and temporal variables in the Schrödinger equation (Sect.1.3).

3.4.1 Statement of the Problem

Let a two-level system, n and k, be illuminated by a clockwise polarized field. According to the selection rules, it is necessary that the magnetic quantum number of level k exceed that of level n by unity for a transition to occur. By choosing the quantization axis z along the direction of propagation of the electromagnetic wave and by assuming that the magnetic quantum number of level n is M and that of level k is M + 1, one can write the exact quantum mechanical equations of motion (2.1) for the stationary-state Hamiltonian (1.18),

$$\hat{H}_{rot} = \hat{H}_0 - Ex - \omega\hat{L}_z \quad , \tag{3.48}$$

in a coordinate frame that rotates with E:

$$i\dot{a}_n = - Ex_{nk}a_k \exp(i\omega_{nk}t) - M\omega a_n \quad ,$$
$$i\dot{a}_k = - Ex_{kn}a_n \exp(i\omega_{kn}t) - (M + 1)\omega a_k \quad . \tag{3.49}$$

It was assumed that the operator x contains no diagonal elements and that the operator $\hat{L}_z$ contains no off-diagonal elements. By introducing new variables

$$a_n = a_n' \exp(iM\omega t) \quad , \quad a_k = a_k' \exp[i(M + 1)\omega t] \quad , \tag{3.50}$$

one obtains

$$i\dot{a}_n' = - Ex_{nk}a_k' \exp(-i\Delta t) \quad ,$$
$$i\dot{a}_k' = - Ex_{kn}a_n' \exp(i\Delta t) \quad , \tag{3.51}$$

where $\Delta = \omega_{kn} - \omega$.

It is readily seen that (3.51), which was found by solving the Schrödinger equation exactly, coincides with (3.3), which was found using the resonance approximation. However, in contrast to (3.3), (3.51) is valid for arbitrary values of E and Δ.

3.4.2 Solution and Discussion

There is no need to solve (3.51). One can simply use the solutions presented in Sect.3.1 and change $V_{nk}^{(1)}/2$ to $-Ex_{nk}$. For example, for the case in which the perturbation is suddenly switched on at time zero and the electron is in the state n, one can use (3.16) as a solution. Returning to the initial laboratory coordinate frame according to (1.16), one finally obtains

$$\Psi(t) = [\cos\Omega t + i \frac{\Delta}{2\Omega} \sin\Omega t] \exp(-i\Delta t/2) \, \Psi_n^{(0)}(t)$$
$$+ \, i(Ex_{nk}/\Omega) \sin\Omega t \, \exp(i\Delta t/2) \, \Psi_k^{(0)}(t) \quad , \tag{3.52}$$

where the Rabi frequency is

$$\Omega = \frac{1}{2} (\Delta^2 + 4E^2|x_{nk}|^2)^{\frac{1}{2}} \quad . \tag{3.53}$$

Thus, for a circularly polarized field the solutions found in the resonance approximation actually coincide with the exact solutions.

In a circularly polarized field there is no Bloch-Siegert frequency shift (Chap.8). Such a shift appears in a linearly polarized field because of the inaccuracy of the resonance approximation.

Also, under the given conditions, no multi-photon resonances similar to those studied in Sect.3.2 are possible. This can be explained by selection rules with respect to the magnetic quantum number. In the case of multi-photon resonance the magnetic quantum number must change by a quantity greater than unity. If the states n and k differ in M by more than unity, one must use other atomic levels to build a multi-photon matrix element. Then the problem, strictly speaking, no longer remains a two-level one. Hence, in contrast to the situation in a linearly polarized field, only one-photon resonance is possible when a circularly polarized field acts on a two-level system.

3.5 A Two-Level System in a Resonance Field: Time-Dependent Parameters

Up to now it has been assumed that the frequency and amplitude of the external field and the separation of the atomic levels are independent of time. Actually, the above parameters often depend on time, usually much more slowly than $\cos\omega t$, i.e., the characteristic times for the variation of these parameters are greater than the perturbation period π/ω. Under these conditions, i.e., for time intervals of the order of π/ω, the fact that the parameters depend on time does not have an effect on the solutions that were · obtained earlier for a monochromatic perturbation. However, over long intervals of time, changes in the populations of the levels as compared with the Rabi solutions accumulate. These changes are greater than the inaccuracy of the solution due to the resonance approximation. What actually happens in practice is that the radiation of a pulsed laser is used, so that one has to take into account the envelope $E(t)$ of the pulse.

3.5.1 The System of Equations

The resonance approximation can be used to write down a set of equations that describes the behavior of a two-level system in an electromagnetic field. The equations should be generalizations of (3.3) and contain a time-dependent detuning, $\Delta(t) = \omega_{kn}(t) - \omega(t)$, and a time-dependent amplitude, $E(t)$, of the perturbation:

$$
\begin{aligned}
i\dot{a}_n &= \frac{1}{2} z_{nk} E(t) a_k \exp\left(-i \int^t \Delta \, dt\right) , \\
i\dot{a}_k &= \frac{1}{2} z_{kn} E(t) a_n \exp\left(i \int^t \Delta \, dt\right) .
\end{aligned}
\tag{3.54}
$$

The fact that the characteristic times for the variation of $\Delta(t)$ and $E(t)$ are considerably greater than π/ω has been used in obtaining (3.54).

Even if a solution to (3.54) can be found, one must be sure that the corrections caused by the anti-resonance terms (which reflect the deviation from the resonance approximation) are small compared to the solution itself.

3.5.2 An Exactly Solvable Problem

Because (3.54) is so complex it is advisable to start with a problem which can be solved exactly (needless to say, in the framework of the resonance approximation). In such a problem it will be assumed that the detuning from resonance, Δ, is constant and that $E(t)$ has the form

$$
E(t) = \frac{E}{\cosh(t/T)} .
\tag{3.55}
$$

Such a function for the amplitude of the perturbation describes a perturbation that is switched on at $t = -\infty$, attains a maximum value, E, at $t = 0$, and vanishes at $t = \infty$, and whose pulse length is equal to T. It is also assumed that $T \gg 1/\omega$.

The solution to (3.54) can be found in an analytical form. If it is assumed that the electron is in the lower level, n, at $t = \infty$, then the probability of a transition to the upper level, k, while the perturbation is being applied, is [3.9]

$$
W_k = \frac{\sin^2(\pi z_{nk} ET/2)}{\cos^2(\pi \Delta T/2)} .
\tag{3.56}
$$

When $\Delta T \gg 1$ this probability is exponentially small and, ultimately, corrections due to the nonresonant nature of the perturbation begin to play a dominant role. The resonance approximation breaks down if

$$e^{-\pi \Delta T} \ll \left| \frac{z_{nk}E}{\omega_{nk}} \right| \quad . \tag{3.57}$$

The above example shows that if $\Delta T \gg 1$, the perturbation is switched on adiabatically. This is in accordance with the results of Sect.3.1. If $\Delta T \ll 1$, then

$$W_k = \sin^2(\pi z_{nk}ET/2) \quad , \tag{3.58}$$

which is a particular case of the general analytical solution to (3.54) for a zero detuning from resonance, $\Delta = 0$:

$$W_k = \sin^2 \left[\frac{1}{2} z_{nk} \int_{-\infty}^{\infty} E(t) \, dt \right] \quad . \tag{3.59}$$

Returning to (3.56), it is evident that the probabilities oscillate, periodically passing through zero at

$$z_{nk}ET = 2K \quad . \tag{3.60}$$

This is a purely quantum mechanical effect.

Moreover, from (3.56) it is evident that the probability only depends on two dimensionless ratios of the three parameters Δ, $z_{nk}E$, and T. This is a general property of (3.54), not only of the present particular problem. It is the result of introducing a dimensionless variable $\tau = t/T$.

The transition probability is greatest when $z_{nk}ET = 2K + 1$, where K is an integer. In particular, when $\Delta T \ll 1$ it may reach 100 percent. In this respect this solution differs substantially from the Rabi solution (Sect. 3.1), for which there was no population inversion. This effect is known as the adiabatic inversion of levels. The adiabatic inversion vanishes if the period of the light pulse, T, is subject to fluctuations. Then it is neces-sary to average (3.56) over T. The same situation arises when the intensity of the field fluctuates so that

$$\delta E > \frac{1}{z_{nk}T} \quad . \tag{3.61}$$

3.5.3 The General Solution

The general solution to (3.54) can be studied using the same method employed for (3.2). Setting

$$
\begin{aligned}
a_n(t) &= A_n(t) \exp\left[i \int^t \left(\Omega - \frac{\Delta}{2} \right) dt \right] \quad , \\
a_k(t) &= A_k(t) \exp\left[i \int^t \left(\Omega + \frac{\Delta}{2} \right) dt \right] \quad ,
\end{aligned}
\tag{3.62}
$$

where $\Omega(t)$ is a time-dependent Rabi frequency and substituting into (3.54) yield the following expression for the Rabi frequency:

$$\Omega_\pm(t) = \pm\frac{1}{2} [\Delta^2(t) + |z_{nk}|^2 E^2(t)]^{\frac{1}{2}} \quad . \tag{3.63}$$

It is a generalization of (3.5) and yields two quasi-energy levels (Sect. 1.3). If the perturbation were switched on and off adiabatically and infinitely slowly, no transitions would take place between these quasi-energy levels. Since the parameters do vary over a finite interval of time, such transitions do take place and one must know how to calculate the probability of their occurrence. This can be done using the Born-Fock theory (Chap.4) which yields

$$a_k(t) = \int_{-\infty}^{t} \frac{1}{\Delta\Omega} \exp\left[i \int_{-\infty}^{t'} \Delta\Omega(t'')dt''\right] \frac{\partial V_{kn}^{(1)}(t')}{\partial t'} dt' \quad . \tag{3.64}$$

Substituting the difference between the quasi-energy levels into (3.64),

$$\Delta\Omega(t) = \Omega_+(t) - \Omega_-(t)$$

$$= 2\Omega(t) \tag{3.65}$$

$$= [\Delta^2(t) + |z_{kn}|^2 E^2(t)]^{\frac{1}{2}} \quad ,$$

yields

$$a_k(t) = \int_{-\infty}^{t} \frac{1}{2\Omega} \exp\left[2i \int_{-\infty}^{t'} \Omega(t'')dt''\right] \frac{z_{kn}}{2} \frac{\partial E(t')}{\partial t'} dt' \quad . \tag{3.66}$$

If one then integrates by parts and sets $t = \infty$, one obtains the final formula for the transition probability throughout the perturbation [3.10]:

$$W_k(\infty) = |a_k(\infty)|^2$$

$$= \left| \int_{-\infty}^{\infty} \frac{z_{nk}}{2} E(t) \exp\left[2i \int_{-\infty}^{t} \Omega(t') dt'\right] dt \right|^2 \quad . \tag{3.67}$$

Since the Born-Fock adiabatic theory leads to the wrong value for the pre-exponential factor in the transition probability, one cannot expect (3.67) give a correct pre-exponential factor. When this probability is close to unity, the formula is in good agreement with the exact solution only for purely numerical reasons.

It is readily seen that when $\Delta = 0$, (3.67) coincides with (3.59). Moreover, for weak fields, (3.67) is a result of perturbation theory (Chap.2):

$$W_k(\infty) = \left| \int_{-\infty}^{\infty} \frac{z_{kn}}{2} E(t) \exp\left[2i \int_{-\infty}^{t} \Delta(t')dt'\right] dt \right|^2 \quad . \tag{3.68}$$

Finally, for a rectangular pulse [$E(t) = E$ when $0 < t < T$ and $E = 0$ outside this interval; $\Delta = $ const], (3.67) yields the Rabi solution (3.17).

When the probability $W_k(\infty)$ falls off exponentially or by a power law, it must be more than the contribution of the anti-resonance terms to the exact transition probability. These terms were neglected in (3.54).

Of course, for (3.67) to be valid, one must make sure that the criteria given in (3.7) are met.

3.6 A Three-Level System in Two Fields

The resonance approximation can not only be used for two-level systems but also for more complex systems which are irradiated by external electromagnetic fields. Consider, for example, the problem of a three-level system in two fields, one of which is in resonance with one transition while the other is in resonance with the second transition. Such a problem can serve as a model for certain physical phenomena, such as stimulated Raman scattering, double optical resonance (Chap.8), and multi-photon optical pumping, that emerge during the interaction of intense light with atoms.

3.6.1 Mixing in a Three-Level System

Consider the three-level system depicted in Fig.3.2. Two external monochromatic fields act on the states n, m, and k. The frequencies of these fields are in resonance with the frequencies of the corresponding transitions, ω_{mn} and ω_{km}. This is also shown in Fig.3.2. Both electromagnetic perturbations are assumed to be small compared to the distances between the atomic levels. Their detuning functions are also assumed to be small, so that the conditions given in (3.7) can be met. The amplitudes of the three levels as func-

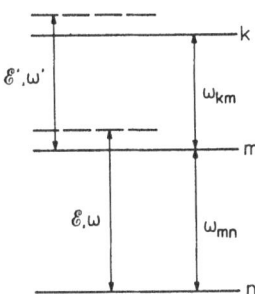

Fig.3.2. A three-level system in two monochromatic fields

tions of time shall be determined as in Sect.3.1. It is assumed that at time zero the electron is in the lower state n and that both perturbations are introduced instantaneously.

In the resonance approximation the system of equations for the probabili-ty amplitudes of the states n, k, and m are

$$i\overset{\bullet}{a}_n = \frac{1}{2} z_{nm} E a_m \exp[i(\omega - \omega_{mn})t] \quad ,$$

$$i\overset{\bullet}{a}_k = \frac{1}{2} z_{km} E' a_m \exp[-i(\omega' - \omega_{km})t] \quad , \qquad\qquad (3.69)$$

$$i\overset{\bullet}{a}_m = \frac{1}{2} z_{mn} E a_n \exp[-i(\omega - \omega_{mn})t] + \frac{1}{2} z_{mk} E' a_k \exp[i(\omega' - \omega_{km})t] \quad .$$

It was assumed that the levels n and k have the same parity while the level m is of opposite parity. All other combinations for the arrangement of the levels and their parities are considered in a similar manner.

In view of what has been said, the initial conditions for (3.69) can be stated as follows:

$$a_n(0) = 1 \quad , \quad a_m(0) = a_k(0) = 0 \quad . \qquad\qquad (3.70)$$

The solution to (3.69) is sought in a standard form, which generalizes what was done in Sect.3.1:

$$a_n(t) = A_n \exp[i(\Omega - \Delta)t] \quad ,$$

$$a_m(t) = A_m \exp(i\Omega t) \quad , \qquad\qquad (3.71)$$

$$a_k(t) = A_k \exp[i(\Omega + \Delta')t] \quad ,$$

where $\Delta = \omega_{mn} - \omega$, and $\Delta' = \omega_{km} - \omega'$. The following cubic equation is obtained for the Rabi frequency:

$$\Omega(\Omega - \Delta)(\Omega + \Delta') - \frac{1}{4} |z_{mn}|^2 E^2 (\Omega + \Delta') - \frac{1}{4} |z_{mk}|^2 E'^2 (\Omega - \Delta) = 0 \quad . \quad (3.72)$$

It is readily seen that each of the three roots of this equation is real. This corresponds to the law of conservation of the number of particles.

Thus, in the general case, each of the three levels is split into three quasi-levels. Their energies are

$$E_m^{(0)} - \omega - \Omega_1 \quad , \qquad\qquad E_m^{(0)} - \omega - \Omega_2 \quad ,$$

$$E_m^{(0)} - \omega - \Omega_3 \quad , \qquad\qquad E_m^{(0)} - \Omega_1 \quad , \qquad E_m^{(0)} - \Omega_2 \quad ,$$

$$E_m^{(0)} - \Omega_3 \quad , \qquad\qquad E_m^{(0)} + \omega' - \Omega_1 \quad ,$$

$$E_m^{(0)} + \omega' - \Omega_2 \quad , \qquad E_m^{(0)} + \omega' - \Omega_3 \quad , \tag{3.73}$$

and are a result of the Floquet theorem (Sect.1.3).

3.6.2 Zero Detuning

To solve (3.72) requires, in general, the use of Cardan's formulas and is thus not straightforward. If one only considers a zero detuning for both resonances, i.e., $\Delta = \Delta' = 0$, the results have a simpler appearance. The three roots are:

$$\Omega_1 = -\Omega_2 = \frac{1}{2} (|z_{mn}|^2 E^2 + |z_{mk}|^2 E'^2)^{\frac{1}{2}} \quad , \tag{3.74}$$

$$\Omega_3 = 0 \quad .$$

In this case the time-averaged probabilities of finding the electron in the states m and k are:

$$W_m = \frac{1}{2} [1 + (z_{mk}E'/z_{mn}E)^2]^{-1} \quad ,$$
$$\tag{3.75}$$
$$W_k = \frac{3}{2} [(z_{mk}E'/z_{mn}E)^2 + (z_{mn}E/z_{mk}E')^2]^{-1} \quad .$$

In particular, when $E' = 0$ the probability $W_m = 1/2$, which was to be expected from the results of Sect.3.1. In this case level k is not populated, while the levels m and n are populated with equal probabilities.

From (3.16) it is evident that the quasi-energy levels of k are equally populated for a two-level system. The situation is different in a three-level system. For example, it is easy to show that the probability W_k, which is determined by the second formula in (3.75), is not the same for each of the quasi-levels of k. The population of two quasi-levels is the same, while that of the third is four times greater.

3.6.3 Population Inversion in a Three-Level System

By varying the field strengths E and E', one can determine the maximum population of state k. At least it can be shown to correspond to a zero detuning for both resonances. Naturally, as this detuning increases, the population of the excited states decreases. From (3.75) it can be seen that the maximum value, equal to 3/8, is attained when $z_{mn}E = z_{mk}E'$. Also, $W_m = 2/8$ and $W_n = 3/8$. Thus, when the average population of level k is a maximum, a population inversion with respect to the lowest level, n, does not exist.

If the population of state k is not at its maximum, an inversion with respect to n can be achieved. Consider the difference

$$W_k - W_n = \frac{[2 - (z_{mn}E/z_{mk}E')^2][(z_{mn}E/z_{mk}E')^2 - 1]}{2[1 + (z_{mn}E/z_{mk}E')^2]} \quad , \tag{3.76}$$

which shows that within the range

$$|z_{mk}E'| < |z_{mn}E| < \sqrt{2}|z_{mk}E'| \quad , \tag{3.77}$$

there is a population inversion of k with respect to n, i.e., $W_k > W_n$. In this respect a three-level system substantially differs from a two-level one. In the latter, a population inversion cannot take place (Sect.3.1). A maximum inversion for a three-level system is achieved when $z_{mn}E = (7/5) z_{mk}E'$ which amounts to about two percent. The population inversion of k with respect to m is about seven percent.

Similarly, one finds that the population inversion of k with respect to m takes place within the range

$$|z_{mn}E| < \sqrt{2}|z_{mk}E'| \quad . \tag{3.78}$$

Thus, the criteria for the amplitude of the perturbation have a common range in which there is population inversion of level k with respect to levels n and m.

4. The Adiabatic Approximation

In quantum mechanics the adiabatic approximation is used for problems in which the perturbation is of an adiabatic nature, i.e., it changes slowly compared to the other variables of the problem. There is no doubt that the adiabatic approximation can be used to describe multi-photon transitions whose frequencies ω_{mn} are much greater than the frequency of the external field, ω:

$$\frac{\omega}{\omega_{mn}} \ll 1 \quad , \tag{4.1}$$

i.e., when the perturbation is a slowly varying quantity.

The crude analogy between the adiabatic and the quasi-classical approximation is well known. It is also known that the problem concerning transitions in a quantized system exposed to an adiabatic perturbation is mathematically equivalent to the problem concerning reflection above a barrier in the quasi-classical approximation [4.1]. To obtain a solution using the adiabatic approximation it is necessary to transfer from the variables momentum and position to those of energy and time.

The primary value of the adiabatic approximation (and the quasi-classical) lies in the possibility of moving into the domain of large perturbations and, hence, into the domain of strong fields, which is not covered by perturbation theory. Obviously, for the limiting case of weak fields, the results obtained using the adiabatic approximation coincide with the results obtained using perturbation theory—to within terms which are small in the adiabaticity parameter ω/ω_{mn}. For resonance processes in the limiting case of weak fields, the results obtained using the adiabatic approximation transform into the results obtained using the resonance approximation.

In this chapter the adiabatic approximation will only be used for the harmonic perturbation $V = zE \cos\omega t$. Bound-bound transitions, degeneracy effects, bound-free transitions, and the criteria for applying the adiabatic approximation to specific problems and for applying perturbation theory to low frequencies will be considered.

93

The adiabatic approximation considerably widens the scope of simple analytical calculations concerning the interaction of an atom and electromagnetic radiation, compared to the scope that can be obtained using perturbation theory or the resonance approximation (Chaps.2 and 3). The sole requirement for the adiabatic approximation to work is that the frequency of the field be small compared to the characteristic differences between the energy levels of the atom. In reality, the word "small" does not mean that the sought quantities must differ by a factor of ten or more. It will be shown that even a factor of three is sufficient for the adiabatic approximation to be used. This ensures a frequent use of the adiabatic approximation in the analysis of the interaction of a strong field with an atom. In certain cases the field strength may even be as high as that inside the atom.

4.1 General Theory

Let us examine the transitions between atomic states under the influence of an adiabatic perturbation. Transitions in two-level systems will mainly be considered because the majority of the calculations can be brought to a final result. Other levels can be taken into account in the framework of the given model.

4.1.1 The Landau-Dykhne Adiabatic Approximation

First of all the general theory concerning transitions between bound (discrete) states will be briefly discussed. Such transitions occur when an adiabatically slow perturbation acts on the system [4.1,2].

In the adiabatic approximation the energy eigenvalues, $E_j(t)$, of the complete Hamiltonian of the system, $H(t)$, are introduced. In these eigenvalues time is considered to be a parameter. In other words, one must solve the Schrödinger equation $i(\partial \Psi_j/\partial t) = H(t)\Psi_j(t) = E_j(t)\Psi_j(t)$. The form of this equation implies that the time dependence of the eigenfunctions Ψ_j is as follows:

$$\Psi_j(t) \propto \exp\left[-i \int^t E_j(t') \, dt'\right] \quad . \tag{4.2}$$

Naturally, the amplitude A_{mn} of the transition from state n to state m under the influence of an adiabatic perturbation is determined by the overlap integral, (Ψ_m^*, Ψ_n). In the general case the adiabatic approximation leads to the following well-known equation for the transition amplitude [4.1]:

$$R \equiv A_{mn} = i \, \exp\left[i \int_{t_1}^{\tau} \omega_{mn}(t) \, dt \right] \quad , \tag{4.3}$$

where t_1 is any point on the real t axis, and τ is a point lying in the upper half-plane of the complex variable t for which

$$E_n(\tau) = E_m(\tau) \quad . \tag{4.4}$$

If there are several such points, one must choose those which lie the closest to the real axis of t, because the contribution of the other roots of (4.2) to R is exponentially small.

According to (4.3) the probability W_{mn} of a transition between the states n and m is

$$W_{mn} = |A_{mn}|^2 = \exp\left[-2 \, \mathrm{Im}\left\{ \int_{t_1}^{\tau} \omega_{mn}(t) \, dt \right\} \right] \quad . \tag{4.5}$$

Equations (4.3-5) correspond to the case where n and m are adjacent levels. If there is a third level, k, between the two, there will be competition between the direct transition from n to m and the cascade transition through the intermediate state k. This case was studied in [4.1]. It will be analyzed later on in this chapter.

4.1.2 The Born-Fock Adiabatic Approximation

The Born-Fock adiabatic approximation [4.3] can be used to find the probability of a transition from the state n to the state m. If it is assumed that at $t = 0$ the electron is in the state n, the probability of a transition to the state m is

$$W_{mn} = \left| \int_0^{\infty} \frac{1}{\omega_{mn}(t)} \exp\left[i \int_0^{t} \omega_{mn}(t') \, dt' \right] \left(\frac{\partial V}{\partial t} \right)_{mn} dt \right|^2 \quad . \tag{4.6}$$

It is known that the higher terms of this approximation, i.e., terms containing powers of the small adiabaticity parameter ω/ω_{mn}, where ω is the frequency of the perturbation and ω_{mn} is the distance between the unperturbed levels, do not form a converging series. This is because the exponential in (4.6) is a rapidly oscillating function (if, of course, $E_m^{(0)}$ is not close to $E_n^{(0)}$). According to the saddle-point method, the value of the integral is determined by the points at which $\omega_{mn}(t) = 0$. It follows from (4.6) that at these points the pre-exponential factor becomes infinite. In the higher terms of the adiabatic approximation, the denominators of expressions similar to (4.6) contain energy differences to powers higher than the first. In this way the pre-exponential singularity is strengthened and compensates for the additional small factors $(\partial V/\partial t)_{mn}$ that appear in each subsequent term of the adiabatic approximation with respect to the preceding term.

As a rule, this means that all of the terms in the Born-Fock adiabatic series have the same order of smallness analytically. Hence, one can hope that by restricting oneself to a single term one is correctly describing the exponential effects. The pre-exponential effects cannot be evaluated using the Born-Fock approximation. With the Landau-Dykhne approximation, they can be evaluated correctly. Herein lies the disadvantage of the Born-Fock approximation. Its one advantage over the Landau-Dykhne approximation is the following: if the perturbation is small, (4.6) turns into an equation that can be obtained using perturbation theory (Chap.2); in the Landau-Dykhne approximation the equations of the adiabatic approximation only approach those of perturbation theory, namely, to within terms of the order of $\omega/\omega_{mn} \ll 1$. For this reason, it is expedient to use the one or the other variant of the adiabatic approximation, especially when constructing approximate solutions. In particular, for weak fields in which the amplitudes vary slowly with time, it is convenient to use the Born-Fock approximation (Chap.3).

4.2 Bound-Bound Resonance Transitions

In this section the Landau-Dykhne adiabatic approximation as applied to a two-level system that is acted upon by a field of frequency ω will be considered. This field is in one- or multi-photon resonance with the energy difference of the two levels n and m. The unperturbed energies of the levels will again be denoted by $E_n^{(0)}$ and $E_m^{(0)}$. Thus, it is indirectly assumed that the difference $E_m^{(0)} - E_n^{(0)}$ is close to an odd multiple of ω.

4.2.1 Transition Probability per Unit Time

The two-level system presently being studied is acted upon by a monochromatic perturbation, $V = V^{(1)} \sin\omega t = zE \sin\omega t$. The Schrödinger equation in the energy representation is:

$$E_n^{(0)} a_n + z_{nm} E a_m \sin\omega t = E(t) a_n \quad ,$$

$$E_m^{(0)} a_m + z_{mn} E a_n \sin\omega t = E(t) a_m \quad . \tag{4.7}$$

To simplify matters, it is assumed that $E_m^{(0)} = -E_n^{(0)} = \omega_{mn}/2 > 0$, so that the distance between the levels is equal to ω_{mn}. As a result (4.7) yields the following expression for the eigenvalues $E(t)$:

$$E_m(t) = -E_n(t) = \left(\frac{1}{4}\omega_{mn}^2 + |z_{mn}|^2 E^2 \sin^2\omega t\right)^{\frac{1}{2}} . \tag{4.8}$$

The points τ, in the upper half-plane of complex time, that satisfy (4.4) are determined by the zeros of (4.8), i.e.,

$$\tau = \frac{\pi N}{\omega} + \frac{i}{\omega}\sinh^{-1}\left(\frac{\omega_{mn}}{2|z_{mn}|E}\right) , \tag{4.9}$$

where N is an integer. All of these points are equidistant from the real axis t and, hence, contribute equally to the transition probability (4.5). Moreover, (4.8) illustrates that, according to the general theory, the τ_N are branch points for the energy $E(t)$ [4.1].

To start with, the contribution of a certain point τ_N, e.g., with N=0, is considered. Substituting (4.8,9) into (4.3), it is found that the transition amplitude is determined by this point τ_0:

$$A_{mn}^{(0)} = \exp\left[i\int_0^{\tau_0}\left(\omega_{mn}^2 + 4|z_{mn}|^2 E^2 \sin^2\omega t\right)^{\frac{1}{2}} dt\right] . \tag{4.10}$$

By changing the variable t to iu, it is easy to evaluate the integral:

$$W_{mn} = |A_{mn}^{(0)}|^2 = \exp\left[-2\frac{\omega_{mn}}{\omega}KD(K)\right] , \tag{4.11}$$

where

$$K = \frac{\omega_{mn}}{(\omega_{mn}^2 + 4|z_{mn}|^2 E^2)^{\frac{1}{2}}} , \tag{4.12}$$

and $D(K)$ is the complete elliptic integral of the third kind [4.3].

Now the other points τ_N are taken into account. The transition amplitude determined by the point τ_N can be written in the following form:

$$A_{mn}^{(N)} = \exp\left[i\int_0^{\tau_0+\pi N/\omega}\left(\omega_{mn}^2 + 4|z_{mn}|^2 E^2 \sin^2\omega t\right)^{\frac{1}{2}} dt\right]$$

$$= A_{mn}^{(0)}\exp(iS_N) , \tag{4.13}$$

where the quantity

$$S_N = \int_0^{\pi N/\omega}[E_m(t) - E_n(t)]\, dt = \int_0^{\pi N/\omega}\left(\omega_{mn}^2 + 4|z_{mn}|^2 E^2 \sin^2\omega t\right)^{\frac{1}{2}} dt \tag{4.14}$$

represents the change in the classical action over N perturbation periods. Evaluating (4.14) one obtains $S_N = NS_1$, where

$$S_1 = \frac{2}{\omega}\left(\omega_{mn}^2 + 4|z_{mn}|^2 E^2\right)^{\frac{1}{2}} E(\sqrt{1 - K^2}) , \tag{4.15}$$

97

i.e., is the same as S_N but over one perturbation period π/ω, and E is the complete elliptic integral of the second kind [4.4]. According to the Feynman principle [4.5], the total transition amplitude over the time T is given by a sum over all possible paths. Of the infinite number of such paths, those that pass through the points τ_N contribute the most [4.2]. Their summation yields

$$A_{mn} = A_{mn}^{(0)} \frac{\exp(iS_1 N) - 1}{\exp(iS_1) - 1} . \tag{4.16}$$

Equation (4.16) determines the probability of a transition per unit time (transition rate) when S_1 is close to an integral multiple of 2π, i.e., if the detuning $S_1 - K\pi$ is small [K is an even integer and $S_1 - K\pi = (\pi/\omega)\Delta_K$]. In squaring the modulus of (4.16) and by using a well-known relationship for the delta function [Ref.4.1, §42] one finds the probability of the transition (n, m) per unit time:

$$w_{mn} = \frac{2\omega^2}{\pi} W_{mn} \delta(\Delta_K) \quad , \tag{4.17}$$

where W_{mn} was defined in (4.11). The argument of the delta function in (4.17) reflects the fact that, with due regard for level shifts in strongly varying fields, energy is conserved in multi-photon processes. The appearance of the delta function in (4.17) also reflects the idealization inherent in the two-level system, i.e., no account is taken of the widths of the levels.

If the equation that determines the transition rate, (4.17), is to be correct, the perturbation time T must not be too large: $w_{mn} T \ll 1$ (Sect.2.1). If this condition is not met, saturation sets in. This effect will be investigated later on.

4.2.2 Criterion of Applicability

Equation (4.11) is true if the exponent is large. In this case the transition probability becomes vanishingly small under the influence of the slowly changing adiabatic perturbation. From (4.11) it is evident that the transition between states n and m must be a multi-photon one, i.e., the criterion in (4.1) must be met. The perturbation can be quite strong, even if $|z_{mn}|E \sim \omega_{mn}$. When the energy of interaction between the perturbation and the atom exceeds the energy separation between the levels, $KD(K) \sim 1$ and (4.11) is still valid. Only for a perturbation

$$|z_{mn}|E \gtrsim \frac{\omega_{mn}^2}{\omega} \gg \omega_{mn} \tag{4.18}$$

is W_{mn} of the order of unity, i.e., the adiabatic approximation breaks down.

4.2.3 The Limiting Transition in Perturbation Theory

Let us consider the above equations for the transition probability and the energy shifts under the influence of periodic adiabatic perturbation in the limiting case of a weak field in terms of perturbation theory [4.6]. Expanding (4.17) in a power series in $K' = (1 - K^2)^{\frac{1}{2}}$ yields [4.7]

$$w_{mn} = \frac{2\omega^2}{\pi} \left(\frac{e|z_{mn}|E}{2\omega_{mn}}\right)^{2K} \delta(\omega_{mn} - K\omega) \tag{4.19}$$

where e is the base of the natural logarithms. This corresponds to a K-photon transition (Chap.2). According to Sect.2.4, a multi-photon matrix element for a two-level system can be written as follows:

$$V_{mn}^{(K)} = \frac{(z_{mn}/2)^K E^K}{[0 - (K - 1)\omega][\omega_{mn} - (K - 2)\omega][0 - (K - 3)\omega]...[\omega_{mn} - \omega]}$$

$$= \left(\frac{z_{mn}}{\omega}\right)^K \frac{\omega}{2^K[(K - 1)!!]^2} E^K \quad . \tag{4.20}$$

Hence, using perturbation theory the transition probability per unit time is

$$w_{mn}^{(K)} = 2\pi |z_{mn}^{(K)}|^2 E^{2K} \delta(\omega_{mn} - K\omega)$$

$$= 2\pi\omega^2 \left(\frac{z_{mn}}{\omega}\right)^{2K} \frac{E^{2K}}{2^{2K}[(K - 1)!!]^4} \delta(\omega_{mn} - K\omega) \quad . \tag{4.21}$$

For the factorials in (4.21) Stirling's formula is used, and one readily sees that (4.21) coincides with (4.19), as expected.

Since Stirling's formula also leads to small errors for small values of K, the adiabatic approximation does not require a very large value of K, e.g., already with K = 3 the error that arises when (4.19) is substituted for (4.21) does not exceed a few percent [4.7].

If the transition probability per unit time given by (4.17) is expanded in a power series in V (one must also take into account the level shift in the variable field), a correction factor to (4.19), which characterizes the deviation from perturbation theory, is obtained. This factor is given in [4.7] and has the form $\exp[-K(z_{mn}E/\omega_{mn})^2]$. To calculate the bound-bound transition probability using perturbation theory, one must make sure that

$$K\left(\frac{z_{mn}E}{\omega_{mn}}\right)^2 \ll 1 \quad . \tag{4.22}$$

99

However, it should be noted that there is a certain range in which the amplitudes of the perturbation are small compared to the separation of the levels. Nevertheless, perturbation theory does not work.

At the same time, (4.15) shows that perturbation theory can be used to calculate the level shift in an adiabatic field if

$$\frac{z_{mn}E}{\omega_{mn}} \ll 1 \quad , \tag{4.23}$$

a condition that is much less stringent than the one in (4.22).

Consequently, the criteria for using perturbation theory in determining the transition probability and the level shifts are different in the multi-photon case. As the field strength grows, perturbation theory first of all can no longer be used to calculate the transition probability and then, finally, cannot be used to calculate the level shifts.

In a weak field the level shifts are quadratic in the perturbation. If one then applies the adiabatic approximation, it is seen that the shifts are half as large as the corresponding shifts in a constant field E (this is because the mean value of $\sin^2\omega t$ is 1/2).

To end this section it should be noted that the solutions ψ_n of the Schrödinger equation were written in the zeroth approximation in the adiabaticity parameter K.

If the next order in the approximation is taken into account in the same way as in the quasi-classical approximation [from $\exp(\pm i \int k \, dx)$ to $k^{-\frac{1}{2}} \exp(\pm i \int k \, dx)$ [4.11]], all of the above-mentioned results are still valid, but the number of absorbed quanta, K, becomes an odd number [4.7]. This is an obvious result since the interaction is a dipole interaction (the perturbation V, as assumed, has zero diagonal matrix elements). The fact that the number of absorbed quanta is odd is also not surprising if the multi-photon matrix elements are studied in the framework of perturbation theory.

4.2.4 Resonance Mixing in a Two-Level System

Up to this point only small perturbation times have been considered, in which the probability of an adiabatic transition is proportional to time (the linear range). For large times the electron begins to oscillate between the resonating states n and m. In describing such oscillations one must take into account the decrease in the number of electrons in state n. The probability of an electron transferring from state n to state m in a time interval π/ω is $W_{mn} = |A_{mn}^{(0)}|^2$ [see (4.11)]. Hence, the probability that the electron remains in the state n is $1 - W_{mn}$. The change in the state vector of a two-level system,

$$\begin{pmatrix} a_m \\ a_n \end{pmatrix} ,$$

in a time interval π/ω is determined by the 2×2 unitary matrix A [4.8]:

$$A = \begin{pmatrix} \mathcal{D} \exp(-iS_1/2) & -A_{mn}^{(0)*} \exp(-iS_1/2) \\ A_{mn}^{(0)} \exp(iS_1/2) & \mathcal{D}^* \exp(iS_1/2) \end{pmatrix} , \tag{4.24}$$

where $\mathcal{D} = (1 - W_{mn})^{\frac{1}{2}}$, S_1 was defined in (4.15), and $-S_1/2$ and $S_1/2$ represent the change in classical phase for states n and m, respectively, in the time π/ω.

The change in the state vector over N periods, i.e., in the time interval $t = \pi N/\omega$, is A^N. To calculate A^N the matrix A must be reduced to a diagonal form. But first it is advisable to simplify matrix A:

$$A \approx \exp(-iS_1/2) \begin{pmatrix} 1 & -A_{mn}^{(0)*} \\ A_{mn}^{(0)} & 1 + i\pi\Delta_K/\omega \end{pmatrix} . \tag{4.25}$$

This is not absolutely necessary; it leaves only the principal nonvanishing term in the resonance width (see below). The detuning from multi-photon resonance, Δ_K, is $\Delta_K = \omega_{mn} - K\omega + \delta\omega_{mn}$, where $\delta\omega_{mn}$ is the change in the difference of the energies of the levels n and m brought on by the variable field.

After diagonalization, the probability of finding the electron in the upper level m [4.9] under the assumption that the perturbation is switched on suddenly is

$$W_{mn}(t) = \frac{W_{mn}}{W_{mn} + (\pi\Delta_K/2\omega)^2} \sin^2\left\{ [W_{mn} + (\pi\Delta_K/2\omega)^2]^{\frac{1}{2}} \frac{\omega t}{\pi} \right\} . \tag{4.26}$$

At exact resonance, i.e., when $\Delta_K = 0$, the two levels are, on the average, equally populated. A similar situation was encountered in the case of one- and multi-photon transitions (Chap.3), in which the condition given in (4.22) was met. When $\Delta_K \neq 0$, the structure of (4.26) and of the corresponding expressions in Chap.3 also prove to be the same.

Hence, resonant transitions and resonant mixing can occur for the multi-photon case, even for perturbations of the order of the distance between the resonating levels. Indeed, W_{mn} remains small compared to unity, and (4.26) preserves its resonant structure.

If the above reasoning is applied to atoms which interact with light, one can say that electromagnetic fields whose strength is of the order of atomic fields can initiate resonant transitions, provided that the frequency of the field is small compared to the distance between the resonating levels. In this case the transitions take place in times that are greater than the characteristic atomic times. All of this is characteristic of multi-photon transitions and is absent in one-photon transitions and in constant fields, in which, when $E \sim E_{at}$, transitions occur in atomic times. It is this fact that always justified the use of fields in which $E \ll E_{at}$. In the case of multi-photon transitions there is no need for such a restriction.

It follows from (4.26) that

$$\Omega_K \equiv \frac{\omega}{\pi} \left[W_{mn} + (\pi \Delta_K/2\omega)^2 \right]^{\frac{1}{2}} , \tag{4.27}$$

whereby Ω_K is called the Rabi frequency for a strong field (Chap.3).

When $W_{mn} \ll (\Delta_K/\omega)^2$, (4.26) becomes equal to (4.17).

4.2.5 Transitions in Multi-Level Systems

If in addition to the two levels n and m, between which transitions can take place due to an adiabatic perturbation, there are other discrete levels in the system, the problem becomes more complex.

Let us first of all study the simpler case in which n and m are adjacent. The energy eigenvalues, E(t), are then determined by equations more complicated than (4.8). However, if these energy eigenvalues are known in principle, the problem can be solved by the above-mentioned method.

When other discrete levels exist between n and m, the problem becomes far more complex. This can be seen using time-dependent perturbation theory (Sect. 2.2). If the distance between one of the levels n or m and the intermediate level is close to an integral number of quanta of the external perturbation, the probabilities of bound-bound transitions are of a resonant nature. Naturally, the quasi-classical expressions for amplitudes of the type given in (4.3) do not reflect the presence of such resonances. But the adiabatic approximation makes it possible, in principle, to describe the intermediate resonance levels. In the neighborhood of such resonances the transitions occur stepwise and have the amplitudes [4.1]

$$A_{m \leftarrow k \leftarrow n} = \exp \left[i \int_{t_1}^{\tau'} \omega_{mk}(t) \, dt + i \int_{t_1}^{\tau''} \omega_{kn}(t) \, dt \right] , \tag{4.28}$$

where τ' and τ'' can be found from equations similar to (4.4):

$$E_m(\tau') = E_k(\tau') \quad , \quad E_k(\tau'') = E_n(\tau'') \quad . \tag{4.29}$$

It was assumed that no intermediate levels other than k exist between n and m. Otherwise these would have contributed to the amplitude. Also, the contribution would be comparable to $A_{m \leftarrow k \leftarrow n}$, even if the levels are nonresonant.

Although the points τ' and τ'' do not exhibit such simple forms as in (4.9) due to the periodicity of the perturbation, they consist of two sets of points in the plane of complex "time". Each set represents a collection of points that are equidistant from the real time axis and that are separated from one another by π/ω. Because there are two sets of points, the summation along the Feynman paths is double compared to the case of two levels (Fig. 4.1). If the point τ'' is fixed, one must sum over all of the points τ' that are within a chosen time interval T to the right of τ''. From the physical point of view such a restriction is an example of the causality principle for a two-step transition [transition (k, m) is followed by (n, k)]. Then one must sum over all of the points τ''. It should be noted that the points τ' and τ'' are branch points for the energies E(t) (as in the case of a two-level system). Through this double summation, the probability of the two-step transition (m, k, n) per unit time is obtained [4.9]:

$$W_{m,k,n} = \frac{\omega^4}{2\pi^3} |A_{m,k,n}|^2 \frac{\delta(\omega_{mn} - K\omega + \delta\omega_{mn})}{(\omega_{kn} - K'\omega + \delta\omega_{kn})^2} \quad , \tag{4.30}$$

where the amplitude $A_{m,k,n}$ has already been defined in (4.28) for points τ' and τ'' arbitrarily chosen from the ones depicted in Fig.4.1; the probability $|A_{m,k,n}|^2$ does not depend on this choice, $\delta\omega_{mn}$ and $\delta\omega_{kn}$ are the changes in the differences between the corresponding energy levels, and K' is the number of perturbation quanta that are in resonance with the transition (n, k). One can see that (4.30) has a resonant structure, as it should have (Sect.2.2) and that the resonance width does not appear in (4.30). The absence of a resonance width of the Breit-Wigner type can be explained by the absence of levels into which the state m can spontaneously decay. The absence of a field broadening similar to the resonance broadening in (4.26) is explained by the

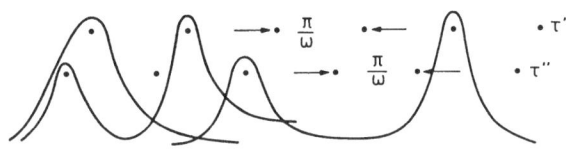

Fig.4.1. Feynman paths for a two-stage transition

fact that only a probability that is linear in time was considered and that there is no resonance mixing in either the transition (n, m) or the transition (n, k).

The situation for long times, during which mixing can occur, can be studied by generalizing the above approach using the matrix in (4.24). In this case the state vector has three elements and its variation during the time interval π/ω is determined by the product of two partial 3×3 matrices. In the first matrix only the transition (k,m) is considered, during which the state n only changes in phase (classical action). In the second matrix the transition (n, k) is considered, during which the state m remains unchanged except for a phase factor. Obviously, the linear regime can be viewed as a particular case in which small perturbation times are used. A detailed study of an intermediate resonance for a three-level system is given in [4.9]. Similar results for intermediate one-photon resonances were discussed in Chap.3.

Similarly, one must take into account every level k, k', ... (of a multi-level system) that lies between n and m. The resulting amplitude $\tilde{A}_{mn}$ has the following form [4.10]:

$$\tilde{A}_{mn} = A_{mn} + \sum_{k} A_{m,k,n} + \sum_{k,k'} A_{m,k,k',n} + \cdots \quad . \tag{4.31}$$

It is readily seen that this amplitude has a structure analogous to that of the multi-photon matrix element obtained using perturbation theory. Hence, in principle, the adiabatic approximation makes it possible to describe all of the intermediate resonances and their interference. This was also done in perturbation theory for weak fields using the multi-photon matrix element.

The first term in (4.31) represents the direct multi-photon transition from state n to state m. All of the terms in (4.31), including the first, effectively take into account the levels k" which do not lie between n and m, since the $E_n(t)$ and $E_m(t)$ depend on these states. Such a dependence is evident in the system of equations that determines the energy eigenvalues for the adiabatic case. The states k" contribute little to the probabilities for multi-step transitions. This becomes evident using perturbation theory, in which these states lead to higher orders of the perturbation parameter compared with the nonzero lowest order.

4.3 A Two-Level System in a Strong Field of Arbitrary Frequency

In Sect.4.2.4 the adiabatic approximation was applied to a case in which the distance between the levels of a two-level system, ω_{mn}, is close to an odd multiple of the frequency ω of the external field. The same method can be used in the general case, which includes nonresonance, with the aid of a proper generalization, because the adiabatic approximation only requires that ω be small compared to ω_{mn}. Nonresonant effects may come into play either when the detuning becomes larger or when the strength of the external field increases so much that the amplitude of the perturbation becomes comparable to the distance between the levels of the system, ω_{mn}. Up until now the approach to the problem in the nonresonant case was solely based on numerical calculations aided by computers [4.11].

The adiabatic approximation makes it possible to take into account analytically the nonresonant terms in the perturbation —terms that were neglected in the resonance approximation but which are significant under the indicated conditions of a strong field or a frequency out of resonance [4.12].

4.3.1 Adiabatic Wave Functions

The equations for the amplitudes a_n and a_m of the lower and upper levels n and m, respectively, are known. The separation of the levels is ω_{mn}. For the sake of simplicity the energy of level n is $-\omega_{mn}/2$ and that of level m is $\omega_{mn}/2$. The perturbation is of the following form:

$$V(t) = V_{mn}^{(1)} \sin\omega t = z_{mn} E \sin\omega t \quad .$$

Hence, the desired equations are

$$i\dot{a}_n = -\frac{1}{2} \omega_{mn} a_n + V(t) a_m \quad , \tag{4.32}$$

$$i\dot{a}_m = \frac{1}{2} \omega_{mn} a_m + V(t) a_n \quad .$$

If the adiabaticity criterion, $\omega \ll \omega_{mn}$, is met, one can search for a solution to (4.32) in an adiabatic form:

$$a_n(t) = A_n(t) \exp\left[i \int_0^t E(t') \, dt' \right] \quad , \tag{4.33}$$

$$a_m(t) = A_m(t) \exp\left[i \int_0^t E(t') \, dt' \right] \quad .$$

Substituting into (4.32), one can find the time-dependent energy values, $E_{m,n}(t) = \pm E(t)$, where

$$E(t) = \left[\frac{1}{4}\,\omega_{mn}^2 + V^2(t)\right]^{\frac{1}{2}} \quad . \tag{4.34}$$

Also the quantities $A_n(t)$ and $A_m(t)$ can be found:

$$A_n(t) = 2^{-\frac{1}{2}}\{1 + [1 + 4V^2(t)/\omega_{mn}^2]^{-\frac{1}{2}}\}^{\frac{1}{2}} \quad ,$$

$$A_m(t) = -2^{-\frac{1}{2}}\{1 - [1 + 4V^2(t)/\omega_{mn}^2]^{-\frac{1}{2}}\}^{\frac{1}{2}} \quad . \tag{4.35}$$

Finally one obtains two adiabatic basis solutions to (4.32):

$$\Psi_n = [A_n(t)|n> + A_m(t)|m>]\,\exp\left[i\int_0^t E(t')\,dt'\right] \quad ,$$

$$\Psi_m = [-A_m(t)|n> + A_n(t)|m>]\,\exp\left[-i\int_0^t E(t')\,dt'\right] \quad , \tag{4.36}$$

where $|n>$ and $|m>$ are the unperturbed wave functions of the states n and m. When the perturbation is switched off, i.e., as $V(t) \to 0$, the adiabatic wave functions of states n and m become the unperturbed wave functions $|n>$ and $|m>$.

The general solution to (4.32) is a linear combination of the two basis solutions given in (4.36). It is convenient to express the general solution at time t,

$$\Psi(t) = a_n(t)|n> + a_m(t)|m> \quad , \tag{4.37}$$

in terms of the solution at another time, $\bar{t}$,

$$\Psi(\bar{t}) = a_n(\bar{t})|n> + a_m(\bar{t})|m> \quad . \tag{4.38}$$

This can be done using a square matrix, S:

$$\begin{pmatrix} a_n(t) \\ a_m(t) \end{pmatrix} = S(t,\ \bar{t}) \begin{pmatrix} a_n(\bar{t}) \\ a_m(\bar{t}) \end{pmatrix} \quad , \tag{4.39}$$

where

$$S(t,\ \bar{t}) = \begin{pmatrix} \tilde{A}^*(\bar{t})\tilde{A}(t) + \tilde{B}^*(\bar{t})\tilde{B}(t) & \tilde{B}(\bar{t})\tilde{A}(t) - \tilde{A}(\bar{t})\tilde{B}(t) \\ \tilde{A}^*(\bar{t})\tilde{B}^*(t) - \tilde{B}^*(\bar{t})\tilde{A}^*(t) & \tilde{A}(\bar{t})\tilde{A}^*(t) + \tilde{B}(\bar{t})\tilde{B}^*(t) \end{pmatrix} \quad . \tag{4.40}$$

The following notations were introduced in (4.40):

$$\tilde{A}(\bar{t}) = A_n(\bar{t}) \exp\left[i \int_0^{\bar{t}} E(t)\, dt \right] \quad ,$$

$$\tilde{B}(\bar{t}) = A_m(\bar{t}) \exp\left[-i \int_0^{\bar{t}} E(t)\, dt \right] . \tag{4.41}$$

Let it be assumed that at some moment, t_0, the electron is in the lower level, i.e.,

$$\Psi(t_0) = \exp\left(i \, \frac{\omega_{mn}}{2}\, t_0 \right) \binom{1}{0} \quad . \tag{4.42}$$

This implies that the perturbation was switched on suddenly. The solution for an adiabatic switch-on will be given later.

As the electron "moves" from t_0 to the "turning point" π/ω, the wave function changes according to the matrix $S(t, \bar{t})$. The result is

$$\Psi\left(\frac{\pi}{\omega} - 0 \right) = S\left(\frac{\pi}{\omega}, t_0 \right) \Psi(t_0) \quad . \tag{4.43}$$

The motion of the electron from $\pi/\omega - 0$ to $\pi/\omega + 0$ is determined by the matrix

$$L = \begin{pmatrix} D & R \\ -R^* & D^* \end{pmatrix} \quad , \tag{4.44}$$

where R is the reflection coefficient:

$$R = i \exp 2i \int_0^\tau \Omega\, dt \quad . \tag{4.45}$$

Evaluating the integral in (4.45) yields:

$$R = i \exp\left[-\frac{\omega_{mn}}{\omega} \, \frac{1}{(q^2 + 1)^{\frac{1}{2}}} \, D\!\left(\frac{1}{(q^2 + 1)^{\frac{1}{2}}} \right) \right] \quad , \tag{4.46}$$

where D is the complete elliptic integral of the third kind and

$$q = \frac{2V_{mn}^{(1)}}{\omega_{mn}} = \frac{2z_{mn}E}{\omega_{mn}} \tag{4.47}$$

is a dimensionless parameter. It is seen that (4.46) coincides with (4.11). Moreover, in (4.44) D is the transmission coefficient, so that $D = (1 - R^2)^{\frac{1}{2}}$.

The motion of the electron from $\pi/\omega + 0$ to $2\pi/\omega - 0$ is determined by the matrix

$$S\left(\frac{2\pi}{\omega}, \frac{\pi}{\omega} \right) = \begin{pmatrix} \exp(iS) & 0 \\ 0 & \exp(-iS) \end{pmatrix} \quad , \tag{4.48}$$

where

$$S = \int_0^{\pi/\omega} \Omega \, dt + \frac{\pi}{2}$$

$$= \frac{\omega_{mn}}{\omega} (1 + q^2)^{\frac{1}{2}} E\left(\frac{q}{(1 + q^2)^{\frac{1}{2}}}\right) + \frac{\pi}{2} \quad , \qquad (4.49)$$

and E is the complete elliptic integral of the second kind. The term $\pi/2$ arises because the next term in the classical approximation has been taken into account.

Hence the motion of the electron from $\pi/\omega - 0$ to $2\pi/\omega - 0$ is the determined by the product of the two matrices:

$$A = S\left(\frac{2\pi}{\omega}, \frac{\pi}{\omega}\right) L$$

$$= \begin{pmatrix} D \exp(iS) & R \exp(iS) \\ -R^* \exp(-iS) & D^* \exp(-iS) \end{pmatrix} \quad . \qquad (4.50)$$

The passage through N turning points is determined by the matrix A^N. Before this matrix is found, A must first be diagonalized (Sect.4.2.4). The eigenvalues of A are

$$A_{\pm} = \exp(\pm i\phi) \quad , \qquad (4.51)$$

where

$$\tan\phi = D^{-1}(|R|^2 + \tan^2 S)^{\frac{1}{2}} \quad .$$

Next, the diagonal elements, $A_{\pm}$, must be raised to the Nth power. The final result is

$$\Psi\left(\frac{(N + 1)\pi}{\omega} - 0\right) = \begin{pmatrix} PA(t_0) \exp\left[-i \int_0^{t_0} E(t) \, dt + iS + i\psi\right] \\ -RB(t_0) \frac{\sin N\phi}{\sin\phi} \exp\left[i \int_0^{t_0} E(t) \, dt\right] \\ -PB(t_0) \exp\left[i \int_0^{t_0} E(t) \, dt - iS - i\psi\right] \\ -R^*A(t_0) \frac{\sin N\phi}{\sin\phi} \exp\left[-i \int_0^{t_0} E(t) \, dt\right] \end{pmatrix} \quad , \qquad (4.52)$$

where

$$P \equiv \left(1 - \frac{|R|^2 \sin^2 N\phi}{\sin^2 S + |R|^2 \cos^2 S}\right)^{\frac{1}{2}} \qquad \text{and} \qquad (4.53)$$

$$\tan\psi \equiv \frac{D \tan N\phi}{(1 + |R|^2 \cot^2 S)^{\frac{1}{2}}} \quad . \tag{4.54}$$

The wave function at time t [which must lie in the interval from $N\pi/\omega$ to $(N+1)\pi/\omega$] is determined by the product

$$\psi(t) = S\left(t, \frac{(N+1)\pi}{\omega}\right)\psi\left(\frac{(N+1)\pi}{\omega} - 0\right) \quad . \tag{4.55}$$

The amplitude of the population of the upper level is:

$$
\begin{aligned}
a_m(t) = \Big(& PA(t_0)B(t) \exp\{i[S(t) - S(t_0) + \psi + S]\} \\
& - RB(t_0)B(t) \exp[iS(t) + iS(t_0)] \frac{\sin N\phi}{\sin\phi} \\
& - PB(t_0)A(t) \exp\{i[-S(t) + S(t_0) - S - \psi]\} \\
& - R^*A(t_0)A(t) \exp[-iS(t) - iS(t_0)] \frac{\sin N\phi}{\sin\phi} \Big) \\
& \times \exp\left(\frac{1}{2} i\omega_{mn}t_0\right) \quad ,
\end{aligned}
\tag{4.56}
$$

where

$$S(t) = \int_0^t E(t) \, dt \quad , \quad S(t_0) = \int_0^{t_0} E(t) \, dt \quad . \tag{4.57}$$

The expression for $a_n(t)$ has a similar structure. Of course, at any moment of time,

$$|a_m(t)|^2 + |a_n(t)|^2 = 1 \quad , \tag{4.58}$$

a relationship that is ensured by the unitarity of the matrices L and $S(t, \bar{t})$.

It should be noted that a_m oscillates with time. In the general nonresonance case there are two periods for such oscillations: $1/\omega_{mn}$ and $1/\omega$. In the neighborhood of a multi-photon resonance there is a third period, i.e., the inverse of the Rabi frequency, $1/\omega R$. The latter oscillation plays a noticeable part when the detuning from multi-photon resonance, $\omega_{mn} - K\omega$, is comparable to the Rabi frequency.

4.3.2 Average Populations

Equation (4.56) determines the population of level m at the time t if the perturbation was switched on at the time t_0. To find the average population, it must be averaged over t, i.e., over the switch-on time of the field t_0, and over N periods.

In an analytical calculation, the averaging process is separated into two independent processes: an averaging over the N turning points (this corresponds to averaging over the Rabi frequency scale) and over the time t between two successive turning points $\pi N/\omega$ and $\pi(N+1)/\omega$.

The small-scale averaging over t and t_0 yields

$$\overline{|a_m(t)|^2}_{t,t_0} \equiv W_m$$

$$= \frac{1}{2} P^2(1 - 4Q^2) + |R|^2\left(2Q^2 + \frac{1}{2}\right)\frac{\sin^2 N\phi}{\sin^2\phi} \quad , \tag{4.59}$$

where

$$Q = \frac{1}{\pi(1 + q^2)^{\frac{1}{2}}} K\left(\frac{q}{(1 + q^2)^{\frac{1}{2}}}\right) \quad , \tag{4.60}$$

and K is the complete elliptic integral of the first kind.

Equation (4.59) can be helpful in finding the populations of levels near a multi-photon resonance which are averaged over short pulses. One need only substitute $\omega t/\pi$ for N.

Averaging (4.59) over N, one finally finds the average population of level m:

$$\overline{|a_m(t)|^2}_{t,t_0,N} \equiv \overline{W}_m = \frac{1}{2} -$$

$$- \frac{2(1 - |R|^2)K^2\left(\frac{q}{(1 + q^2)^{\frac{1}{2}}}\right)}{\pi^2(1 + q^2)\left\{1 + |R|^2\tan^2\left[\frac{\omega_{mn}}{\omega}(1 + q^2)^{\frac{1}{2}}E\left(\frac{q}{(1 + q^2)^{\frac{1}{2}}}\right)\right]\right\}} \quad . \tag{4.61}$$

The population probability for the lower level is obviously $\overline{W}_n = 1 - \overline{W}_m$.

4.3.3 Results, Limiting Cases, and a Comparison with Numerical Calculations

If only weak fields and regions that are not too close to resonances corresponding to multi-photon transitions are studied, (4.61) yields

$$\overline{W}_m = \left(\frac{z_{mn}}{\omega_{mn}}\right)^2 E^2 \quad , \tag{4.62}$$

which coincides with the equation that was obtained using first-order perturbation theory (Chap.2).

From (4.61) it also follows that, at the points where

$$\frac{\omega_{mn}}{\omega}(1 + q^2)^{\frac{1}{2}} E\left(\frac{q}{(1 + q^2)^{\frac{1}{2}}}\right) = \frac{(2K + 1)\pi}{2} \quad , \tag{4.63}$$

the average population of each level is 1/2. Equation (4.63) describes the positions of the multi-photon resonances in a strong field.

For a point of exact resonance (4.59) can be simplified:

$$W_m = \frac{1}{2} - \frac{2}{\pi^2(1 + q^2)} K^2\left(\frac{q}{(1 + q^2)^{\frac{1}{2}}}\right) \cos\left(\frac{2\omega t}{\pi} \sin^{-1}R\right) \quad . \tag{4.64}$$

This equation implies that W_m undergoes large-scale oscillations similar to the Rabi oscillations in a strong field (Sect.3.1). The amplitude of these oscillations is

$$A = \frac{2}{\pi^2(1 + q^2)} K^2\left(\frac{q}{(1 + q^2)^{\frac{1}{2}}}\right) \quad . \tag{4.65}$$

This amplitude does not depend on the multiplicity of the resonance, i.e., on the frequency ω. The frequency of these oscillations is

$$\Omega = \frac{2\omega}{\pi} \sin^{-1}R \quad . \tag{4.66}$$

It depends strongly on the multiplicity of the resonance. Of course, this frequency is very small compared to ω, and even more so compared to $\omega_{mn} = 5\omega$.

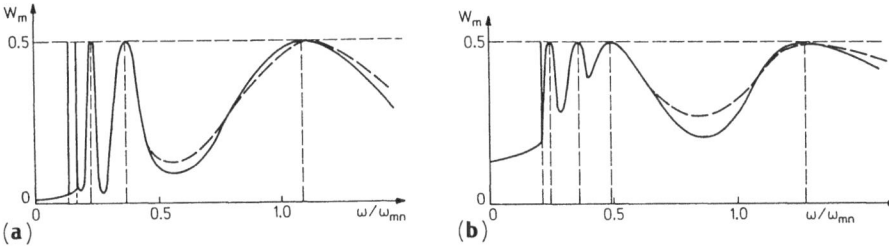

(a)　　　　　　　　　　　　　　　　　(b)

Fig.4.2a,b. The population of the upper level of a two-level system calculated using the adiabatic approximation (*solid line*) and a computer (*dashed line*): (a) q = 0.5, and (b) q = 1.54

The average population of the upper level m and the switch-on phase of the field is illustrated in Fig.4.2. It was calculated with the help of (4.61) [4.12], and by using a computer [4.11]. It is seen that, starting with a three-photon resonance, the adiabatic solution is in very good agreement with the exact solution. Moreover, the agreement is better in the neighborhood of resonances than in the nonresonant region. As the frequency ω increases, the discrepancy between the two types of solution becomes larger. However, it is not too large even in the neighborhood of the resonance on the fundamental harmonic (K = 1), especially if the intensity of the perturbation, q, is not too great. At low frequencies, i.e., outside the region of three-photon resonance, the analytical equation (4.61) coincides, for all practical purposes, with the exact solution over the entire frequency range, including the nonresonant intervals, in which the transitions between the levels n and m are not determined by the multi-photon matrix elements, which is the case in the neighborhood of resonances, but are determined by the usual non-resonant matrix elements (which do not require that the energy be conserved). For example, in weak fields, the populations inside the nonresonant intervals are determined by the first-order term in q. These populations are defined from the expansion in (4.61) of the complete elliptic integral K in powers of q at $q \ll 1$.

At the resonance points the average maximum population of level m never exceeds 1/2. For a two-level system this is true without the adiabatic approximation.

Finally, a solution to (4.61) is sought for the case in which the perturbation is switched on adiabatically and slowly. Just as in Chap.3, it can be found by using one of the adiabatic basis solutions in (4.36), not both of them, as was done for the sudden switch-on mode. When the perturbation is switched on adiabatically and slowly, the population probabilities do not oscillate as they do for the sudden switch-on regime. Hence, they do not require an averaging over time and over the switch-on phase. The final expression for W_m has a form similar to that in (4.61):

$$W_m = \frac{1}{2}\left[1 - \frac{\frac{2}{\pi}\left(\frac{1 - |R|^2}{1 + q^2}\right)^{\frac{1}{2}} K\left(\frac{q}{(1 + q^2)^{\frac{1}{2}}}\right)}{\left\{1 + |R|^2\tan^2\left[\frac{\omega_{mn}}{\omega}(1 + q^2)^{\frac{1}{2}}E\left(\frac{q}{(1 + q^2)^{\frac{1}{2}}}\right)\right]\right\}^{\frac{1}{2}}}\right] . \tag{4.67}$$

4.4 Transitions Between Degenerate States

Up until now adiabatic transitions between nondegenerate levels in a two-level system have been studied. However, the states in real atoms are always degenerate. Degeneracy effects have already been studied both in the framework of perturbation theory (Chap.2) and in the framework of the resonance approximation (Chap.3). This section is devoted to the study of the role that degenerate states play in multi-photon transitions between atomic levels. Of course, one is not dealing with cases in which the system splits into a number of two-level systems in an external field and in which these systems, for one reason or another (e.g., selection rules), are not coupled. Of interest are those cases in which the perturbation matrix elements between the degenerate states are nonzero. This leads to a mixing of all the degenerate states. Theory must provide the means by which one can calculate the transition probabilities with due regard for the degeneracy of these states.

4.4.1 Reducing the Problem Concerning Degenerate States to One Involving a System that Has a Constant Dipole Moment

In general, degenerate states make the problem concerning resonance mixing in an external monochromatic field much more difficult (Chap.3). However, resonance mixing usually occurs in such a way that the degenerate sublevels are evenly populated. This fact simplifies the problem considerably.

Let us examine the transitions between the degenerate levels n and m under the influence of a monochromatic field, $V^{(1)}\sin\omega t$, with the frequency $\omega \ll \omega_{mn}$. The sublevels of n are denoted by n' and those of m by m'. The equation that governs the variation in the population amplitude for state m is

$$i\dot{a}_m = \sum_{m'} V^{(1)}_{mm'} a_{m'} \sin\omega t + \sum_{n'} V^{(1)}_{mn'} a_{n'} \sin\omega t + \frac{1}{2} \omega_{mn} a_m \quad . \tag{4.68}$$

The equations for the variation with time of the other amplitudes for n, m', and n' are all similar.

In view of what has been said, it can be assumed that the populations of all of the sublevels of m are the same, i.e., $a_m = a_{m'}$. It can also be assumed that $a_n = a_{n'}$. Then, instead of (4.68) one has

$$i\dot{a}_m = V^{(1)}_{mm} a_m \sin\omega t + V^{(1)}_{mn} a_n \sin\omega t + \frac{1}{2} \omega_{mn} a_m \quad , \tag{4.69}$$

where

$$V^{(1)}_{mm} = \sum_{m'} V^{(1)}_{mm'} \quad , \quad V^{(1)}_{mn} = \sum_{n'} V^{(1)}_{mn'} \quad . \tag{4.70}$$

The second equation can be written in a similar manner

$$i\dot{a}_n = V^{(1)}_{nn} a_n \sin\omega t + V^{(1)}_{nm} a_m \sin\omega t - \frac{1}{2} \omega_{mn} a_n \quad , \tag{4.71}$$

where

$$V^{(1)}_{nn} = \sum_{n'} V^{(1)}_{nn'} \quad . \tag{4.72}$$

The system of equations (4.69,71) resembles the one for nondegenerate levels. However, the levels m and n have constant dipole moments equal to $V^{(1)}_{mm}$ and $V^{(1)}_{nn}$, respectively. Hence, the system of equations for two degenerate levels has been reduced to a problem concerning two nondegenerate levels with constant dipole moments. It has also been shown how these dipole moments are related to the matrix elements that connect the degenerate sublevels. In conclusion, it can be said that our above approximation is based on the assumption that the sublevels of the degenerate states are equally populated, i.e., that $a_m = a_{m'}$ and $a_n = a_{n'}$ (but, of course, $a_m \neq a_n$), and is realized either if we assume that at the initial moment of time one of the levels is equally populated or if we are interested only in quantities averaged over the scale $1/\omega$ (but not over the scale of the Rabi frequencies). The simplification of the problem is justified if the dipole moment of the state m does not depend on the quantum numbers that characterize the sublevels of this degenerate level. The same is true for the state n.

4.4.2 Degeneracy and Transition Probability

If one compares the system of equations (4.69,71) with (4.32), one notices that the diagonal matrix elements $V_{nn}^{(1)}$ and $V_{mm}^{(1)}$ appeared in (4.69,71). The time-dependent energy $E(t)$ is no longer given by (4.34). Substituting (4.33) into (4.69,71) yields

$$E_{m,n}(t) = \frac{1}{2} (V_{mm}^{(1)} + V_{nn}^{(1)}) \sin\omega t$$

$$\pm \left[\left(\frac{\omega_{mn}}{2} + \frac{V_{mm}^{(1)} - V_{nn}^{(1)}}{2} \sin\omega t \right)^2 + |V_{nm}^{(1)}|^2 \sin^2\omega t \right]^{\frac{1}{2}} . \qquad (4.73)$$

Using the general equation (4.3), one obtains the following expression for the transition amplitude R, instead of (4.46) [4.12]:

$$R = i \, \exp\left(i \int_{t_1}^{\tau} \left\{ \left[\omega_{mn} + (V_{mm}^{(1)} - V_{nn}^{(1)}) \sin\omega t \right]^2 + 4|V_{nm}^{(1)}|^2 \sin^2\omega t \right\}^{\frac{1}{2}} dt \right) , \qquad (4.74)$$

where τ is a point in the upper half-plane of complex time determined by the equation

$$\omega_{mn} = (V_{nn}^{(1)} - V_{mm}^{(1)}) \sin\omega\tau - 2i|V_{nm}^{(1)}| \sin\omega\tau . \qquad (4.75)$$

At this point the integrand in (4.74) vanishes. Also, t_1 is any point on the real time axis. If the diagonal elements vanish, τ is determined by (4.9).

From (4.74) one obtains the transition probability,

$$W_{mn} = |R|^2$$

$$= \exp\left(-\frac{2\omega_{mn}}{\omega} \, \mathrm{Im} \left\{ \int_{t_1}^{\tau} [(1 + p \sin\phi)^2 + q^2 \sin^2\phi]^{\frac{1}{2}} \, d\phi \right\} \right) \qquad (4.76)$$

which generalizes (4.11) to the case of $p \neq 0$. Here, the notations

$$p = \frac{V_{mm}^{(1)} - V_{nn}^{(1)}}{\omega_{mn}} \quad , \quad q = \frac{2|V_{nm}^{(1)}|}{\omega_{mn}} \qquad (4.77)$$

have been introduced. The integral in (4.76), which is a function of p and q, is tabulated in [4.13].

Similarly, one can generalize the condition in (4.63) for the Stark shift of the positions of the multi-photon resonances:

$$\frac{\omega_{mn}}{\omega} \int_0^{\pi/2} [(1 + p \sin\phi)^2 + q^2 \sin^2\phi]^{\frac{1}{2}} d\phi = \frac{2K + 1}{2} \pi \quad . \qquad (4.78)$$

If the stated modifications of the transition probability and of the Stark shift are made, then the rest of the equations in Sects.4.1 and 4.2 can be applied to a system with a constant dipole moment.

If the integrals in (4.76,78) are studied carefully, it can be seen that W_{mn} increases with the dipole moment, p. Thus, in general, the degeneracy of levels brings about an increase in the probability of a transition. However, for this to occur, the dipole moments $V_{mm}^{(1)}$ and $V_{nn}^{(1)}$ must differ considerably. If $V_{mm}^{(1)} = V_{nn}^{(1)}$, the degeneracy of the levels does not have an influence on the transition probabilities or the Stark shifts.

4.5 Bound-Free Transitions

An atomic system that is ionized adiabatically and slowly by a varying monochromatic perturbation can easily be analyzed if the field vanishes at infinity at any moment of time. However, in reality,

$$V(\mathbf{r},t) = zE \sin\omega t \xrightarrow[z \to \infty]{} \infty \qquad . \qquad (4.79)$$

This is the reason why, when the field is switched on, the edge of the continuous spectrum is shifted, and all of the solutions to the adiabatic Schrödinger equation,

$$H(t)\psi_n(\mathbf{r},t) = E_n(t)\psi_n(\mathbf{r},t) \quad , \qquad (4.80)$$

where time is considered a slowly varying parameter, become quasi-stationary. The goal of this section is to be able to describe the ionization probability as a function of the amplitude, E, and frequency, ω, of the field.

4.5.1 Comparison of the Perturbation in Discrete and Continuous States

Let us consider the simple case in which the variables in (4.80) are separable. The coordinate is denoted by z. A typical dependence of the effective potential $U(z,t)$ on time is shown in Fig.4.3.

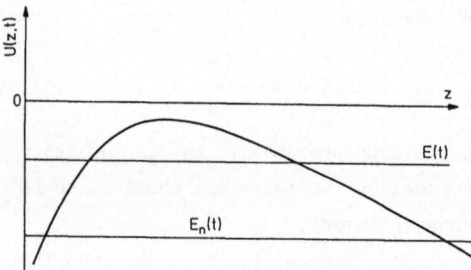

Fig.4.3. A typical interaction potential of a low-frequency field with an atom

It is assumed that the widths of the quasi-stationary levels from which ionization takes place are small compared to their energies. This will be true if the amplitude of the perturbation is sufficiently small compared to the amplitude of the atomic field: $E \ll E_{at}$.

How will the initial and final states of an atom change under the influence of an external field? The final state is usually perturbed much more than the initial state because the electron in the initial state is bound more strongly to the nucleus. For this reason one can consider the initial state in (4.5) to be unperturbed, i.e., one can assume that $E_n(t) = E_n^{(0)}$, and only consider the perturbation of the final state (a state in the continuum).

The potential energy of the electron in the final state will also be neglected. This is certainly justified for the ionization of an anion, whose core is a neutral atom and whose potential is short-range. But, if one is dealing with the ionization of a neutral atom, whose core is a cation, the Coulomb field of the core makes it impossible to neglect the potential energy of the electron a priori. However, if exponential accuracy is sufficient, the field of the nucleus can always be neglected.

4.5.2 Probability of a Transition in a Low-Frequency Field

To calculate the probability of ionizing an atom, one can use (4.5) of the Landau-Dykhne adiabatic approximation. The final state will be taken to be the state of a free electron in an electromagnetic linearly polarized field. In such a field the electron will be forced to oscillate at the frequency ω of the external field. The energy of such oscillations is

$$E(t) = \frac{E^2}{2\omega^2} \sin^2 \omega t \quad . \tag{4.81}$$

This is an elementary consequence of classical physics. Thus, the probability of ionization is

$$W_{nE} = \exp\left(-2 \; \text{Im}\left\{\int_0^\tau [E(t) - E_n^{(0)}] \; dt\right\}\right) \quad , \tag{4.82}$$

where τ is a point in the upper half-plane of complex time that is determined by the condition that $E_n^{(0)} = E(t)$. Hence,

$$\tau = \frac{i}{\omega} \sinh^{-1} \left(\frac{\omega(2|E_n^{(0)}|)^{\frac{1}{2}}}{E}\right) . \tag{4.83}$$

It is important to note that since the motion of the electron is periodic, τ is not the only turning point. There are other points the same distance from the real time axis but separated from one another by $N\pi/\omega$, where N is an integer. In Sect.4.2 it was shown that this leads to a transition from absolute probability to a probability per unit time. Here, this difference is not of interest.

Using (4.81) and (4.83) to evaluate the integral in (4.82) yields [4.14]

$$W_{nE} = \exp\left[-\frac{2|E_n^{(0)}|}{\omega} f(\gamma)\right] \quad , \tag{4.84}$$

where (for a linearly polarized field)

$$f(\gamma) = \left(1 + \frac{1}{2\gamma^2}\right) \sinh^{-1}\gamma - \frac{1}{2\gamma}(1 + \gamma)^{\frac{1}{2}} \tag{4.85}$$

and

$$\gamma \equiv \frac{(2|E_n^{(0)}|)^{\frac{1}{2}}\omega}{E} \quad . \tag{4.86}$$

The parameter γ is known as the adiabaticity parameter. It determines how an atom is ionized by a low-frequency field. The condition $|E_n^{(0)}| \gg \omega$ guarantees that the probability W_{nE} remains exponentially small.

Originally (4.84) was derived from the exact equation for the transition probability [4.14]:

$$W_{nE} = \left|\int_{-\infty}^\infty (\psi_n^{(0)}, V\Psi_E) \; dt\right|^2 \quad , \tag{4.87}$$

where $\psi_n^{(0)}$ is the unperturbed wave function of the initial state, V is the perturbation due to the electromagnetic field, and Ψ_E is the wave function of the final state, which is perturbed by the field. It is easily verifiable that, in the adiabatic approximation for Ψ_E, (4.82) and (4.87) are equivalent to within exponential accuracy. Naturally, after substituting for Ψ_E the known wave function of a free electron with momentum **p** in an electromagnetic field [4.15],

$$\Psi_E(\mathbf{r},t) = \Psi_{\mathbf{p}}^{(0)}(\mathbf{r},t) \times \exp\left[i\, \frac{\mathbf{p} \cdot E}{\omega^2}\, \sin\omega t - i\, \frac{E^2}{4\omega^2}\left(t + \frac{\sin 2\omega t}{2\omega} \right) \right] \quad , \qquad (4.88)$$

where

$$\Psi_{\mathbf{p}}^{(0)}(\mathbf{r},t) = \exp\left(i\mathbf{p} \cdot \mathbf{r} - \frac{1}{2}\, ip^2 t \right) \quad , \qquad E = \frac{1}{2}\, p^2 \qquad\qquad (4.89)$$

are the wave function and the energy of the free electron, one obtains (4.84).

In [4.16] another method of obtaining (4.84) was studied. It is called the "imaginary time method" and is based on an equation similar to (4.82), in which the classical Hamiltonian functions are introduced instead of the quantum-mechanical energies $E(t)$ and $E_n^{(0)}$. These functions are written as sums of the kinetic and potential energies, which are given in terms of the classical coordinates and velocities of the electron as functions of time. It is easily verifiable that the result coincides with (4.82) to within exponential accuracy. Naturally, after neglecting the atomic potential of the final state, one arrives at (4.84).

4.5.3 Limiting Cases

Let us discuss the various limiting cases of (4.84). The limiting expressions for $f(\gamma)$ are:

$$f(\gamma) = \frac{2}{3}\gamma \quad , \qquad f(\gamma) = \ln(2\gamma/e^{\frac{1}{2}}) \quad . \qquad\qquad (4.90)$$
$$\gamma^2 \ll 1 \qquad\qquad\quad \gamma^2 \gg 1$$

In the first case,

$$W_{nE} = \exp\left| -\frac{4\sqrt{2}\,|E_n^{(0)}|^{3/2}}{3E} \right| \quad . \qquad\qquad (4.91)$$

This is nothing more than the probability, calculated to within exponential accuracy, of extracting an electron from a potential well with a constant electric field E. A comparison of the solution for the ionization from a short-range potential well and that for the ionization from a Coulomb potential [Ref.4.1, §77, Problems 1 and 2] shows that the exponent of the exponential function is the same for both cases. Differences between the types of potentials only influence the pre-exponential factor. This justifies neglecting the atomic potential when $\gamma \ll 1$. Also, the condition $\gamma \ll 1$ does not contradict the condition $E \ll E_{at}$, because $|E_n^{(0)}| \gg \omega$. Equation (4.91) corresponds to the tunneling of the electron through the potential barrier depicted in Fig.4.3.

In the opposite limiting case, when $\gamma^2 \gg 1$,

$$W_{nE}_{\gamma^2 \gg 1} = \left(\frac{e^{\frac{1}{2}}E}{2(2|E_n^{(0)}|)^{\frac{1}{2}}\omega}\right)^{2|E_n^{(0)}|/\omega} . \tag{4.92}$$

This is nothing more than the probability, calculated within the framework of perturbation theory to the $|E_n^{(0)}|/\omega^{th}$ order, of an electron being extracted. Also, $K_0 = |E_n^{(0)}|/\omega$ is the number of photons needed to transfer the electron from the initial state n to the continuum.

Equation (4.92) coincides with a similar result obtained in Chap.2 with an arbitrary potential. Hence, when $\gamma \gg 1$, the role that the atomic potential plays in the motion of the electron in the continuum is small. The only exceptions are those in which the differences $|E_n^{(0)}| - K\omega$ (with $1 \leqslant K < K_0$) coincide with the energy of an excited discrete state. One then is dealing with an intermediate resonance, which greatly complicates the problem. In this case the adiabatic theory under consideration cannot be applied. However, such resonances may become weaker in strong fields because of the broadening of the intermediate states. This is an additional argument for the usefulness of (4.92). Hence, (4.92) describes the multi-photon limit in which the electron in the potential well continues to absorb photons from the external field until its energy becomes positive.

On the basis of what has been said, one can conclude that the atomic potential only plays a significant role in finding the pre-exponential factor to the ionization probability in the intermediate case in which $\gamma \sim 1$. Summing up, one can say that (4.84) is an accurate (to within exponential accuracy) description of the ionization process for any value of γ and for any atomic potential.

5. Laser Radiation

When studying lasers as sources of light, the four basic parameters that are of interest are: the intensity, the frequency, the polarization, and the degree of monochromaticity. In reality the situation is much more complex, since it is necessary to take into account not only the absolute value of the intensity but also the spatial-temporal distribution of the beam over the atomic target; not only the absolute value of the frequency but also the possibility of a smooth variation in the frequency and the spectral width of the radiation; not only the degree of polarization but also the degree of ellipticity.

Does the radiation of modern lasers satisfy the specifications that theory imposes on electromagnetic fields? Can perturbing fields be set up that are as strong as atomic fields in a frequency range from the ultraviolet to the infrared, that exhibit linewidths smaller than those of atomic levels, and that show complete polarization and a given degree of ellipticity? At present this is impossible, although no doubt it will become possible in the near future. At present, such optimal values can only be realized for one or several parameters and only with certain lasers. For this reason the choice of a laser must be made on the basis of the features of the particular experiment. Although there are no lower limits for the values of the basic parameters that characterize strong electromagnetic fields, theoretical estimates predict that these fields are of the order of 10 V/cm, have wavelengths that range from several tenths of a micrometer to several micrometers, and that their radiation is completely polarized and has a given degree of ellipticity. This lower limit on the field strength corresponds to a mixing of levels in the event of a one-photon resonance in a two-level system. For multi-photon or nonresonant transitions, the lower limit can increase to 10^4V/cm, sometimes even higher. Consequently, if in the first case one uses continuous-wave (cw) lasers, one will likely use Q-switched pulsed lasers in the second [5.1]. The short duration of the pulses in the second type of laser (about 10^{-8}s) and their high energies produce the needed power. Since the yield of nonlinear, optical effects largely depends on the strength of the perturbing field, such effects can be used to transform frequencies only if the initial radiation is powerful enough, i.e., practically speaking, only with the radiation from Q-switched lasers.

Although lasers are known as sources of unidirectional, monochromatic radiation, this radiation is at best only quasi-monochromatic ($\Delta\omega/\omega \gtrsim 10^{-7}$). The wavefront has a very large but finite radius of curvature, and the amplitude of the field varies strongly near the wavefront. This must be kept in mind whenever experimental and theoretical findings are compared. In theory it is always assumed that the wave is planar and monochromatic ($\Delta\omega/\omega \equiv 0$), and that the amplitude of the field near the wavefront is a constant.

5.1 Intensity of Radiation

The intensity is defined as the number of photons that pass through a unit area per unit time. The following formula relates the intensity of the radiation p (photons per square centimeter per second) to the field strength E (volts per centimeter) and the power density P (gigawatts per square centimeter):

$$F \approx \frac{0.6}{\hbar\omega} 10^{28} P \approx \frac{8.3}{\hbar\omega} 10^{15} E^2 \quad , \tag{5.1}$$

where $\hbar\omega$ is the energy of the radiation in electron volts. In the following sections all three characteristics will be used in accordance with current nomenclature.

5.1.1 Spatial-Temporal Distribution of the Intensity of Laser Radiation

When studying nonlinear effects, it is obvious that the laser radiation must be characterized by its intensity, and not by the energy in a single pulse, as is the case with linear effects. One must bear in mind that the spatial-temporal distribution of the radiation is essentially nonuniform. For cw lasers this spatial nonuniformity is of utmost importance, whereas for pulsed lasers both effects play an important role. Pulsed lasers will be discussed later. The method of taking into account the nonuniformity of cw laser radiation is a part of the problem which will now be dealt with.

The nonuniformity in the distribution of laser radiation is a typical phenomenon. Spatial nonuniformity is due to the fact that as a source of light a laser is very close to a point source. Consequently, the spatial distribution of the field is determined by diffraction. Temporal nonuniformity is due to the fact that the duration of the wavefront and the length of the pulse are of the same order of magnitude. In a good laser the envelopes of both distributions are close to Gaussian. Obviously, the nonuniformity does not vanish when the laser radiation is transformed by linear or nonlinear optical methods.

The root of the difficulties in characterizing the distribution of the laser radiation lies in its spatial-temporal nonuniformity. Hence, one must use averaged parameters. When the interaction is linear, one uses arithmetic means, such as the well-known concepts of half-width and half-height. Such an approach is not justified in nonlinear optics, since the result of the action of a laser beam is a nonlinear function of its intensity. When the effects are nonlinear, the effective value of the intensity of the light can be taken as the maximum value. Such a choice reflects the real situation the better, the higher the degree of nonlinearity of the process and the greater the nonuniformity of the distribution.

From what has been said one can conclude that it is desirable to write the intensity of the laser radiation $F(x,y,z,t)$ in the form $F = F_{max} f(x,y,z,t)$. Let it be assumed that the spatial and temporal variables in f can be separated, i.e., $f = \phi(x,y,z)\psi(t)$. Such a possibility is very important for the actual measurement of the intensity of the laser radiation. It is not realizable in all cases because it is dependent upon the mode in which a particular laser is operating. Let it also be assumed that the variation in the intensity along the direction of the beam (the z axis) can be neglected compared to the variation in a plane that is normal to the direction of propagation (whether this assumption is true or not depends on the mode of operation). Then, by substituting $\eta(x,y)$ for $\phi(x,y,z)$ one can write an equation that relates the intensity of the radiation to the energy E of the pulse:

$$F_{max} = \frac{E}{S \tau \hbar \omega} , \qquad (5.2)$$

where S is the effective cross section of the beam, obtained as a result of normalizing the distribution function $\eta(x,y)$ to the maximum value of this function:

$$S = \frac{\int \eta(x,y)\, ds}{\eta_{max}} , \qquad (5.3)$$

and τ is the effective length of the pulse:

$$\tau = \frac{\int \psi(t)\, dt}{\psi_{max}} , \qquad \eta_{max}\psi_{max} = 1 . \qquad (5.4)$$

From (5.2) it follows that the greater the energy, E, of the pulse, the smaller the cross section, S, through which this energy is transported, and the shorter the pulse length, τ, the higher the intensity of the radiation. The measurement of the intensity can be reduced to a measurement of the pulse energy, the illumination η in the xy plane, and the energy distribution function over the pulse length ψ.

5.1.2 Dependence of the Intensity on the Type, the Design, and the Mode of Operation of a Laser

The values of E, S, and τ strongly depend on the choice of the active material, the method by which a population inversion is created (the method of pumping), the design of the cavity, and the method of Q-switching.

One can distinguish between four time domains of laser operation, which depend on the pumping method and on the Q-factor of the cavity: i) continuous-wave operation (continuous pumping, Q-factor of the cavity is fixed); ii) millisecond, or normal mode (0.1 ms $\leqslant \tau \leqslant$ 1 ms, pulsed pumping, Q-factor of the cavity is fixed, free-running operation, the intensity of the radiation of a pulse changes randomly but with a characteristic time of the order 1 μs, the so-called spiking); iii) nanosecond operation (1 ns $\leqslant \tau \leqslant$ 100 ns, pulsed pumping, Q-switching, length of the leading edge of the pulse 1-10 ns, the "giant pulsing"); iv) picosecond operation ($\tau \leqslant 10^{-11}$s, pulsed pumping, Q-switching, length of the leading of the pulse is of the order of 10^{-11}s, transverse-mode locking).

In (iii) and (iv), the repetition rate is about one pulse per minute. Obviously, for a fixed pulse energy, the intensity of the radiation in picosecond operation (iv) is about eight orders of magnitude greater than that in millisecond operation (iii). In going from the cw operation to Q-switching, (i) to (iii), the pulse energy is only reduced by a few times, whereas the length of the pulse is reduced by several orders of magnitude. Hence, Q-switching is a very effective method of increasing the intensity of the radiation.

Although the intensity of cw lasers is much weaker than that of pulsed lasers, it is sufficient to allow the study of a great variety of processes involving small numbers of photons or the study of resonance processes. This is because the probability of their occurrence is relatively high.

Depending on the active material, a variety of lasers can work in one of the above modes of operation. Some lasers can even be used in all of them (e.g., the ruby laser). To a great extent, the time domain of lasing depends both on the pumping and on the method of Q-switching.

An electromagnetic field in an open cavity in which an active material is located between reflecting elements exhibits definite, stable configurations called modes of oscillation [5.1]. It is standard practice to denote modes by a symbol such as TEM_{mnq}, in which the subscripts correspondto functions that characterize the distribution of the field in a plane normal to the axis of the cavity (labels m and n) and along the axis (label q). The field can have a transverse structure if diffraction causes the light to travel at an angle to the cavity's axis. A longitudinal structure results when an integral number of half-waves fits into the separation of the reflecting elements, and some of the corresponding frequencies fall in the finite width of the luminescence spectrum of the active material. At optical wavelengths (λ) this number $N \approx \ell/\lambda$ is very large (about 10^6), ℓ being the distance between the reflecting elements. The modes whose frequencies satisfy this condition and experience gain are known as axial or longitudinal modes. They correspond to a standing wave with an integral number of half-waves in the resonator and with nodes on the surface of the reflecting elements. For every axial mode there exists a group of nonaxial (transverse) modes. The distribution of the field along the wavefront has been calculated for various cases. For simplicity, the axial mode of the lowest transverse order (TEM_{00q}) is assumed to have a Gaussian distribution: $E^2(r) = \exp(-2r^2/r_0^2)$, r_0 being a measure for the spatial width [5.2].

The mode composition of the radiation plays a very important role whenever there arises the need to measure the intensity of the laser light exactly and to maintian this intensity over a large number of successive pulses. In a large number of lasers suitable for experiments (solid-state lasers [5.3], dye lasers [5.4], excimer lasers [5.5]), the initial inhomogeneity of the active material, the distortion of the configuration of the active material by the pumping light, and the inhomogeneity of the pumping light cause a spread in the lengths of the optical paths of the rays in the cavity. It is for this reason that the distribution of the intensity along the wavefront is not only essentially inhomogeneous but also irregular and is not necessarily reproduced in successive pulses.

A complexer spatial-temporal distribution is found in Q-switching. The finite time required for the Q-factor of the resonator to increase, the large number of modes for which the threshold conditions are satisfied, and the above-mentioned inhomogeneities all lead to differences in the temporal development of the modes. The changes in the divergence of the beam and in the spatial distribution of the radiation that take place during a long pulse suggest that, when there is Q-switching in multi-mode lasing, the intensity is not exactly described by (5.2), because the spatial and temporal variables cannot be separated. This relationship can only be used as an approximation.

Single-mode lasing has considerable advantages over multi-mode operation. The intensity distribution over the wavefront is close to a Gaussian function and is repeated over a large number of successive pulses. Moreover, one can use (5.1) to calculate the intensity. Different transverse-order modes TEM_{mnq} result from differences in diffraction losses at the mirrors; the lowest-order mode TEM_{00q} has the lowest losses. To isolate a single mode one can place an aperture perpendicular to the cavity's axis. This reduces the pulse energy E, and can be offset by a decrease in beam cross section due to a smaller divergence of the resulting mode of lower transverse order.

5.1.3 Increasing the Intensity of Laser Radiation

There are two ways of increasing the intensity of laser radiation: i) using a quantum amplifier, and ii) focusing the beam, i.e., reducing its cross section. Amplification of light is achieved as it passes through an inversely populated medium due to coherent de-excitation of the excited states. The linear amplification of a traveling wave is restricted by, for example, saturation in the resonantly amplifying medium. Such nonlinear effects change both the temporal and the spatial distributions of the radiation [5.6]. The amplifier gain is limited by the lifetime of the population inversion in the active medium of the amplifier under the influence of laser radiation. Since the gain depends on the intensity, there is no way of finding numerical parameters of a sufficiently general nature. Usually one speaks of a gain of several units to several tens of units per gain stage.

The exceptional properties of laser radiation make it possible, in principle, to focus the beam to an area of one wavelength. Indeed, if the laser generates a TEM_{00q} mode and if the focusing lens is far enough away from the laser, then the wavefront near the surface of the lens may be very close to being planar and the distribution of the intensity over the entrance pupil may almost be homogeneous. In this case, the intensity distribution in the focal plane is described by Airy's formula for Fraunhofer diffraction for a circular aperture [5.6]. The first minimum of the Bessel function is found at $r_1 \approx 1.2 \, f/d$ (the radius of the Airy disc), where f is the focal length of the lens and d is the diameter of the entrance pupil. If the relative aperture of the objective, d/f, is of the order of unity, the light can be focused into a circle of radius λ, which is roughly 1 μm at optical wavelengths. In reality everything is less than ideal because of the large spherical aberrations that accompany relatively large apertures and because the entrance pupil is not uniformly illuminated. However, the general picture of a distribution in the form of a Bessel function with alternating maxima and minima is still valid. The characteristic distribution of light along the axis of the focusing system consists of periodically spaced minima which are due to interference, between axial and aperture rays. A good estimate for the Airy disc can be made, even if the beam does not cover the entrance pupil uniformly, using the diameter of the beam.

Even when the laser operates in the TEM_{00q} mode, calculations of any kind can only serve to estimate the order of magnitude of the intensity of the focused light. It is also always important to determine the distribution not only in the focal plane but also in a small region near the focus of the lens, where the observation takes place. Even for the ideal case (with a plane wave, a uniform distribution of the intensity over the wavefront, a circular entrance pupil, an ideal lens) the spatial distribution has a very complex configuration [5.8]. Since real cases are far from ideal (almost always when the intensity of the radiation must be measured exactly), one must first determine the spatial-temporal distribution.

When using short-focus lenses, one must always bear in mind that, in reality, the wave is not planar. The distance between the plane of focus and the focal plane may be comparable to the distance along the axis of the focusing system over which the intensity changes considerably.

5.1.4 Varying the Intensity of Laser Radiation

In the study of most nonlinear phenomena, one of the main objectives is to measure the dependence of the various parameters of a particular effect on the beam intensity. The very fact that a nonlinear phenomenon is observed follows from observation of the nonlinear dependence of the probability of the phenomenon (its yield) on the intensity. For this reason the linear variation of beam intensity (due to noise or fluctuations) is a typical problem of most experiments. In highly nonlinear processes, the variation of intensity must be kept small. This results in preciser measurements.

To change the beam intensity, one never changes the operation mode of the laser (though in principle this is possible), because too many parameters that must be monitored are changed. The usual practice is to thoroughly stabilize the lasing so as to obtain the needed maximum intensity and then to place an attenuator into the path of the beam. This makes it possible to reduce the energy E within required limits without changing the spatial-temporal distribution (S and τ). The main practical difficulty lies in keeping the attenuation of the intensity linear. If absorbers are used, one must be sure that no nonlinear phenomena appear. For example, if filters made of colored optical glass are to be used, one must choose those whose absorption coefficient is a slowly varying function of frequency. Identical requirements are imposed on spatial-temporal attenuators which are solutions of various substances of varying concentrations. Another convenient method of attenuation is to reflect light from a glass plate placed at an angle α to the direction of the beam. Here, the radiation to be attenuated must be linearly polarized, no interference should exist between the transmitted and the reflected light, and a second plate placed at the angle $(180^{\circ} - \alpha)$ must restore the initial direction of the beam. For elliptically or circularly polarized light an attenuator of this kind will change the polarization.

5.1.5 Measuring the Intensity of Radiation

To measure the intensity of radiation one must measure the pulse energy E, the spatial distribution of the illumination $\eta(x,y)$ in a plane perpendicular to the beam's axis, and the temporal distribution of the energy $\psi(t)$. Obviously, both the intensity in the beam and at the focus of the lens can be found once these parameters are known.

To measure the energy of a pulse, it is common practice to use different types of calorimeters. These make it possible to measure energies ranging from 10^{-3} to 10^{2}J with an absolute accuracy of several percent [5.10].

The spatial distribution of beam intensity integrated over the length of the pulse can be determined using the photographic process [5.11]. The radiation is attenuated in such a way that the illumination of the photographic film placed in the xy plane corresponds to the linear part of the curve film blackening versus illumination. One must bear in mind that for nanosecond (even more so for picosecond) pulses the law of the interchangeability of the intensity and the duration breaks down [5.12]. For this reason one must measure the gamma of the photographic emulsion (which correlates the blackening of the film and the exposure time) using laser radiation rather than use the data given by the manufacturer, which is intended for standard exposures of about 1 s. To find $\eta(x,y)$ and S the light which falls on the film must be measured photometrically. By moving the film along the axis of the beam, one can obtain, with a sequence of laser pulses, data on the spatial distribution of the laser radiation.

If the radiation is focused on the target in such a way that the distribution of illumination in the focal region cannot be directly measured, the measurements are made in the focal region of an identical simulating lens. This lens is placed the same distance from the laser (i.e., the virtual light source) as the main, focusing lens in the path of a simulating, low-intensity beam.

If, according to the focusing conditions, S is very small, then a non-aberrational system (lens or microscope), which projects the xy plane onto the photographic film with a definite magnification, is placed on the axis of the focusing system to diminish measuring errors. It is essential that the depth of focus of this optical system be smaller than the characteristic distance along the axis of the focusing system, over which the intensity changes considerably.

The length of the laser pulse is usually measured with the help of a photodiode and an oscilloscope. Depending on the aim of the experiment, either the whole light beam or only a fraction is directed towards the photodiode. The standard resolution of such a system is of the order of 10^{-10} s. Nowadays, it is possible to obtain higher resolution (up to 5×10^{-13} s) with the help of image converters, photorecorders, or streak cameras [5.12,13].

If the laser operates in transverse modes with different indices, the measurement of the intensity becomes very complex. In any case, one can easily over- or underestimate the intensity by several orders of magnitude. In the case of cw lasers, the main source of error, i.e., the inhomogeneity in the spatial distribution, still prevails. For an automatic measurement of the spatial-temporal distribution of pulse energy a scheme was proposed in [5.14].

5.2 Frequency of the Radiation

At present intensities light beams can be produced for a wide range of frequencies, from the ultraviolet to the infrared, by an appropriate selection of the active material (and, hence, laser type). The width of the laser line not only depends on the active material but also on the design of the laser. In principle, it can vary from several hundred reciprocal centimeters (for spontaneous transitions such as luminescence and fluorescence) to $\sim 10^{-3}$ cm^{-1} (for a single longitudinal mode). Methods of nonlinear optics can also be used to vary the frequency of intense laser radiation.

The possibility of generating one longitudinal mode is very important for the study of nonlinear optical phenomena. Although the spectral widths of a small number of modes may satisfy the above requirements, multi-mode lasing significantly influences the experimental results because there are such experimentally unaccountable parameters as the phase and the amplitude of a mode. With regard to the phase, one can distinguish between types of multi-mode generation: random phase distribution, partial mode locking, and total mode locking. The advantage of these types lies both in the reproducibility of the results and in their clear theoretical description. By applying complete mode locking the experimenter can produce ultrashort pulses in the picosecond range [5.15] and [Ref.5.12, Chap.2]. In most cases a smooth variation of the lasing frequencies is associated with the use of an active material with a wide spontaneous absorption or emission spectrum and a cavity the Q-factor of which can change within a range of frequencies.

From the frequency point of view it is most interesting to utilize the visible and ultraviolet regions when studying nonlinear optical phenoma in atoms and in atomic spectra. The criteria for the spectral width are more complex because they largely depend on the specific problem and the experimental arrangement. Typical problems in laser spectroscopy demand that the spectral widths be as small as possible [5.16].

5.2.1 Changes in the Tunability and Spectral Narrowing

When the experimental conditions demand a smooth change in the lasing frequency, the experimenter chooses an active material with a sufficiently broad spectral width and uses a cavity with a dispersive element [5.4]. One essential difficulty that arises with the use of dispersive elements is the usual contradiction when one wishes to increase the region of dispersion (an interval designed for waves that do not overlap in different orders) and the resolving power (which determines the linewidth). To obtain the narrowest possible lasing line for the widest spontaneous emission line, one must use several dispersive elements, e.g., multiplexers (complex Fabry-Perot interferometers). Methods which can bring about a change in frequency but which do not cause a change in the spatial-temporal characteristics of the laser radiation (e.g., the geometry of the cavity) have great practical advantages. An example of this is the Wood filter, whose operation is based on the change that takes place in the length of a doubly refracting crystal as the temperature changes [5.17]. In multi-mode lasing a decrease in spectral width corresponds to a decrease in the number of generated modes. This causes a wider spread in the yield of the nonlinear phenomenon due to fluctuations in the phase of the mode. Obviously, the smallest spectral width is obtained with a single-mode laser.

5.2.2 The Single-Mode Laser

With a sufficiently inhomogeneous spatial distribution in the inversely populated lasing medium one can generate in any laser the simplest transverse
mode (TEM_{00q}) and suppress the higher modes by introducing diffraction losses,
for example, by changing the aperture size. The frequency width is determined
only by the number of longitudinal orders excited. The best method of isolating a longitudinal order (single-frequency operation) involves the use of
a complex cavity that consists of two or more coupled cavities. In contrast
to a simple two-mirror cavity, such a system generates nonequidistant frequency
lines [5.3]. The relative gain of modes in such a cavity depends on the tuning of the cavity and on the amount of pumping above threshold. The use of
coupled cavities makes single-frequency lasing possible. Since the width of
the single-frequency line is determined by the pulse length ($\Delta\omega T \sim 1$), a typical value for $\Delta\omega$ in the giant pulse operation is $10^{-3} cm^{-1}$ which corresponds
to a typical value of 10^{-8}s for T [Ref.5.12, Chap.2].

5.2.3 The Measurement of Laser Frequency

Three distinct problems always arise in conducting experiments: i) the measurement of the absolute frequency of the laser, ii) the measurement of the
relative variation of this frequency, and iii) the measurement of the frequency distribution over the spectrum (the shape of the laser line and its
half-width). All three problems can be solved by standard spectroscopic
methods [5.4].

Nowadays, spectral measurements are usually much simpler than those that
were made in pre-laser optics because laser light is highly monochromatic and
its wavelength can be measured with considerable precision. Difficulties
arise when the spectrum is measured using a single-mode laser. The resolving
power of standard spectral instruments is restricted to values of about
$10^{-2} cm^{-1}$. For this reason the width of the spectrum is usually determined by
measuring the length of the emitted pulse. In fact, one studies the temporal
distribution of the laser radiation to be assured that single-frequency lasing is taking place. With a few adjacent modes whose phases are nonlocked,
beats can take place between the modes, influencing the temporal distribution of the radiation. The shape of the envelope is not a Gaussian function;
it exhibits several maxima and minima, which are usually not resolved by
conventional photodiodes.

5.3   The Polarization of the Radiation

Since laser radiation is not perfectly monochromatic but only quasi-monochromatic, it is usually characterized by its degree of polarization (for
partially polarized light) and by its degree of ellipticity (for fully polarized light) [5.16]. The high intensity of the non-monochromatic pump light
used to produce population inversions leads to an anisotropy in the active
medium of solid-state lasers. This destroys the natural anisotropy of the
crystal and the isotropy of glass. Thus, whenever a laser must produce
fully polarized light with a given degree of ellipticity, one must introduce an element into the cavity which makes the Q-factor dependent on the
polarization of the light. The simplest way to do this is to postion one or
several plane-parallel glass plates at the Brewster angle to the direction of
the light. The absence of losses for light that is linearly polarized creates favorable conditions for the generation of linearly polarized light.
Since the development of the lasing process is associated with a multiple
propagation of light along the axis of the cavity, the small differences
in the losses for light of different polarizations are increased several orders of magnitude. To change the degree of ellipticity of light linearly
polarized in two perpendicular planes and, in particular, to obtain circularly polarized light, one uses the phase-plate method [5.16]. The phase
plate divides the impinging light into two components, shifts the phase of
one relative to the other, and then combines the two into one beam. Since
the path length of the light in the phase plate may change with the temperature of the plate, different degrees of ellipticity for different wavelengths can be achieved for a constant spatial distribution of the light.

The degrees of polarization and ellipticity are measured using standard
methods [5.8] with the help of a $\lambda/4$ phase plate, an analyzer (polarizing
prism), and a calorimeter. Two types of analyzers, the Foucault prism and
the Glan-Thompson prism have an advantage over others because neither has a
glued surface and the radiation resistance is sufficiently high. Since the
measurement of the degree of polarization is complex (one must independently change the orientations of the phase plate and the analyzer), it is expedient to use compensators for some investigations (e.g., the Babinet com-
pensator or the Soleil plate [5.8]). They allow the determination of the
phase shift in elliptically polarized light in a single measurement. Such
methods should be used when working with high-power pulsed lasers which emit
about one pulse per minute. The accuracy of the measurements depends, primarily, on the relative precision with which the pulse energy can be measured.

5.4 The Monochromaticity of the Laser Radiation

It has already been mentioned in the introduction to this chapter that laser radiation is always quasi-monochromatic. The degree of monochromaticity depends on the operation of the particular laser and on the stability of the particular mode excited. It is highest in single-frequency lasing, where the width of the laser line is 10^{-3} to $10^{-4} cm^{-1}$, i.e., of the order of the natural linewidth of excited atomic states, and lowest in multi-mode lasing, where the width of the laser line may reach hundreds of cm^{-1}.

In addition to the linewidth, which is an integral characteristic of a spectrum, the radiation also differs considerably in the distribution $F(\omega)$ and $F(t)$. In single-mode operation when longitudinal and transverse indices are fixed, the distributions $F(\omega)$ and $F(t)$ are smooth bell-shaped functions having widths of $\Delta\omega T \sim 1$. In multi-mode lasing, the distribution $\Delta\omega$ and $F(t)$ are not usually smooth functions; their specific forms are determined by mode locking. The most interesting case from a practical point of view is when the modes are unlocked and their phases are distributed randomly. This occurs when no special steps are taken to select the modes and no nonlinear elements are introduced into the laser cavity. The radiation contains multiple frequencies (many longitudinal orders with a fixed transverse configuration) as well as multiple modes (many modes with different transverse orders).

Without mode locking, the radiation field depends on time in a random manner and the distribution $F(t)$ exhibits pronounced fluctuations so that the intensity drops to zero and is restricted by the energy per axial period (the time it takes for the light to pass through the cavity). The inverse width of the spectrum is determined by the time scale of the fluctuations, i.e., the correlation times $\tau_{cor} \sim 1/\Delta\omega$. For times $t < \tau_{cor}$, the function $F(t)$ changes only slightly so that it may be assumed that $F(t) \approx F(0)$. When $t > \tau_{cor}$ the function $F(t)$ can take on all possible values which are given by the distribution $P(F)$, —one of the integral characteristics of the radiation. Typical linewidths of multi-mode radiation are less than $100\ cm^{-1}$. The correlation time $\tau_{cor} \sim 10^{-12}$ s, so that the intensity of the radiation fluctuates very rapidly. By using the standard method (a photodiode and a high-speed oscillograph) it is impossible to register the fluctuations in the intensity with the present available instruments which can determine time intervals of up to $\sim 10^{-10}$ s. Hence, one can usually only measure integral characteristics of multi-mode radiation. In this case one must use correlalation techniques to determine the instantaneous values of the different parameters that characterizes the radiation [Ref.5.12, Chap.3].

The dependence of the distribution function P(F) on the number of lasing modes is given by [5.15]

$$P(F) = \frac{1}{<F>} \left(\frac{N-1}{N}\right)\left(1 - \frac{F}{N<F>}\right)^{N-2} \quad , \tag{5.5}$$

where <F> is the ensemble average of F. This equation is valid for a real Gaussian form of $F(\omega)$ if it is assumed that the number of modes is equal to twice the width of the distribution $F(\omega)$ [5.15]. As $N \to \infty$, P(F) turns into a well-known exponential distribution which describes a source of heat:

$$P(F) = \frac{1}{<F>} \exp\left(- \frac{F}{<F>}\right) \quad . \tag{5.6}$$

For simplicity and reliability in the interpretation of experimental data the situation in which the properties of multi-mode laser radiation are equivalent to those of a heat source exhibits obvious advantages. The closeness of the distributions given in (5.5,6) depends on the degree of nonlinearity of the process that takes place due to the laser radiation. This will be discussed in Sect.9.1.

The other limiting case occurs when all of the phases of the modes are completely locked. An ultrashort pulse of picosecond duration is emitted with a width of $\Delta\omega \sim 1/T \sim 10^2 cm^{-1}$. There are two distribution functions, $F(\omega)$ and $F(t)$. To determine $F(\omega)$ streak cameras must be used [5.13].

The intermediate cases involve partially interlocked modes. They are the least suitable for practical use due to the obvious difficulties that arise in measuring the parameters that characterize the radiation, in keeping these parameters stable, and in describing the radiation mathematically.

The radiation from a single-frequency pulsed laser operating in a single (axial) mode with constant phase ϕ and amplitude that changes with time according to the shape $\psi(t)$ of the pulse is given by

$$E(t) = E_0[\psi(t)]^{\frac{1}{2}} \exp[i(\omega_0 t + \phi)] \quad . \tag{5.7}$$

The frequency ω_0 may undergo smooth variations during the pulse due to changes in the refractive index of the active medium. This leads to a broadening of the radiation spectrum as compared with the inverse time of the pulse length.

The field of a single-frequency cw laser is weak and varies slowly with time due to natural fluctuations. These, in turn, are due to the contribution to the radiation of spontaneous noise. The width of the spectrum is determined by the fluctuations. Such radiation may be described by the phase-diffusion model:

$$E(t) = E_0 \exp[i\omega_0 t + i\phi(t)] \quad , \tag{5.8}$$

where E_0 is a constant and the phase is dependent on time.

The field of a multi-frequency pulsed laser with unlocked modes is simply the sum of fields with fixed (for a given momentum) phases ϕ_n and amplitudes $[\psi(t)]^{\frac{1}{2}} A_n$ that change according to the shape $\psi(t)$ of the pulse:

$$E(t) = [\psi(t)]^{\frac{1}{2}} \sum_n A_n \exp(i\omega_n t + i\phi_n) \quad , \tag{5.9}$$

where $\omega_n = \omega_0 + 2\pi n/T$ and n is the index of the mode. The phases and amplitudes of the different modes change from one pulse to another independently of each other. The shape of the spectrum is close to a Gaussian function, and the distribution $P(F)$ is described by (5.6).

In conclusion, one can say that the differences of the influences produced by single-frequency and multi-frequency radiation are qualitative rather than quantitative in a number of cases. These are cases in which the instantaneous, not the averaged, characteristics of the radiation play a significant role. Section 9.1 is devoted to these cases.

6. Experimental Aspects

The diversity of the physical phenomena that are observed when intense electromagnetic radiation interacts with atoms is the reason for the great variety
of experimental methods in this field. However, only those features that are
common to all of these experimental methods will be discussed in this chapter.

The unique conditions in which the nonlinear phenomena are observed, namely, the small volume in which the strong electromagnetic field is created,
the small time interval in which the field interacts with the atom, the high
intensity of the field which allows the electron to be transferred from one
state to another, and the coherent nature of the interaction all lead to
experimental conditions which are qualitatively different from the classical
methods of linear optics.

The requirements imposed on the target and the electromagnetic field are
determined by the aim of the experiment, which is to study the phenomena that
follow the nonlinear interaction of electromagnetic radiation of a fixed frequency with an isolated atom. In linear optics it is only important to con-
serve the product of the number of photons and the number of atoms in the
target. No requirements are imposed on the target or the field. When the interaction is nonlinear, classical experimental aspects such as background and
competing processes acquire new meaning. In addition, processes which are absent in linear optics, such as the self-induced distortion of intense light
passing through an atomic medium and the coherent scattering of the light,
can be observed. The effect of the self-induced distortion is to change the
spatial-temporal distribution of the light beam when it interacts with the
medium. The coherent scattering leads to the appearance of a second electromagnetic wave with a different frequency in the dispersive medium, so that
the classical field approximation cannot be applied. Of course, with coherent
scattering the incident wave is also attenuated. However, this only changes
the wave quantitatively, whereas the new wave changes the incident field
qualitatively. The requirements imposed on the target and the electromagnetic

field are also related to the method by which the reaction products are detected. The small yield of the process is often a limiting factor because the
region in which the field is created is very small.

6.1 Atomic Target

The design of the atomic target is determined by the aim of the experiment,
the type of atoms to be used, and their physical and chemical properties. It
goes without saying that the targets must be gases. Otherwise, if the substance of the target is in another state, one cannot speak of isolated atoms
or atomic beams.

The special features of experiments in this field are a result of the
presence (or absence) of resonances with bound electronic states. If such
resonances are present, they play a significant role or are the subject of
the research. In this case a target in the form of an atomic beam has obvious
advantages over a gaseous target. If two orthogonal intersecting beams (electromagnetic and atomic) are used, a linear Doppler effect cannot be observed
and the resolution of the frequency is determined only by the width of the
laser line, provided that the field does not perturb the atomic spectrum.
The with Γ_D of the Doppler absorption line of a collection of atoms whose
velocity distribution is Maxwellian is determined by the frequency ω_D at
which the absorption is e times less than at the peak of the distribution,
ω_0. One then has the following dependence of Γ_D on the parameters of the gas:

$$\Gamma_D = 2|\omega_D - \omega_0| = \omega_0 v/c = \omega_0 (2kT/mc^2)^{\frac{1}{2}} \tag{6.1}$$

where v is the speed of the atom, T is the temperature [K], and m is the
mass of the atom. Since at room temperature $v/c \approx 10^{-5}$, $\Gamma_D \approx 0.1$ cm^{-1} for visible light. The Doppler width Γ_D is several orders of magnitude greater than
the width of a single-mode laser line (Γ_D closely approximates the width of
the atomic levels in the absence of a perturbation). This comparison shows
the advantages of atomic beams.

The advantage of a target chamber filled with gas over an atomic beam is
the high density of the target. In the case of atomic beams, technological
aspects limit the density of the beam (see below), whereas for a gaseous
target the density of the gas is only limited by competing effects (Sect.
6.2) and certain other physical phenomena which shall be considered in this
chapter. It must be noted that the pressure of the gas in the target chamber
may be as high as 1 mm Hg, which corresponds to a density of about 10^{16}cm^{-3}.

It is possible to obtain a narrow directional pattern, a large cross section, and a high density of atoms by using multi-channel collimators or Laval nozzles [6.1]. It is easy to produce beams several millimeters wide with densities up to $10^{12} cm^{-3}$ and divergences of about 1°. The temperature at which the vapor pressure of a given substance is sufficiently high determines the feasibility of working with noble gases (room temperature), alkali metals (temperatures between ten and a thousand degrees Celsius), and certain other elements whose melting points are not very high (not more than several hundred degrees Celsius), such as bismuth, cadmium, gallium, indium, iodine, mercury, phosphorus, selenium, and sulfur. However, experiments have also been carried out with elements with much higher melting points, e.g., with titanium which has a melting point of about $1800°C$.

It is possible to increase the density of the atoms in a beam to $10^{12} cm^{-3}$. However, $10^{14} cm^{-3}$ is the limit, which is two orders of magnitude less than the density attainable with a target chamber filled with gas.

To better understand the importance of the target's density, let us further examine typical experimental conditions. If, to obtain a strong electromagnetic field, the light is focused onto a disc $10^{-2} cm$ in diameter and the depth of the focusing region, ℓ, is assumed to be about 10 d, where d is the diameter of the disc, then the strong field is produced in a volume V of about $10^{-5} cm^3$. With a density $N \sim 10^{12} cm^{-3}$ and a 100 percent efficiency, the yield is of the order 10^7 photons. If these are scattered isotropically into a solid angle of 4π steradians, then only about 10^4 will reach the detector (if a photocell with a solid angle of less than $10^{-2} sr$ is used). This exceeds the value of the standard noise of a photocathode by no more than one order of magnitude.

The conditions (temperature and pressure) under which a given substance can exist in an atomic or molecular state are also of importance. At room temperature the elements hydrogen, oxygen, and nitrogen exist as molecules. The interest in a target of atomic hydrogen is obvious. However, the traditional method of dissociating the hydrogen molecules in an oven and forming an atomic beam using diaphragms cannot be applied in view of the extremely low density of such a beam. It is possible to attain fairly high densities (up to $10^{12} cm^{-3}$) by carrying out the dissociation in a gas discharge and by using multichannel collimators. The main disadvantage of this method is that the beam contains small amounts of excited hydrogen atoms.

Whenever it is necessary to produce a target of excited atoms, the plasma of a gas discharge or of a discharge afterglow is used. The latter is especially suitable if the target consists of atoms in metastable states. To a large extent the maximum densities of such targets depend on the characteristics of the discharge (e.g., with helium the density may be as high as 10^{12}cm^{-3} [6.2]).

In producing atomic targets special attention must be paid to their purity. Foreign atoms or molecules of the atoms being investigated can cause background effects. If the phenomena under study are nonlinear, the entire experimental results may be due to impurities. When ions are being registered, the presence of pump oil in the apparatus is very undesirable. As a rule, the ionization potentials of such organic molecules are not very high [6.3], and the ionization process involves relatively few photons. This results in a high yield of electrons and ions [6.4]. The recording of ions rather than electrons and their mass analysis makes it possible to reduce this background effect [6.5].

6.2 Competing Processes

With respect to the process under investigation, i.e., the interaction of a strong electromagnetic field with an isolated atom, competing processes encompass all those processes that involve the interaction of a given atom with other particles. These particles may be atoms identical to the given atom, or other atoms, molecules, ions, or electrons present in the target or produced in it by the electromagnetic field. They may also be photons of a different frequency which can be produced in the target by the strong electromagnetic field. Since the creation of photons of a different frequency can be a coherent process, the target can be acted upon not only by the external perturbing field but also by an induced field of another frequency. Then, some or all of the target atoms will be under the influence of two or more fields, i.e., the classical field approximation is no longer valid.

Obviously, if the density and the size of the target are reduced, one can bring about conditions under which competing processes do not play a significant role. However, the reduction of these quantities is limited by the need to be able to clearly observe the investigated effect. In Sect.6.1 the lower limit of the density of a target was estimated. To estimate the upper limit specific competing processes must be taken into account.

In the following section two types of targets consisting of a rarified gas or an atomic beam will be examined. Targets which employ plasmas, including the cold plasma of a discharge afterglow, will not be discussed because their field of application is too narrow and also because the basic

138

phenomena which take place in such targets are of a specifically plasmatic
nature. It will also be assumed that both the gas and the atomic beam consist of identical atoms and that the admixtures, including the molecular
phase, are small enough to be ignored.

6.2.1 Collisions Between Identical Atoms

Collisions between the atoms of a gas have been studied in detail. Such collisions lead to a broadening of absorption lines because the field in the
interior of an atom is perturbed by the field of a passing atom [6.6]. The
linewidth Γ_{col} is related to the frequency $1/\tau_{col}$ and the effective cross
section σ_{col} as follows:

$$\Gamma_{col} = 2/\tau_{col} = 2Nv\sigma_{col} = 2\pi Nv\rho^2_{col} \quad , \tag{6.2}$$

where N is the density and v is the speed of the atoms in the gas. A question
that must be answered is: What is the value of the effective cross section
(or the effective radius ρ_{col}) of such collisions? To answer this question
it must first of all be noted that σ_{col} is not given by the gas-kinetic
cross section. In the classical theory of collisional broadening it is assumed that the electromagnetic field is weak [6.7]. It can thus be assumed
that between two consecutive collisions an atom does not undergo any transitions due to the external field, i.e., the field does not have an influence
on the collision. Hence only two atoms are involved in a collision, and light
is absorbed by atoms whose levels are broadened by single collisions. The
collisional cross section greatly depends on the type of colliding particles.
For identical particles the cross sections may be extremely large due to the
resonant nature of the transfer of energy from an excited atom to an atom in
its ground state.

The resonance cross section is greater than the gas-kinetic cross section
by several orders of magnitude. At $T \approx 300$ K and $N \lesssim 10^{16}$ cm^{-3}, the linewidth
$\Gamma_{col} \approx 10^{-16}$N, where Γ_{col} is measured in reciprocal centimeters. When the interaction between atoms in their ground states is of a van der Waals nature,
the effect is smaller by two or three orders of magnitude.

If the electromagnetic field is strong, the two above-mentioned assumptions may not hold and nonlinear phenomena may occur. In its most general
form the nonlinear theory of collisional broadening is based on the assumption that an atom, A_1, interacts with a system consisting of an atom, A_2,
and a field and that transitions that take place as a result of collisions between A_1 and A_2 determine the absorption process [6.8].

A strong field differs markedly from a weak field in that the collisional cross section in a strong field strongly depends on the frequency and the intensity of the external field.

The nonlinear phenomena that were predicted by the theory of collisions in strong fields, in particular the departure from the law ($W \propto E^2$) governing the probability of light absorption, have been observed in experiments [6.9]. The critical field strength beyond which nonlinear effects are observed is not very large: $E_{cr} \sim 10^5$-10^6V/cm. Up to now no effects are known which drastically increase the broadening in a strong field. The broadening is always decreased in a strong field.

It can thus be said that if a sufficiently pure gas at room temperature and with an atomic density $N \lesssim 10^{15}$cm^{-3} is used as the target, the collision broadening ($\Gamma_{col} \lesssim 10^{-1}$cm^{-1}) will not exceed the Doppler broadening irrespective of the strength of the electromagnetic field.

6.2.2 Collisions with Electrons

It may seem at first that when one uses a gas at room temperature as a target, there should be no need to take into account the free electrons because they cannot change the state of the atoms under study. However, for a gas under the influence of a strong electromagnetic field, such a conclusion is not valid. Electrons can acquire energy from the electromagnetic field and pass it on to an atom through collisions, and hence excite or ionize the latter. A free electron can acquire energy from an electromagnetic field as a result of inverse bremsstrahlung or as a result of an induced Coulomb effect. There are many reasons, mainly related to the technical aspects of an experiment, for the appearance of free electrons in a gas, for example, the ionization of organic molecules, excited atoms, or of molecules with low ionization potentials by the electromagnetic field. Even a small number of free electrons in the gas can lead to macroscopic effects. If these initial electrons acquire enough energy from the electromagnetic field to ionize atoms, then each ionization step will double the number of electrons. Hence, an electron avalanche is initiated. As soon as the plasma which is formed in the process ceases to be transparent to the laser radiation, the light will be strongly absorbed by the target and cause the plasma to heat up. The result is the formation of a laser spark in the focusing region of the laser radiation.

Let us consider the conditions under which the free electrons can absorb energy from the electromagnetic field. It is a well-known fact that a free electron in the field of a planar, monochromatic electromagnetic wave cannot absorb any amount of energy from this wave because neither energy nor momentum can be simultaneously conserved. However, the wavefront is not exactly planar (the virtual source is situated at a great but finite distance) and the radiation is only quasi-monochromatic, $\Delta\omega/\omega \sim 10^{-3}$-$10^{-7}$. In such a field a free electron can acquire energy by absorbing a photon of frequency ω_1 (wave vector $\mathbf{k}_1$) and emitting a photon of frequency ω_2 (wave vector $\mathbf{k}_2$). Such a process is known as the induced Compton effect.

With the conservation of energy and momentum in mind, one can formulate the criterion for induced Compton scattering as follows [6.10]:

$$\frac{\Delta\omega}{\omega} > \left(\frac{kT_e}{mc^2} \Delta\Omega\right)^{\frac{1}{2}} \quad , \tag{6.3}$$

where T_e is the electron temperature and $\Delta\Omega = [4\pi\sin^2(\theta/2)]$ is the solid angle of the beam with a divergence of θ. The rate at which an electron acquires energy is described by the following equation [6.11]:

$$\frac{dE_e}{dt} = \frac{2\pi r_0^2}{m\omega^2} \frac{\Delta\Omega}{\Delta\omega} E^4 \quad . \tag{6.4}$$

Estimates have shown that for an optimal relationship between $\Delta\Omega$ and $\Delta\omega$ the rate at which an electron acquires energy, dE_e/dt, is related to the field strength of the laser radiation by the following equation:

$$\frac{dE_e}{dt} \lesssim 10^{-20} E^4 \quad , \tag{6.5}$$

from which it follows that the electron acquires an energy of one electron volt if $TE^4 > 10^{24}$, where T [s] is the length of time the field acts on the electron, and E [V/cm] is the strength of the field. For a typical high-powered laser pulse $T \approx 10^{-8}$s, while the critical field strength, E_{cr}, is about 10^8V/cm. It is obvious that, because of the Compton effect, the target density does not play a role in the acquisition of energy by the free electron. However, the efficiency of the excitation (or ionization) of the target atoms by the accelerated electrons does depend on the density of these atoms and the effective cross section of the collision [6.12].

Finally, it must be noted that the optimal values for $\Delta\omega$ and $\Delta\Omega$, which were used to obtain the last criterion, are never realized in experiments involving the use of high-powered laser radiation. For example, if it is assumed that $\Delta\Omega \sim 1$ sr (which is the case only if the relative aperture of the focussing objective is close to unity), in practice $\Delta\Omega \sim 10^{-2}$ sr. Hence, for a standard experimental setup, the critical field E is approximately 10^9V/cm.

Another process by which a free electron can acquire energy from an electromagnetic field is called the inverse bremsstrahlung absorption process, in which an electron collides with an atom of the gas. Its name is an apt description, i.e., this process is the inverse of the well-known bremsstrahlung for a charged particle moving in the field of an atom [6.13]. In the presence of an external field, an electron in the field of an atom can absorb a photon and hence acquire additional energy. In this case energy and momentum are conserved because a third particle is present, i.e., the atom with which the electron collides. If the density of the gas is not too high, the frequency of collisions is much less than the frequency of the external field ($\nu \ll \omega$) and the rate at which the electron acquires energy is determined by the following equation [6.14]:

$$
\begin{aligned}
\frac{dE_e}{dt} &= \frac{4\pi e^2}{mc^2} \frac{\nu}{\omega^2} E^2 \\
&= \frac{4\pi e^2}{mc^2\omega^2} N\sigma_{tr} E^2 \quad ,
\end{aligned}
\tag{6.6}
$$

where $\nu = N\sigma_{tr}$ is the frequency of collisions and σ_{tr} is the transport scattering cross section, $\sigma_{tr} = \sigma(1 - \cos\theta)$, σ being the cross section and θ the scattering angle. In contrast to stimulated Compton scattering, the increase in the energy of the electron in inverse bremsstrahlung absorption is directly proportional to the density, N, of target atoms. Numerical estimates have led to the following relation, which guarantees that an electron has acquired one electron volt: $\tau N E^2 > 10^{23}$. For a typical laser pulse length of 10^{-8}s and a gas pressure $p \sim 1$ mm Hg ($N \sim 10^{16} cm^{-3}$), this condition is for fields greater than 10^7V/cm. Obviously, for a fixed pulse length and a sufficiently high field strength one can always use a gas of such low density that the electrons do not acquire enough energy to excite the atoms. In fact, there are many factors that have been ignored (the size of the region in which the strong field is localized, the spectrum of the target atoms, etc.) [6.15] which can change the value of the estimate but not its order of magnitude. In the literature dealing with inverse bremsstrahlung absorption, the estimates were usually made for conditions involving the development of an electron avalanche, which leads to the production of a plasma with a density exceeding the critical density ($> 10^{15} cm^{-3}$) for visible light, i.e., for conditions conducive to the development of a laser spark. However, the above estimates involve conditions under which an electron acquires a specific amount of energy from the field.

The aforementioned considerations allow one to find conditions under which one can neglect the competing processes that arise due to the action of free electrons. These relations illustrate the prevailing role of the inverse bremsstrahlung absorption. With a laser pulse length of 10^{-8} s and with optical focussing systems whose numerical apertures exceed 1/10, one can conduct experiments with fields up to 10^7 V/cm and gas densities up to 1 mm Hg. Stronger fields require shorter pulse lengths, which can be attained using picosecond pulses with mode locking [6.16].

6.3 The Self-Induced Distortion of Intense Light in Atomic Targets

When intense light passes through a target, certain self-action effects that change the spatial-temporal distribution of the light may be observed. The macroscopic cause of such effects is the nonlinear dependence of the refractive index of the medium on the light intensity. The fact that the refractive index is not uniform in an initially homogeneous medium leads to a nonlinear refraction of light rays [6.17,18]; a phenomenon that is undesirable, not so much because there is need to monitor the variations in the dimensions of the beam, but that these variations may be qualitative. For example, if refraction leads to a self-focussing of the beam, the field will sharply increase in certain small regions inside the target, and hence, drastically change the nature of the nonlinear phenomena being studied.

For a quantitative description of these self-induced effects the concepts of critical field strength E_{cr} and self-induced length L are used. Both correspond to conditions under which the diffractive divergence of the laser beam compensates for the divergence due to nonlinear refraction: i.e., $\theta_{dif} = \theta_{ref}$.

For the simplest model, i.e., a cylindrical beam of radius r with a planar wavefront and a uniform amplitude distribution over the wavefront and a medium with quadratic nonlinearity,

$$n = n_0 + n_2(\omega)E^2 \quad , \tag{6.7}$$

the values of E_{cr} and L are determined by the condition that the angle of refraction,

$$\theta_{ref} \approx [n_2(\omega)]^{\frac{1}{2}}E \quad , \tag{6.8}$$

and the diffractive angle of divergence, $\theta_{dif} \approx \lambda/r$, be equal:

$$E_{cr} \approx \frac{\lambda}{r} \left[n_2(\omega) \right]^{-\frac{1}{2}} \quad , \qquad L \approx \frac{r}{\theta_{ref}} \approx \frac{r}{E} \left[n_2(\omega) \right]^{-\frac{1}{2}} \quad . \tag{6.9}$$

Although a real beam and a real medium do not correspond to this simple model, estimates based on this model adequately describe the experimental results [6.19].

The main problem lies in the calculation of the refractive index, which in this case is a nonlinear function of the intensity of the radiation. The microscopic phenomena that lead to the nonlinearity depend a great deal on the properties of the medium.

The classical causes of self-focussing — the high-frequency Kerr effect, striction, and the heating of a rarefied atomic medium by the absorbed light — play a significant role.

In a rarefied atomic medium the main effects are due to the occurrence of a resonance between the frequency of the external field and the frequency of a transition from the ground state to an excited state. The electromagnetic wave can change the populations of the ground and excited states, mix these states (Chap.8), and detune (or tune) the resonance according to the perturbation of the atomic spectrum (Chaps.7 and 8). Since the sign of the dynamic polarizability of an atom is dependent upon whether $\omega > \omega_{mn}$ or $\omega < \omega_{mn}$ (where ω is the frequency of the external field and ω_{mn} is the frequency of the transition: Chap.7), the nonlinear phenomenon can lead to both a focussing and a defocussing of the incident beam. It must also be borne in mind that in a highly nonlinear medium a considerable broadening of the spectral lines of the incident radiation occurs in the resonant frequency region due to frequency modulation [6.20] when there is resonance. The expressions which nonlinearly correct the refraction index at resonance [6.21,22] adequately describe the results of experiments in which focussing and defocussing at a one-photon resonance [6.23] and focussing at a two-photon resonance [6.24] were observed. However, the great variety of experimental conditions does not allow quantitative estimates of a sufficiently general nature to be made. This can be seen if only from the expression for the general form of the refractive index near resonance:

$$n_0 = 1 + \frac{2\pi N |d_{nm}|^2}{\hbar(\omega_{nm} - \omega)} \quad , \qquad n_2 = |d_{nm}|^2 \frac{n_0 - 1}{4(\omega_{nm} - \omega)^2 \hbar^2} \quad , \tag{6.10}$$

where the dipole matrix element d_{nm} for the resonance transition (n,m) can assume significantly different values depending on the kind of atom and the transition. For a two-photon resonance between the frequency of the external field, ω, and the frequency of the atomic transition, ω_{mn}, the matrix element

144

d_{nm} is much smaller than for a one-photon resonance. In an atomic medium whose density $N \sim 10^{14} \mathrm{cm}^{-3}$ ($P \sim 10^{-2}$mm Hg) the critical length L was observed to be $10-10^2$cm both in the case of a one-photon resonance ($E_{cr} \approx 10^4$V/cm) and in the case of a two-photon resonance ($E_{cr} \approx 10^5$V/cm). If these values are compared with the restrictions imposed on the density of the target and the field strength by other effects, it is evident that with resonance the self-action effects can play a significant role for sufficiently dense and extended targets.

Other phenomena may result in a change of the temporal distribution of the radiation and the pulse length. In a nonlinear medium, self-frequency modulation of the pulses can occur [6.20]. The length of the frequency-modulated pulses may change if the medium is dispersive. Depending on the sign of the nonlinear term in the expression for the refractive index, either high ($n_2 < 0$) or low frequencies ($n_2 > 0$) are found at the pulse front. The pulse length will either increase or decrease depending on the sign of the derivative $dv/d\omega$ (where v is the group velocity of the pulses moving through the nonlinear medium) and the distribution of frequencies over the pulse. With an incident wave whose frequency is close to the frequency of a transition, both the nonlinear term and the dispersion may be significant. In a nonlinear medium, the critical length over which the length of a pulse changes considerably is given by [6.25]

$$L_T \sim cT\left(- \frac{n_2 E^2 c_\omega}{d^2 n_0/d\lambda^2}\right)^{-\frac{1}{2}} , \tag{6.11}$$

where the refractive indices n_0 and n_2 have been previously defined. Numerical estimates based on the above relationships have shown that for $N \sim 10^{16} \mathrm{cm}^{-3}$ and $|d_{nm}|^2 \sim 10^{-35}$esu (a standard value for resonance transitions in the spectra of alkali atoms), the effective length $L_T \sim cT$, i.e., for a nanosecond pulse the length L_T is about 10 cm.

It should be noted that the above estimate yields the upper limit of the phenomenon, since the matrix elements for all other transitions are smaller. Hence, in contrast to spatial effects, any temporal effects may only be observed for ultrashort pulses in a very dense target of alkali atoms, as has been experimentally verified [6.26].

6.4 Experimental Methods for Studying the Phenomena Produced by Strong Electromagnetic Fields Interacting with Atoms

When a strong electromagnetic field interacts with an atom, the final state of the atom may be any one of the following three: the ground state, an excited state, or an ionized state (a positive ion and a free electron). In addition, the spectrum of the bound electronic states in an atom may be disturbed by the electromagnetic field, the transition frequencies may be changed, or the states themselves may be broadened. The light that interacts with an atom can change its direction or its frequency. These special phenomena that arise when an atom and light interact determine the range of experimental methods that can be used in studying such an interaction. Obviously, one can detect an interaction directly if light of a different frequency, or light that moves in a direction different from that of the exciting light is registered in a target consisting of neutral atoms or ions (electrons). Using indirect methods, e.g., an auxiliary light source, one can observe a change in the absorption of such light in a target. A method widely used in classical linear optics involves the study of the absorption of the exciting light by a target. However, as a rule, this method cannot be used in nonlinear optics because the number of atoms in a target is usually many orders of magnitude smaller than the number of photons passing through the target. In a certain sense, using laser radiation for the exciting light simplifies an experiment in that the high degree of monochromaticity and the small angular divergence of laser radiation makes it possible to register small variations in the frequency or the direction of the scattered light. The high intensity of the exciting light, however, is a disadvantage because it creates a strong background of scattered light.

6.4.1 Detecting Ions

The positive ions and free electrons that are produced inside a target are separated, then accelerated by a constant or pulsed field, and finally recorded by a detector (Fig.6.1). Detecting ions, and not electrons, has a definite advantage from the standpoint of reliability. If the masses of the ions are monitored by measuring the time of flight to the detector (or by deflecting them in a magnetic field), one can reliably separate the true signal from the background which is due to easily ionized impurities [6.4]. When only the total number of ions produced in the target must be measured, standard electron-optical methods are used to focus the ions. If it is unnecessary to measure the absolute number of ions, electron multipliers are

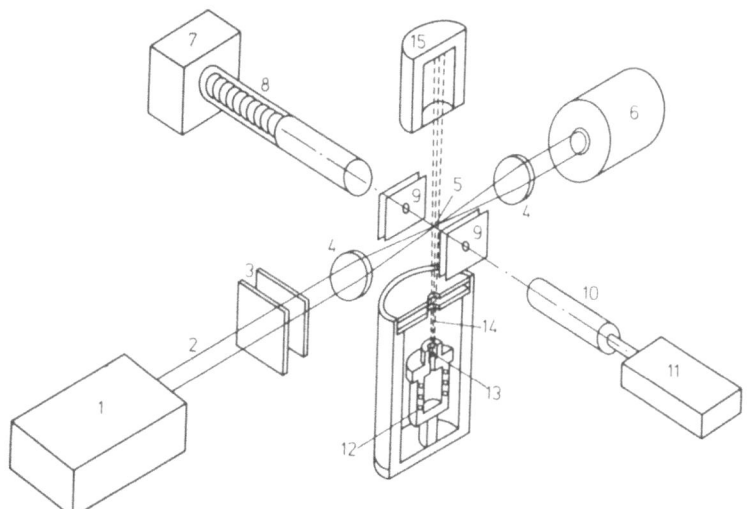

Fig.6.1. Experimental setup for detecting the ions produced in the ionization of atoms by laser radiation (the method of intersecting beams) (1: laser, 2: laser beam, 3: attenuators, 4: focussing lens, 5: focussing region, 6: calorimeter, 7: monitors of the signal from the electron multiplier, 8: electron multiplier, 9: system that accelerates and focusses the ions, 10: Faraday cup, 11: monitor of the signal from the probe, 12: source of the atomic beam, 13: collimator, 14: atomic beam, 15: condenser)

employed as detectors. In this case the sensitivity of the experimental apparatus is such that it is possible to register single ions. However, the amplification factor of an electron multiplier depends a great deal on the conditions under which the multiplier operates. When the absolute number of ions must be measured, e.g., in measuring cross sections, it is best to use as a detector a single probe in the form of a Faraday cup. The sensitivity of such a probe is much lower than that of an electron multiplier (about 10^3 ions). To measure an absolute number of ions, conditions must be maintained that rule out any losses due to collisions in the flight-path interval. For a typical mean free path of about 10 cm, it is necessary to reduce the pressure in the chamber to about 10^{-3} mm Hg or to carry out a fractional evacuation of it. When an atomic beam represents the target, one can always achieve the required degree of evacuation.

If there is a need to measure the angular distribution of the electrons produced in the process of ionizing the atoms, one can use a cylindrical capacitor to accelerate them [6.27]. For the detection of the electrons or ions produced in a gaseous target, the leading edge of a voltage pulse at the output of a recording instrument has a duration of about 10^{-8} s. This is due to: 1) the high density and degree of ionization of the plasma in the focusing region of the light, 2) a divergence in the lengths of the trajectories of the charged particles from the place where they are produced to the detector, and 3) the transmission band of the electron multiplier and the amplifier (when a probe is used). Hence, when detecting ions (or electrons) the experimenter can only measure the overall effect produced by a single high-power laser pulse with an average duration of about 10^{-8}s.

6.4.2 Absorption of Light

In one way or another a study of the absorption of light in an atomic medium always involves spectra of bound electronic states. Indeed, the change in absorption with the frequency of the light is due to the occurrence of a resonance between the frequency of the light and the frequency of a definite atomic transition. The absorption method is used in various modifications, the majority of which use a laser beam as the source of light. The absorbed light may originate from an auxiliary source and can be used to analyze the effect the main intense beam has in perturbing the atomic spectrum. In this case the field E^* of the auxiliary light must be sufficiently weak so as not to perturb the spectrum, i.e., $\delta E_i(E)$, $\delta\Gamma_i(E) < \Gamma$, where $\delta E_i(E)$ is the shift in energy, $\delta\Gamma_i(E)$ is the variation of the linewidth, and Γ the spontaneous width of state i in experiments using beams or the Doppler width in experiments using gaseous targets. The absorbed light can also originate from the main intense light that perturbs the atomic spectrum. The number of photons that are in resonance with an atomic transition can also vary.

Several methods can be applied to study the absorption of light. Among these are: 1) the attenuation at certain frequencies of the light impinging on the target; 2) the detection of the light produced by the spontaneous relaxation of an excited state; and 3) the detection of ions when the probability of transition of an electron from an excited state to the continuum is greater than the probability of the relaxation of the excited state.

In the case of multi-photon absorption, the light may consist of photons that travel in different directions, and that have different frequencies or degrees of ellipticity. The aims of a particular experiment determine the choice of the optimal modification of an absorption method. For example, the two-photon absorption method as applied to counter-propagating beams [6.28,29] is used for the spectroscopic measurements of bound atomic states that are not very energetic. However, this method compensates for the Doppler effect in a target which consists of a rarefied gas (the absorption-saturation method [6.25] is also used for this purpose). The radio-optical double resonance method [6.31,32] is employed to study the perturbation of the ground states of atoms. Finally, the multi-photon ionization method makes it possible to study the behavior of highly excited states in strong fields where ionization predominates [6.31-34].

Hence, the most important methods used to investigate the effects of a strong electromagnetic field on an atom are: 1) the one-photon absorption method, 2) the two-photon absorption method as applied to counter-propagating beams, 3) the multi-photon ionization method, and 4) the relaxation of excited states.

6.4.3 One-Photon Absorption

Although in the one-photon case the absorption probability is relatively high, both the short duration of the strong electromagnetic field and the small number of atoms on which this field acts place very stringent demands on the intensity of the auxiliary light. A typical experiment is illustrated in Fig.6.2 [6.35]. Its aim was to study the changes in the transition energy of the principal doublet of potassium produced by a strong electromagnetic field. The source of the strong field was a ruby laser, which was also used to pump a dye laser. The absorption of the dye-laser radiation (with a relatively broad spectrum) by the potassium vapor was observed. To measure the absorption spectrum during laser emission (10^{-8}s), it was necessary to use a source of auxiliary light with a field strength of about 10^3V/cm and an optical length of the order of 1 mm, and to have a vapor density of about 10^{16}cm^{-3}.

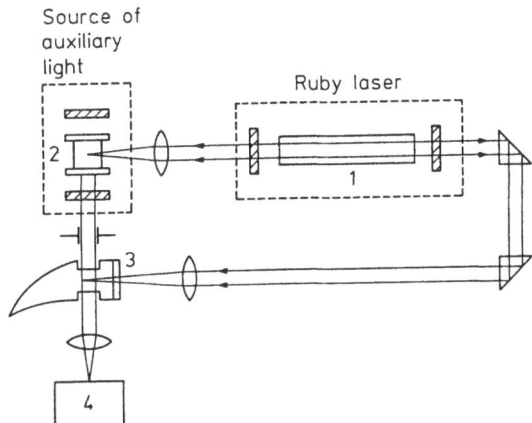

Fig.6.2. Experimental setup for one-photon absorption (1: high-powered laser produces a beam that irradiates a gaseous target 3 and pumps the dye laser 2, whose light is absorbed in the target and is registered by the spectrograph 4)

6.4.4 Two-Photon Absorption

In this section the two-photon absorption method as applied to counter-propagating beams will be studied [6.28,29]. If a gas consisting of atoms with a thermal velocity v is irradiated by two beams of light which travel in opposite directions with the frequencies ω_1 and ω_2, then, in a reference frame with respect to a target atom, $\omega_1' = \omega_1(1 - v/c)$ and $\omega_2' = \omega_2(1 + v/c)$ (the Doppler effect). When $\omega_1 = \omega_2$, the sum $\omega = \omega_1 + \omega_2 + (v/c)(\omega_2 - \omega_1) = \omega_1 + \omega_2$ does not depend on v. The Doppler effect is compensated by the absorption of light of a fixed frequency in the form of two counter-propagating beams. Of course, it is assumed that the light is perfectly monochromatic (or, at any rate, the spectral width of the incident light Γ is much less than the Doppler width Γ_D) and that the wavefront is planar. Both assumptions are valid in practice. Indeed, for a gas at room temperature, Γ_D typically lies between 0.1-1 cm^{-1}, the spectral width of a single-mode pulsed laser, Γ, is about 10^{-4} cm^{-1}, and the beam divergence, θ, from the axial mode is of the order of 10^{-4} radian. If the transition energy of an atom, E_{nm}, is $\hbar\omega$, $\omega_1 = \omega_2 = \omega/2$, and the transition takes place in accordance with the selection rules for the absorption of two photons of a given polarization, the target atoms in state k will shift to state m when the target is irradiated by two counter-propagating beams of frequency $\omega_{1,2}$. The width of the absorption spectrum will then be determined by the spectral width Γ of the perturbing light or the line-width of a transition, $\Gamma_{n,m}$ (depending on which of the two quantities is greater). By changing the frequency of the laser radiation one can measure E_{nm} to an accuracy of Γ, $\Gamma_{n,m}$. With typical values for E_{nm} (10^4 cm^{-1}) and for Γ (10^{-4} cm^{-1}), the relative accuracy $\Delta E/E$ can be as high as one part in a hundred million).

The main disadvantage of this method is that in general $\omega = 2\omega_{1,2}$, so that the transition (n,m) can take place by the absorption of two photons from two different beams or two photons from the same beam. As always, the absorption line for the latter is broadened by the Doppler effect. Superimposed on the broad Doppler contour can be seen a narrow band that corresponds to the absorption of two photons from two different beams. Indeed, if the number of photons in each beam is the same, i.e., $N(\omega_1) = N(\omega_2) = N(\omega)$, then the number of two-photon transitions in which both photons are from the same beam is proportional to $N^2(\omega_1) + N^2(\omega_2) = 2N^2(\omega)$, whereas the number of two-photon transitions in which the photons are from different beams is proportional to $[N(\omega_1) + N(\omega_2)]^2 = 4N^2(\omega)$, i.e., is twice as large as the former. However, as a rule, this factor of two is not sufficient for reliable measurements to be made. Two methods can be used to increase this factor.

The first of these methods uses frequencies that are only approximately equal: $\omega_1 \sim \omega_2$. Moreover, these frequencies are such that $|\omega_1 + \omega_2 - \omega_{nm}| < \Gamma_{n,m}$ and $|2\omega_{1,2} - \omega_{nm}| > \Gamma_D$. It is easy to see that these conditions put limits on the values of $(\omega_1 - \omega_2)$:

$$\omega_{nm}\left(\frac{v}{c}\right)^2 \ll (\omega_1 - \omega_2)\frac{v}{c} \lesssim \Gamma_{n,m} \quad . \tag{6.12}$$

If this is so, the transition (n,m) can only take place if the photons are absorbed from different beams. Of course, since the frequencies are only approximately equal the Doppler effect is not completely compensated for. However, if the linewidth of the atomic levels $\Gamma_{n,m}$ is not that great, the ratio $\Delta E/E$ is sufficiently large.

Figure 6.3 shows the theoretical dependence of the efficiency of the absorption of two photons from two counter-propagating beams, η, and of the two-photon absorption linewidth $\Gamma_{1,2}$ on the Doppler width Γ_D and on the difference of frequencies of the light in the beams, namely on $\Delta\omega = |\omega_1 - \omega_2|$ [6.26]. For large values of $\Delta\omega$ (when $Y \gg 1$), $\Gamma_{1,2} \to \Gamma_D$; for small values of $\Delta\omega$ (when $Y \ll 1$), $\Gamma_{1,2} \to \Gamma_{n,m}$ and $\eta \to 1$.

The second method can only be used for transitions between the states n and m that take place without a change in the value of the orbital quantum number. However, it eliminates the Doppler effect. Circularly polarized laser radiation is split into two beams, which are then focussed on the target from opposite directions. Two-photon transitions can take place if two photons of opposite helicities are absorbed from the counter-propagating beams, since in this case the net transfer of angular momentum to the atom is zero. Transitions cannot take place if photons of the same beam are involved, since

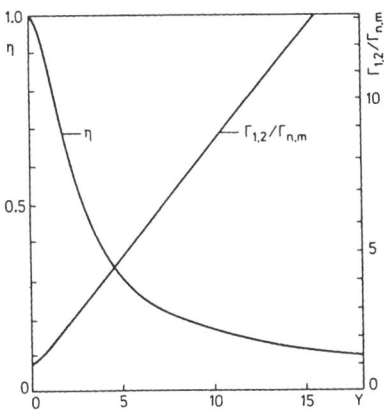

Fig.6.3. Calculated efficiency η of the absorption of two photons from counter-propagating beams and the two-photon absorption linewidth, $\Gamma_{1,2}$ (in terms of the natural linewidth $\Gamma_{n,m}$ of the two-level system), as functions of the relative difference in the frequencies of the photons; $Y = (\Gamma_D/\Gamma_{n,m})\delta\omega$ and $\delta\omega = \Delta\omega/\omega_{nm} = (\omega_1 - \omega_2)/\omega_{nm}$

the angular momentum must then change by two units of ħ. Hence, it may be that $\omega_1 = \omega_2 = \omega_{nm}/2$ but still no background of photons from one beam can be observed.

In the method of counter-propagating beams the experimental arrangement dictates the necessity not to record the absorption of light but to measure the fluorescence resulting from the relaxation of an excited state. The principle of absorption of photons from two counter-propagating beams is illustrated in Fig.6.4.

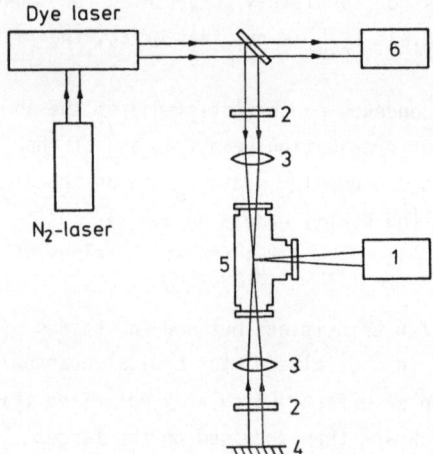

Fig.6.4. Experimental setup for two-photon absorption in two counter-propagating beams

The counter-propagating beams that meet at the target 5 are produced by a dye laser whose light passes through the target before being reflected by the mirror 4. The degree of polarization of the transmitted and the reflected beams is changed by polarizers 2. To increase its intensity the light is focussed by two lenses 3. The fluorescence produced through the resonance absorption of two photons by the target is detected by the photomultiplier 1 (6 represents the equipment for monitoring the parameters of the laser radiation).

6.4.5 Multi-Photon Ionization Spectroscopy

If the exciting field is strong or the principal quantum number of an excited state is large, then the multi-photon excitation of an atom can lead to its ionization. By changing the frequency of the exciting light and by measuring the ion yield as a function of this frequency, one can observe a resonance between several quanta and an atomic transition by noting a resonance increase in the yield. An optimum experimental arrangement is one in which two lasers or a constant ionizing field are used to bring about excitation and ionization. Both processes, excitation and ionization, are realized with the highest efficiency when the fields are separated. The sensitivity of the multi-photon ionization method is considerably higher than that of the fluorescence method. Each ion that is generated can be detected. The frequency resolution can also be very high. Due to the high detection efficiency, an atomic beam (whose density is lower and whose dimensions are smaller than those of a gaseous target) can be used as a target. In this case, if the atomic and light beams are orthogonal, the linear Doppler effect is absent. If the perturbation of the atomic levels is less than their natural width, the multi-photon ionization method can be used to determine the energy of highly excited states. To this end a tunable laser with a minimum spectral width is used (Sect.6.4.1).

The spectral results of experiments with cesium are typical for highly excited atoms [6.33]. A dye laser with a wavelength that can be varied from 655 to 695 nm and a spectral linewidth $\Gamma \sim (6-8) \times 10^{-3}$nm was used. In the presence of an intermediate resonance involving bound electronic states with principal quantum numbers from 9 to 19 the three-photon ionization of an atom was observed. The resonance width of the ion yield is of the order of the laser linewidth, so that the characteristic doublet structure for alkali atoms is well resolved.

If the field substantially perturbs the spectrum, multi-photon ionization spectroscopy can be used to measure the degree of perturbation. It must be borne in mind that the spatial and temporal distributions of the intensity are not uniform in the region where the ions are produced. The ion yield integrated over the volume of the target in which the field is strong, and over the length of the pulse, is recorded. This produces two effects. First of all, nonlinear broadening is observed, which leads to a wider maximum in the ion yield compared to the resonance width. Secondly, the frequency that corresponds to the maximum ion yield must be related to a certain effective field strength. Obviously, both the large volume of the target in which the field is weak and the very small volume in which it is a maximum contribute very little to the integral ion yield compared to the effective volume in which the field is less than the maximum. The larger the degree of nonlinearity of the multi-photon resonance, the smaller the effect of the nonlinearity.

To obtain qualitative results concerning the perturbation of the atomic levels (by using the data on the resonance of the ion yield), it is necessary to determine the functions $\phi(x,y,z)$ and $\psi(t)$, which characterize the spatial-temporal distribution of the radiation (Sect.5.1). Moreover, one can draw certain conclusions about the nature of the transitions from the ground state to an excited state and from an excited state to the continuum [6.32] using theory and the experimental findings.

Saturation with respect to any one of these transitions (as a rule, saturation is attained sooner with respect to the latter transition, whose degree of nonlinearity is smaller) leads to a change of nonlinearity. If the field strength is varied and a resonance in the ion yield is observed, the change in the frequency that corresponds to a maximum yield can be monitored. Hence, one can determine the energy of the transition as a function of the strength of the perturbing electromagnetic field.

6.4.6 Detecting the Light Produced in the Target

The most sensitive light detector, regardless of the nature of the light, is the photomultiplier. Its sensitivity is determined by shot noise and the quantum yield of the photocathode. In the visible range the sensitivity threshold is about 10^3 photons. Other detectors have a much lower sensitivity. For example, photographic films require a flux of about 10^{12} photons/ cm^2. When the perturbing light is generated by a laser, its high uni-directionality and monochromaticity simplify an experiment. If the light that emerges from a target and the laser radiation that impinges on it are not coherent, the light to be measured will be distributed fairly evenly in all directions, so that the detector must be placed at large angles (approximately 90°) to the laser beam. Then the detector only registers a small fraction of the light produced in the target (as a rule, less than 10^{-3}-10^{-4}). Consequently, the lower limit on the number of photons produced in the target after one laser pulse is about 10^6-10^7 (taking into account the sensitivity of the multiplier. In fact, this limit is actually higher, since a dispersing agent (spectrograph, monochromator, or Fabry-Perot interferometer) must be placed between the target and the multiplier. As a result the solid angle which encloses the light to be registered becomes smaller and additional losses are introduced. The advantage of an experimental arrangement in which the detector is placed at a large angle to the laser beam lies in the relatively low level of laser radiation at such angles.

If, however, the light that emerges from the target and the laser radiation that impinges on it are coherent, then, on the one hand, the detector can practically record all of the photons produced in the target but, on the other hand, the effect of the incident light must be reduced and the emerging light must be measured against a strong background of laser light. Separating the actual quantity to be measured from the background is carried out using conventional spectroscopic methods and is not difficult if the frequencies differ considerably. But, reducing the effect of the perturbing light is always a complex problem, since an attenuating filter that can withstand high intensities but yet not be a cause of secondary radiation is required. Solutions of various inorganic compounds with suitable absorption spectra are often used as such filters. Their large volume ensures that no damage occurs (an effect that is common in colored glass filters). With organic compounds Raman scattering usually appears, which changes the spectrum of the impinging light.

An extremely large number of photons is present in one pulse of a high-power laser. To observe the light produced in the target, the intensity of the laser light must be reduced using all of the available means (i.e., changing the geometry of the experimental arrangement or introducing dispersing agents or filters) by a factor of enormous magnitude — as high as 10^{12}-10^{15}. It is for this reason that the customary pre-laser concepts of light diffusion through the structural elements of the experimental setup, the transparency of various filters, and light scattering on dust and sludge are totally inadequate when laser light is used.

Figure 6.5 depicts an experiment in which both the coherent light emerging from a target (the third harmonic of the incident laser light) and the incoherent light (fluorescence) are recorded [6.31].

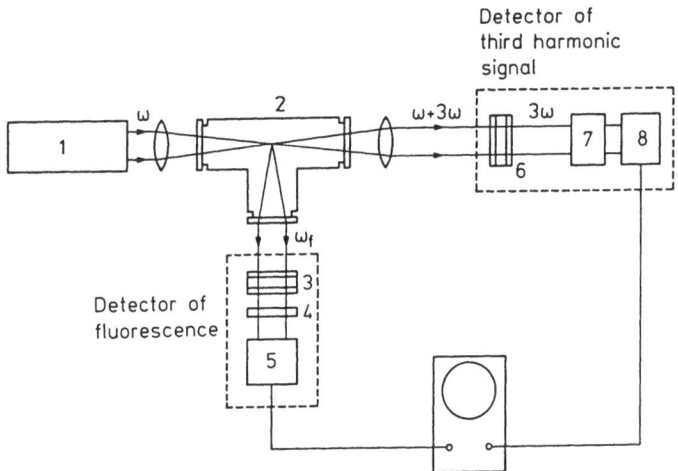

Fig.6.5. Experimental setup used to register the light produced in an atomic target by laser radiation (1: laser that emits light of frequency ω; 2: atomic gaseous target; 3 and 4: filters; 5: photomultiplier that registers the fluorescence of frequency ω_f ; 6: filters; 7: monochromator; 8: photomultiplier that registers the third harmonic of the laser radiation at 3ω)

6.5 The Measurement of the Main Parameters that Characterize the Interaction Between Intense Electromagnetic Radiation and an Atom

The large number of parameters that characterize the interaction between intense electromagnetic radiation and an atom and the various phenomena that occur in a strong electromagnetic field are the reasons why it is necessary to measure so many quantities in such experiments. In this section the methods of measuring two quantities of a sufficiently broad nature will be discussed: the degree of nonlinearity and the multi-photon cross section for power-law processes.

6.5.1 Degree of Nonlinearity

If the interaction between intense electromagnetic radiation and an atom involves the absorption of K photons and if the absorption probability is described by the power law $w \propto F^K$, F being the intensity of the light, then K is usually called the degree of nonlinearity of the process. As a result of the absorption an excited atom (multi-photon excitation), a pair of ions (multi-photon ionization of the atom), or a photon of a different frequency (generation of harmonics, Raman scattering) may be produced. If the degree of nonlinearity of a process is to be found for a pulsed laser, when the spatial and temporal distribution functions do not change for a sequence of pulses (Sect.5.1), it is sufficient to measure the dependence of the amplitude of the yield, A, on the energy of the pulse, E. It is not necessary to determine w and F, since $A \propto E^K$. The dependence of w on F (and thereby the dependence of A on E) may deviate from the ideal dependence due to a scattering effect if the fraction of atoms that are involved in the interaction is not negligible. Saturation is reached when there is a breakdown in the condition,

$$\int_0^{T_K} w \, dt \ll 1 \quad , \tag{6.13}$$

where w is the absorption probability per unit time and T_K is the effective pulse length.

A specific feature of experiments with pulsed lasers is the reduction of saturation. This is due to the increase in the effective target volume as the energy of the laser beam that is focussed on the target increases. This effect is explained by the nonuniform spatial distribution of light over the target (Sect.5.1). One can determine the degree of nonlinearity by reducing, in accordance with the linear law over the cross section of the beam, the amount

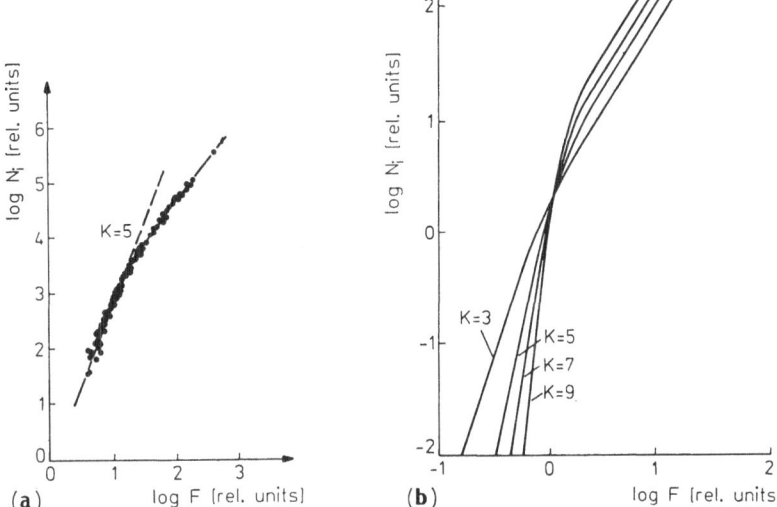

Fig.6.6a,b. Number of ions, N_i, as a function of the intensity of the laser radiation, F, in the presence of saturation: (a) the experimental data for the direct five-photon ionization of sodium atoms: (b) the results of calculations for processes with different values of K

of energy that falls on the target in a series of successive laser pulses and by monitoring the yield of the process in these pulses.

Results of such an experiment are presented in Fig.6.6a. A dependence of the yield of sodium ions (the ionization potential $I \approx 5.2$ eV) produced in a five-photon ionization process on the pulse energy of a neodymium glass laser ($\hbar\omega \approx 1.2$ eV) was observed [6.38]. The spread in the ion signal for a fixed energy is due to the nonuniformity of the spatial-temporal distribution of the light over the target. This leads to fluctuations in F (for a fixed value of E). Due to saturation a reduction of the growth rate of A with respect to E can also be observed. In the region where saturation is absent, the dependence of A on E can be approximated by a power law using the method of least squares.

Calculations of similar (normalized) functions for different values of K are shown in Fig.6.6b [6.39].

6.5.2 Multi-Photon Cross Sections

If the power law $w \propto F^K$ holds, the proportionality factor does not depend on the intensity of the radiation, so that the following relationship holds:

$$w = \alpha^{(K)} F^K \quad . \tag{6.14}$$

The proportionality factor plays the role of a multi-photon cross section whose dimensions are $cm^{2K}s^{K-1}$. When $K = 1$ the dimensions are square centimeters, which are the generally accepted dimensions of the effective cross section for processes in which one photon is absorbed in an elementary act. In (6.14) w is the absorption probability in s^{-1} and F is the intensity of the light in $cm^{-2}s^{-1}$.

Just as for one-photon processes, the multi-photon cross sections depend on the frequency ω and the polarization ρ of the radiation, i.e., $\alpha^{(K)}$ = $\alpha^{(K)}(\omega,\rho)$. A multi-photon cross section, just as a one-photon cross section, is a main quantitative parameter of a power-law process.

The measurement of multi-photon cross sections is hindered by the fact that the spatial-temporal distribution of the light over the target is not uniform. Consider a target with an atomic density N. Let it be assumed that the atoms undergo multi-photon ionization. The rate at which the density decreases is proportional to the probability of ionization, i.e., $dN/dt = -wN$. After one pulse, the initial density N_0 decreases by the following amount:

$$\Delta N = N_0 - N = N_0\left[1 - \exp\left(-\int_0^{\tau_K} w \, dt\right)\right] \,, \tag{6.15}$$

where τ_K is the effective pulse length for the absorption of K photons. The number of ions produced can be found by integrating (6.15) over the effective volume of the target. In doing so one must allow for the nonuniformity of the spatial distribution of the light over the target and for the nonlinear nature of the elementary act of absorption. Both of these factors determine the effective target volume V_K for a process involving the absorption of K photons. Hence, the number of ions produced is given by

$$N_i = N_0 \int_0^{V_K} \left[1 - \exp\left(-\int_0^{\tau_K} w \, dt\right)\right] dv \,. \tag{6.16}$$

If it is assumed that the ionization probability and the intensity of the light are related by a power law, one finally arrives at the following formula for N_i:

$$N_i = N_0 \int_0^{V_K} \left[\int_0^{\tau_K} \alpha^{(K)}F^K \, dt\left(1 - \frac{B}{2!} + \frac{B^2}{3!} - \ldots\right)\right] dv \,, \tag{6.17}$$

where

$$B = \int_0^{\tau_K} \alpha^{(K)}F^K \, dt \,. \tag{6.18}$$

Two methods can be used to measure cross sections:

1) Let $\int_0^{\tau_K} w\ dt \ll 1$; then there is no saturation, and in (6.17) all of the terms except unity can be neglected. Then,

$$\alpha^{(K)} = \frac{N_i}{N_0} \left(\int_0^{V_K} \int_0^{\tau_K} F^K\ dv\ dt \right)^{-1} . \tag{6.19}$$

Separating the temporal and spatial variables in $F(x,y,z)$ and normalizing to the maximum intensity (Sect.5.1.1), one can find a formula for $\alpha^{(K)}$ that incorporates measurable quantities:

$$\alpha^{(K)} = \frac{N_i}{N_0} \left(\frac{\hbar\omega}{E} \right)^K \frac{S^K \tau^K}{V_K \tau_K} , \tag{6.20}$$

where $V_K = \int \phi^K\ dv/\phi_{max}^K$ and $\tau_K = \int \psi^K\ dt/\psi_{max}^K$ are the effective target volume and pulse length, respectively, for a process involving the absorption of K photons in one step, $S = \int \eta(x,y)\ ds/\eta_{max}$, and $\tau = \int \psi(t)\ dt/\psi_{max}$, see (5.1). The measurable quantities are N_i, N_0, E, $\phi(x,y,z)$, and $\psi(t)$. Since the determination of $\alpha^{(K)}$ involves the absolute measurement of all of these quantities, the accuracy of such an absolute method of measuring cross sections is relatively low, especially for large values of K [6.40].

2) Let $\int_0^{\tau_K} w\ dt \sim 1$; then saturation sets in. If in this case one obtains experimental data for the region in which there is no saturation and for the region with saturation; then through extrapolating one can determine the ratio of the ideal value N_i, see (6.17), to the extrapolated value N_i, see (6.20):

$$\frac{N_i}{N_i^*} = 1 + \sum_{n=1}^{\infty} P_n \alpha_n^{(K)} F_n^K , \qquad \text{where} \tag{6.21}$$

$$P_n = \frac{(-1)^n \tau_K^n V_K^{n+1}}{(n+1)!\ V_1} . \tag{6.22}$$

This ratio does not incorporate the atomic density N_0. It incorporates the ratios of ion signals and of effective volumes, but not their absolute values. This greatly increases the accuracy of the measurements. Equation (6.21) can be fitted to the experimental data by choosing an appropriate value for $\alpha_n^{(K)}$. The accuracy of such a relative method of measuring cross sections is much higher than that of the absolute method [6.38,41].

In all of the reasoning so far, it has been assumed that the radiation is strictly monochromatic, that the distribution of intensity over the laser pulse is a smooth function, and that the distribution can be measured fairly accurately with standard equipment with a resolution of about 1 ns. If the radiation is generated by a multi-mode laser, considerably shorter fluctuations in intensity are observed due to fluctuations in the phases of different modes. The envelope curve which is measured by standard equipment does not reflect these fluctuations. If the laser contains nonlinear elements in its cavity, the phases of the modes are unlocked. If the number of such modes is large enough (the higher the degree of the nonlinearity of the process, the greater the number), and the average intensity of the multi-mode and single-mode radiation is fixed, the ratio of the yields of a nonlinear process of the Kth degree is (Sect.9.1) [6.42,43]

$$\frac{N_i(\text{many modes})}{N_i(\text{one mode})} = K! \quad . \tag{6.23}$$

The factor $K!$ must be taken into account when cross sections are measured in the field of a multi-mode laser, especially when K is very large.

7. Nonresonant Phenomena

Certain phenomena are called nonresonant if the frequency of the external field or of its higher harmonics is not close to the frequency of an allowed transition between bound atomic states. As a result, such transitions do not take place. Nonresonant phenomena can, therefore, only be described using the concept of virtual transitions.

The difference between virtual and real transitions lies in the value of the lifetime of an electron in an intermediate state. For real transitions (which, in fact, occur in the presence of a resonance; see Chap.8), the lifetime can be determined from the width of a given state, Γ, i.e., from the probability of a spontaneous or stimulated transition from a particular state. For virtual transitions, the concept of lifetime, τ, is a result of the Heisenberg uncertainty principle, $\Delta E \times \tau \sim \hbar$, in which the detuning from resonance, $\Delta E \equiv \Delta = \omega_{nk} - \omega$, represents the uncertainty in the energy of the virtual state. Correspondingly, virtual transitions can take place in violation of the energy conservation law, and the intermediate virtual states through which these transitions pass cannot be defined by their populations as real states can. Obviously, τ is always much less than $1/\Gamma$. Virtual transitions become real transitions when Δ decreases and approaches a value of the order of Γ.

In a real atom, which has an infinite number of electronic states, all of the states determine the amplitudes of the nonresonant phenomena with statistical weights that are inversely proportional to the detuning from the corresponding resonance. Naturally, this is only true for those states between which transitions are allowed by certain selection rules. Examples of nonresonant processes were discussed in Chap.2 in which the matrix elements that describe the higher orders of perturbation theory were studied. The structure of such matrix elements, see (2.13) and (2.21), i.e., an infinite sum over all of the virtual states in which each term contains the detuning from the corresponding resonance in the denominator, is characteristic of virtual transitions.

If the equations for the compound matrix elements are examined closely, it may seem that the classification of processes as being resonant or non-resonant is an arbitrary one and that the real difference lies in the fact that with resonance the value of the corresponding matrix elements increases sharply. Actually, this is not the case. With resonance the respective intermediate state becomes the final state in the quantum mechanical sense and the various competing transitions may start from this state. In this way the transition to this state can change drastically.

In this chapter phenomena which are essentially nonresonant will be studied. A well-known example is the linear nonresonant scattering of light from an atom. In this process the atom absorbs a quantum of light, passes over to a virtual state, and then spontaneously emits a photon in proceeding to its final state. If the final state coincides with the initial state, the process is known as Rayleigh scattering. If the two do not coincide, it is known as Raman scattering. These processes will be briefly examined before the nonlinear scattering of light from an atom is discussed. In this case the atom absorbs and/or emits not just one but several photons.

Naturally, nonlinear scattering occurs, in principle, at any intensity of the scattered light. But it is significant only at a sufficiently high intensity of the incident light, when the amplitude of the particular process is sufficiently great. For this reason, nonlinear scattering is usually realized when the frequency of the incident light, ω, or of its higher harmonics, $K\omega$, is close to the frequency of one of the atomic transitions. However, it will not be assumed that the frequencies are so close to each other that significant deviations from the predictions of perturbation theory occur (perturbation theory was considered in Chap.2; the deviations will be dealt with in Chap.9). Although processes which involve nonlinear scattering are often called resonance processes in the literature, it has been found advisable to include their consideration in this chapter because they can be described by a power series in terms of the intensity of the external field (which is in agreement with time-dependent perturbation theory). On the other hand, if the frequency of the incident light lies far from the frequencies of atomic transitions, then with fields that are weak compared to atomic fields the nonlinear scattering of light is negligible compared to the linear scattering.

Both linear and nonlinear scattering can be described in a unified way if one is interested in the polarization of an atom produced by one or more external monochromatic fields. The polarization vector $\mathbf{P}(t)$ is defined to be the induced dipole moment, $\mathbf{d}$, averaged over the state of the atom-field system:

$$P(t) = <\Psi(\mathbf{r},t,E,\omega)|\mathbf{d}|\Psi(\mathbf{r},t,E,\omega)> \quad . \tag{7.1}$$

In the absence of a field it will be assumed that the atomic system does not have a constant dipole moment. This is true for all cases except those which involve the hydrogen atom or the highly excited (hence hydrogen-like) states of complex atoms. In a weak field the polarization is linearly proportional to the field E:

$$P_i = \sum_j \chi_{ij}^{(1)} E_j \quad . \tag{7.2}$$

This is the first term in the perturbation series. The proportionality factor $\chi^{(1)}$ is called the linear susceptibility (polarizability) of the atom. In a strong field the following terms in the expansion are very important. The coefficients are known as the nonlinear susceptibilities (hyper-polarizabilities) of the atom. The linear susceptibilities determines the cross section of the linear scattering of light, while the nonlinear susceptibilities determine the cross section of the nonlinear scattering. Obviously, there is no way one can write a general formula for the nonlinear susceptibilities. This is because the different physical processes are described by different terms in the Taylor series for $\mathbf{P}(t)$, and the number of such terms varies considerably depending on the nature of the absorption or the emission of light from the external field (or several fields).

It is obvious that polarization is time-dependent. If it is expanded in a Fourier series in t, one arrives at the concept of polarization, $\mathbf{P}_\omega$, at a frequency ω' (which in nonlinear processes does not necessarily coincide with the frequency of the external field, ω). It is this quantity that is connected with the experimental observables so characteristic of the nonresonant interaction between electromagnetic radiation and atoms.

In addition to the nonresonant scattering of light, there is also the nonresonant absorption of light, i.e., the process of nonresonant ionization. In Sect.2.4 the theoretical ways of describing the transition of an electron to the continuum were studied. Now, the phenomenon of nonlinear nonresonant ionization will be studied.

7.1 Nonlinear Atomic Susceptibilities

The aim of this section is to present a unified approach to the problem of describing the scattering of light from an atom and the perturbation of the atomic spectrum in terms of the linear and nonlinear atomic susceptibilities.

When the interaction is nonresonant, perturbation theory (developed in Chap.2) can be used if the amplitude of the electric field strength E of the perturbing light is small compared to the characteristic atomic field strength E_{at}. It will be assumed that this condition holds.

First of all it will be shown that the linear atomic susceptibilities can be related to the multi-photon matrix elements (introduced in Chap.2) using the diagram language of Feynman (Sect.2.3). In this way the cross sections of various processes involving linear scattering will be expressed in terms of the linear susceptibilities. Then, the perturbation of the atomic levels will also be expressed in terms of the susceptibilities.

7.1.1 The Scattering of Light and the Linear Susceptibility

The linear scattering of light from an atom (elastic or inelastic) was dealt with in Chap.2. In this process an atom in the initial state n absorbs a photon of the external field $E \cos\omega t$, goes over to the state m, then spontaneously emits a photon and passes on to the final state k. If $k = n$, one has Rayleigh scattering, otherwise one is dealing with Raman scattering. The matrix element that describes this process consists of two diagrams (Sect.2.3):

where a dashed line denotes the absorption of a photon from the external field $E \cos\omega t$, and a wavy line denotes the spontaneous emission of a photon of frequency ν. If the field is sufficiently weak, the above diagrams will contribute the most because diagrams with two dashed lines (corresponding to the nonlinear scattering of light), for example, are quadratic in E. This process is called linear because the matrix element, $V_{kn}^{(2)}$, is linear in E. Energy conservation ensures that $E_k^{(0)} = E_n^{(0)} + \omega - \nu$. In the event of Rayleigh scattering, $E_k^{(0)} = E_n^{(0)}$. Hence, $\nu = \omega$.

Let us find the relation between $V_{kn}^{(2)}$ and the atomic polarization. According to the Golden Rule (Sect.2.1), the probability of scattering is

$$dw_{kn} = 2\pi |V_{kn}^{(2)}|^2 \rho = \frac{\nu^3}{2\pi c^3} |V_{kn}^{(2)}|^2 d\Omega \quad , \tag{7.3}$$

where ρ is the density of the final states of the emitted photon, and $d\Omega$ is the solid angle into which the photon is emitted. According to the well-known semiclassical theory of dipole radiation, the probability of spontaneous emission by a dipole with a moment $\mathbf{d} = \mathbf{d}_\nu\, e^{i\nu t}$ (oscillating with a frequency ν) is:

$$dw_{kn} = \frac{\nu^3}{2\pi c^3}\, |\mathbf{e} \cdot \bar{\mathbf{d}}_\nu|^2 d\Omega = \frac{\nu^3}{2\pi c^3}\, |\mathbf{e} \cdot \mathbf{P}_\nu|^2 d\Omega \quad , \qquad (7.4)$$

where P_ν is the atomic polarization at the frequency ν (defined in the introduction to this chapter), $\mathbf{e}$ is the polarization of the emitted photon, and the bar over $\mathbf{d}_\nu$ signifies that a quantum mechanical averaging was performed. Comparing (7.3,4), one finds a relation between the two-photon matrix element $V_{kn}^{(2)}$ and the atomic polarization P_ν at the frequency of the emitted photon ν:

$$V_{kn}^{(2)} = \mathbf{e} \cdot \mathbf{P}_\nu \quad . \qquad (7.5)$$

Writing out the above diagrams for $V_{kn}^{(2)}$ explicitly in accordance with the rules given in Sect.2.3 yields

$$P_\nu = \sum_m \left(\frac{r_{km}(r_{mn} \cdot E)}{\omega_{mn} - \omega} + \frac{r_{mn}(r_{km} \cdot E)}{\omega_{mn} + \nu} \right) \quad . \qquad (7.6)$$

If the dependence of the polarization on the external field $\mathbf{E}$ in (7.6) is isolated, one arrives at the concept of a linear susceptibility (or polarizability) tensor, $\chi_{ij}^{(1)}$:

$$P_{\nu i} = \sum_j \chi_{ij}^{(1)} E_j \quad . \qquad (7.7)$$

The quantity $\chi_{ij}^{(1)}$ is also known as the scattering tensor, where the labels $i,j = 1,2,3$ denote the projections on the x,y, and z axes [7.1]. Hence, the following expression describes the linear susceptibility tensor [7.1]:

$$\chi_{ij}^{(1)} = \sum_m \left(\frac{r_{km}^i r_{mn}^j}{\omega_{mn} - \omega} + \frac{r_{mn}^i r_{km}^j}{\omega_{mn} + \nu} \right) \quad , \qquad (7.8)$$

where r^i, r^j are the projections of $\mathbf{r}$ on the respective axes. One can say that the linear susceptibility can be represented by the same diagrams that represent the multi-photon matrix element, only that the vertices in the diagrams now correspond to r^i instead of $\mathbf{r} \cdot \mathbf{E} = \sum_i r^i E_i$. Equation (7.8) defines the linear susceptibility of an atom at a frequency ν in an external field of frequency ω. Obviously, in Rayleigh scattering $\nu = \omega$; in Raman scattering ν is also uniquely determined in terms of ω due to the conservation of energy. Hence, one can write $\chi_{ij}^{(1)}$ in the form $\chi_{ij}^{(1)}(\nu;\omega)$. It follows from (7.8) that

the susceptibility may be diagonal in the quantum numbers, i.e., so that $k = n$.
One then speaks of the susceptibility of an atom in the state n. On the other
hand, the susceptibility can also have nonzero off-diagonal elements (when
$k \neq n$).

7.1.2 The Nonlinear Scattering of Light and Nonlinear Susceptibility

If the polarization vector **P** of an atom in an external field is expressed
as a power series in E, one obtains terms which are nonlinear in the perturbation E. For example, the quadratic term in the Taylor series is

$$P_i = \frac{1}{2} \sum_{j\ell} \chi^{(2)}_{ij\ell} E_j E_\ell \quad , \tag{7.9}$$

where the quantity $\chi^{(2)}_{ij\ell}$ represents one of the nonlinear susceptibilities
(or hyper-polarizabilities) of an atom [7.2]. In the given case it is the
second-order polarizability in the external field E.

A typical diagram for $\chi^{(2)}_{ij\ell}$ has the following form (of course, this is
not the only possible form):

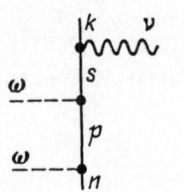

This diagram represents the nonlinear scattering of light accompanied by the
absorption of two photons from the external field and the spontaneous emission of a single photon. According to the conservation of energy, $\nu = \omega_{nk} + 2\omega$.
The dipole matrix elements r^i_{np}, r^j_{ps}, and r^ℓ_{sk} are associated with the vertices. If one recalls Sect.2.3, one realizes that such a diagram can be
written in an analytical form:

$$\sum_{ps} \frac{r^i_{np} r^j_{ps} r^\ell_{sk}}{(\omega_{pn} - \omega)(\omega_{sn} - 2\omega)} \quad .$$

By analogy with (7.4) and using (7.9) one obtains the following expression
for the probability of the corresponding nonlinear scattering:

$$dw_{kn} = \frac{\nu^3}{2\pi c^3} \left| \sum_{ij\ell} e_i \chi^{(2)}_{ij\ell} E_j E_\ell \right|^2 d\Omega \quad . \tag{7.10}$$

Understandably, the probability ω_{kn} dominates the usual linear scattering only when $\chi_{ij\ell}^{(2)}$ has one or several small denominators. If one turns to the above diagram, one can see that this is the case when ω_{pn} is close to ω or ω_{sn} is close to 2ω. However, these frequencies cannot be so similar that perturbation theory breaks down, i.e., the condition $\Delta \gg \Gamma$ must still be valid.

The above diagram represents one of the terms of the nonlinear suscepti- bility $\chi_{ij\ell}^{(2)}$ $(\nu;\omega, \omega)$ at the frequency ν. Obviously, for the same order of perturbation theory there is also a nonlinear susceptibility (hyper-polariz- ability) $\chi_{ij\ell}^{(2)}$ $(\nu'; \omega, -\omega)$, where $\nu' = \omega_{nk}$, etc.

Since it has been assumed that the parity of state k is opposite to that of state n, Rayleigh scattering ($k = n$) cannot be described by second-order per- turbation theory. The process is called hyper-Raman scattering because $k \neq n$. Rayleigh scattering can only be described by the next order of perturbation theory, and by the following diagram:

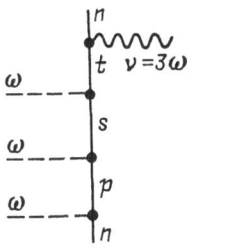

Since $\nu = 3\omega$, this diagram describes the generation of the third harmonic of the incident radiation. It corresponds to $\chi_{ij\ell o}^{(3)}$ $(3\omega; \omega,\omega,\omega)$. Recalling once more Sect.2.3, one realizes this diagram can be written in the following ana- lytical form:

$$\sum_{pst} \frac{r_{np}^i r_{ps}^j r_{st}^\ell r_{tn}^o}{(\omega_{pn} - \omega)(\omega_{sn} - 2\omega)(\omega_{tn} - 3\omega)} \quad .$$

Another diagram of the same nonlinear susceptibility $\chi_{ij\ell o}^{(3)}$ $(3\omega; \omega,\omega,\omega)$, which also represents the generation of the third harmonic, has the follow- ing form:

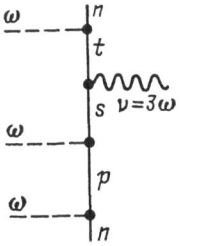

167

Analytically ($\nu = 3\omega$), the diagram is represented thus:

$$\sum_{pst} \frac{r^i_{np} r^j_{ps} r^\ell_{st} r^0_{tn}}{(\omega_{pn} - \omega)(\omega_{sn} - 2\omega)(\omega_{tn} + \omega)} .$$

If the final atomic state is not n but some other state k, the scattering is called hyper-Raman.

Finally, the diagram for the third-order nonlinear susceptibility that describes the nonlinear elastic scattering of light has the following form:

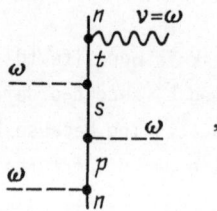

where $\nu = \omega$. The diagram represents one of the terms of the hyper-polarizability $\chi^{(3)}_{ij\ell 0}$ (ω; ω, $-\omega$, ω). Analytically this diagram can be written thus:

$$\sum_{pst} \frac{r^i_{np} r^j_{ps} r^\ell_{st} r^0_{tn}}{(\omega_{pn} - \omega)\omega_{sn}(\omega_{tn} - \omega)} .$$

7.1.3 Linear and Nonlinear Susceptibilities and the Perturbation of Atomic Spectra

In this section it will be shown how the susceptibilities that describe the scattering cross section of electromagnetic radiation also determine the shifts in the levels of an atom in a monochromatic field, $E \cos\omega t$.

But first it must be clarified what is meant by the energy of a state when the perturbation is time-dependent. In Sect.1.3 it was found that the wave function of an atomic state n, when acted upon by a monochromatic field of frequency ω, has the following form, see (1.26):

$$\Psi_n(\mathbf{r},t) = e^{-iE_n t} \sum_{k=-\infty}^{\infty} C^n_k(\mathbf{r}) e^{-ik\omega t} , \tag{7.11}$$

where E_n is the quasi-energy of the perturbed atomic state n.

It is evident that in general there is a linear combination of an infinite number of states with energies $E_n + k\omega$ ($-\infty < k < \infty$), so that no definite value of energy can be ascribed to the perturbed state.

If, however, certain conditions are satisfied such that only one of the terms in (7.11) is significant (e.g., the term with $k = 0$), a stable quantum mechanical state with an energy E_n is obtained. Hence, in a monochromatic field the shift of the energy level is equal to $\delta E_n = E_n - E_n^{(0)} \propto E^2$.

Let us see, first of all, under what conditions one can neglect all of the terms with $k \neq 0$. A two-fold degenerate level (n,p) which is acted upon by a field $E \cos\omega t$ can be introduced into the model studied in Sect.2.5.1. The population amplitude, see (2.23), of the state n for the initial condition, $a_n(0) = 1$, is

$$a_n(t) = \cos\left(\frac{z_{np}E}{\omega} \sin\omega t\right) \quad . \tag{7.12}$$

Expanding $a_n(t)$ in a Fourier series yields a wave function whose form is similar to the one in (7.11) and whose expansion coefficients are

$$c_k^n(\mathbf{r}) \underset{k \text{ even}}{=} J_k\left(\frac{z_{np}E}{\omega}\right)\psi_n^{(0)}(\mathbf{r}) \quad , \quad E_n = E_n^{(0)} \quad . \tag{7.13}$$

This implies that if

$$\frac{z_{np}E}{\omega} \ll 1 \quad , \tag{7.14}$$

then all of the harmonics except the zeroth ($k = 0$) are negligible. The criterion given in (7.14) is thus sufficient for the appearance of a stable state with an energy E_n. It is obvious that this criterion is of a general nature and is not limited to the above example. In the optical spectral range, $\omega \sim \omega_{pn}$ and (7.14) implies that E is much smaller than E_{at}.

On the other hand, for small ω's,

$$\frac{z_{np}E}{\omega} \gg 1 \quad , \tag{7.15}$$

and, according to the well-known properties of the Bessel functions in (7.13), all of the terms will be small except those for which the order, $|k|$, of the function is approximately equal to the argument of the function, $z_{np}E/\omega$. Such points correspond to the energies $E_n^{(0)} + k\omega = E_n^{(0)} \pm |z_{np}E|$. Hence, it is obvious that the energy shift is linear in E and that in comparison the shift quadratic in E may be neglected [7.3]. Such a linear shift is qualitatively analogous to the linear Stark effect, which can be observed when hydrogen is placed in a constant electric field.

Let us now turn to the relationship between energy shifts and linear and nonlinear susceptibilities under the condition that (7.14) is met. An approach will be used that is based on time-independent perturbation theory and that was already examined in Sect.1.3.5. The essence of this method is as follows. The formula for an energy shift in a time-independent field E is used, but the energies $E_n^{(0)}$ are replaced by the quasi-energies $E_n^{(0)} + k\omega$ and the constant field E is replaced by the varying field $E\cos\omega t$. Then, in calculating the dipole matrix elements, the integration over the coordinates is averaged over time.

As a result, the well-known formula of time-independent perturbation theory in the lowest (second) order yields the following formula:

$$\delta E_n^{(2)} = E_n - E_n^{(2)}$$

$$= \frac{1}{4} \sum_m |\mathbf{r}_{nm} \cdot E|^2 \left(\frac{1}{\omega_{nm} - \omega} + \frac{1}{\omega_{nm} + \omega} \right) \quad . \tag{7.16}$$

If (7.16) is compared with (7.8), one obtains an expression that relates the energy shift to the atomic polarizability:

$$\delta E_n^{(2)} = -\frac{1}{4} \sum_{ij} \chi_{ij}^{(1)} E_i E_j \quad , \tag{7.17}$$

where $\chi_{ij}^{(1)}$ is the linear susceptibility (polarizability) tensor of an atom in the state n [a tensor defined in (7.8) for $\nu = \omega$ and $k = n$].

If perturbation theory is used for the fourth order in E, a correction to the energy appears which is proportional to the hyper-polarizability $\chi_{ij\ell o}^{(3)}$. Naturally, this process can be continued. Hence, one can see how the linear and nonlinear susceptibilities (polarizabilities) are related to the energy shift, δE_n, of the corresponding states and what the criteria are for the realization of a shift that is linear or quadratic in E.

7.2 Perturbation of Isolated Atomic States

This section shall be devoted to nondegenerate, isolated, bound electronic states that are perturbed by a monochromatic field. It will be shown how the energy shift of the states depends on the frequency and the polarization of the electromagnetic radiation and also on the structure of the atomic spectrum.

What is an isolated state? In the absence of an external perturbation, the bound electronic states are well defined and isolated from each other. Indeed, the probability curve for an electron with an energy E_n has a natural width γ that is much less than the difference between E_n and the energy of the nearest state, E_p. The situation is quite different in the presence of an external perturbation. The high intensity of the field can lead to a mixing of adjacent states (the term "mixing" was defined in Chap.3). Mixing can also be a result of the polarization of the field. For example, in an elliptically polarized field the bound electronic states are linear combinations of states with different magnetic quantum numbers [7.4].

The remainder of this section deals with states that stay isolated after the external field has been switched on. Effects that are associated with a mixing of states are discussed in Sect.7.3.

7.2.1 Dynamic Polarizability of an Isolated Nondegenerate Atomic State

Equation (7.17) shows how the variation in the energy of a discrete atomic state, n, is determined by the linear susceptibility (polarizability) $\chi_{ij}^{(1)}$. This quantity is called the dynamic polarizability to distinguish it from the static polarizability which corresponds to the limit $\omega \to 0$.

In the limiting case, as $\omega \to 0$, (7.16) yields

$$\delta E_n^{(2)} = \frac{1}{2} \sum_m \frac{1}{\omega_{nm}} \left| \mathbf{r}_{nm} \cdot \mathbf{E} \right|^2 \quad . \tag{7.18}$$

As expected, this coincides with the results of ordinary time-independent perturbation theory if E is replaced by $E \cos\omega t$ and the time-dependent part of the perturbation is averaged over time, i.e., $\cos^2\omega t = 1/2$.

If it is assumed that $\omega > |E_n^{(0)}|$, then one of the terms in (7.8) (with $\nu = \omega$ and $k = n$) will have a nonzero imaginary part, since one must substitute $\omega_{mn} - i\lambda$ for ω_{mn} ($\lambda = +0$) in the neighborhood of a singularity (Sect.2.2). Thus,

$$\text{Im} \{\delta E_n^{(2)}\} = -\frac{\pi}{4} \left| \mathbf{r}_{nm} \cdot \mathbf{E} \right|^2 \delta(\omega_{mn} - |\omega|)$$

$$= -\frac{1}{2} w_{mn} \quad . \tag{7.19}$$

The quantity w_{mn}, which was defined in (2.7), is the ionization width of level n and characterizes the decay of the level due to the escape of the electron into the continuum.

From (7.8) it can be seen that the dynamic polarizability $\chi_{ij}^{(1)}$ is a tensor of rank 2. For linearly polarized light (e.g., along the z axis) this tensor has a nonzero diagonal element $\chi_{zz}^{(1)}$.

For very high frequencies, i.e., $\omega \gg \omega_{mn}$, the general expression for the dynamic polarizability,

$$\chi_{ij}^{(1)} = \sum_m \left(\frac{r_{nm}^i r_{mn}^j}{\omega_{mn} - \omega} + \frac{r_{mn}^i r_{nm}^j}{\omega_{mn} + \omega} \right) \quad , \tag{7.20}$$

and the dipole sum rule [7.5] yield

$$\chi_{ij}^{(1)} = -\frac{Z}{\omega^2} \delta_{ij} \quad , \tag{7.21}$$

where Z is the number of electrons in the atom.

Equation (7.21) is valid for one-electron atoms. In reality one must take into account the Pauli principle, which states that the closed shells of an atom are open to the valence electrons. For this reason, Z must be replaced by the number of valence electrons in the atom (e.g., $Z = 1$ for alkali atoms). In addition, the summation with respect to m in (7.20) must be restricted to vacant discrete states (and all of the states of the continuum) because the Pauli principle does not allow valence electrons (virtual electrons included) to "enter" closed shells. It is assumed that $\omega_{n_0 m_0} \gg \omega \gg \omega_{mn}$, where the ω_{nm} are the characteristic binding energy differences for the valence electrons, and the $\omega_{n_0 m_0}$ are similar differences for the electrons of the closed shells. Thus, the electrons of the core contribute to the energy shift. Due to the fact that $\omega \ll \omega_{n_0 m_0}$, this contribution is similar to the Stark effect in a constant field:

$$\delta E_{n_0}^{(2)} \sim \frac{1}{\omega_{n_0 m_0}} \left| r_{n_0 m_0} \cdot E \right|^2 \quad . \tag{7.22}$$

The contribution is small compared to that of the valence electrons, which according to (7.21) is

$$\delta E = \frac{Z^2}{4\omega^2} E^2 \quad , \tag{7.23}$$

where it has been assumed that $r_{n_0 m_0}^2 \sim \omega_{n_0 m_0}^{-1}$.

One can also arrive at (7.23) without referring to the general equation (7.20), which is nothing more than the contribution to the energy from the natural oscillations of Z valence electrons in a monochromatic field, $E \cos \omega t$. Since $|z_{nm} E| \ll \omega_{mn} \ll \omega$, the high-frequency effect given by (7.23) is very small compared to the Stark effect in a constant field.

The diagonal elements of the dynamic polarizability $\chi_{ij}^{(1)}$ are even functions of ω. For example, (7.20) yields

$$\alpha_n \equiv \chi_{zz}^{(1)} = 2 \sum_m \frac{\omega_{mn} |z_{nm}|^2}{\omega_{mn}^2 - \omega^2} \quad . \tag{7.24}$$

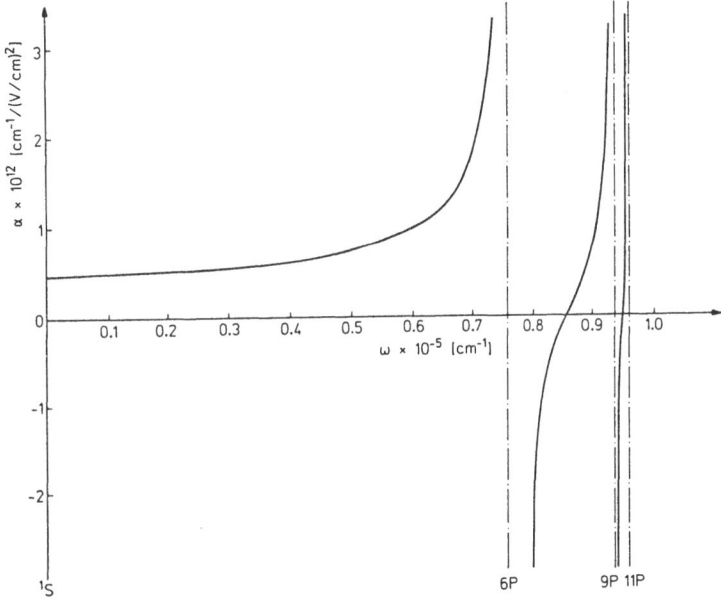

Fig.7.1. Dynamic polarizability α of the ground state of xenon as a function of the frequency of the external field, ω

A preciser criterion for the dynamic polarizability $\chi_{zz}^{(1)}$ to reach the asymptotic value (7.21) (for $i = j = z$) is $\omega^2 \gg \omega_{mn}^2$. Hence, the asymptotic value is reached fairly rapidly as ω increases. It is also reached much faster, the greater the principal quantum number n of the particular spectral term. This is accompanied by a rapid decrease in the characteristic energy differences for the valence terms, ω_{mn}. Numerical calculations confirm these statements [7.6].

The transition to static polarizability occurs when $\omega^2 \ll \omega_{mn}^2$. This explains the rapid change-over from resonance to the static limit (Fig.7.1).

Equation (7.24) illustrates how, if the state n is the ground state and the light is linearly polarized, in all of the intervals between the resonances the dynamic polarizability, α_n, will vanish. If the above two conditions are not met, such a straightforward statement cannot be made.

At frequencies for which $\chi_{ij}^{(1)} \rightarrow 0$, one must take into account the following terms in the Taylor series expansion, i.e., the nonlinear susceptibilities (or dynamic hyper-polarizabilities) come into play. This is also important when the hyper-polarizability has a sharp resonance (at values of ω_{mn} close to 2ω; see Sect.7.2.2). In this case the linear polarizability does not experience a resonance.

7.2.2 Dynamic Hyper-Polarizability

The third-order term in E does not contribute anything to δE_n for the same reason as the first term, i.e., in the double bracket method one averages over the period $2\pi/\omega$ (Sect.1.3.5), so that the third-order term vanishes. Hence, our attention shall be turned to the fourth-order term.

Applying the results of time-independent perturbation theory to $\delta E_n^{(4)}$ and using the double bracket method yields [7.7]

$$\delta E_n^{(4)} = -\frac{1}{24} \sum_{ij\ell o} \chi_{ij\ell o}^{(3)} E_i E_j E_\ell E_o \quad , \qquad \text{where} \tag{7.25}$$

$$\chi_{ij\ell o}^{(3)} = \chi_{ij\ell o}^{(3)} (\omega; -\omega, \omega, \omega)$$

$$+ 2\chi_{ij}^{(1)} (\omega; -\omega) \frac{d}{d\omega} \left[\sum_m \left(\frac{r_{mn}^\ell r_{nm}^o}{\omega_{mn} - \omega} - \frac{r_{mn}^o r_{nm}^\ell}{\omega_{mn} + \omega} \right) \right] \quad . \tag{7.26}$$

The nonlinear susceptibility $\chi_{ij\ell o}^{(3)}(\omega; -\omega, \omega, \omega)$ can be represented by the following sum of diagrams:

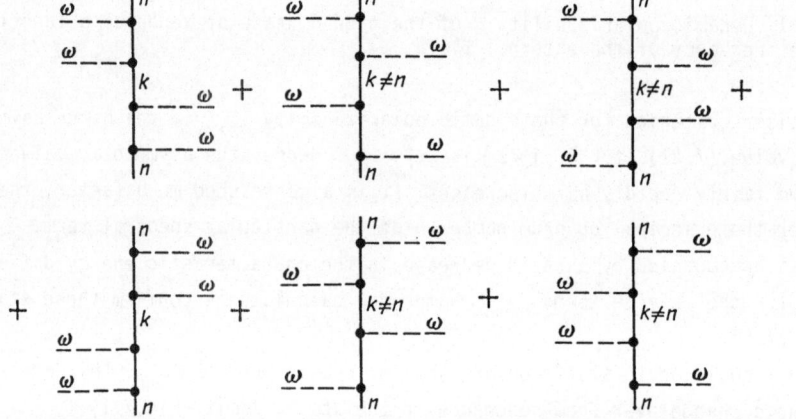

The dynamic polarizability, defined in (7.26), is a tensor of rank 4. The last term in (7.26), which cannot be reduced to a nonlinear susceptibility, is the result of the secular terms for the energies of spectral terms.

It is easy to see that as $\omega \to 0$, (7.26) approaches the formula of time-independent perturbation theory if, in the latter formula, E $\cos\omega t$ is substituted for E and one uses the fact that $\cos^4 \omega t = 3/8$.

If $|\omega| < |E_n^{(0)}| < 2|\omega|$, the dynamic hyper-polarizability exhibits an imaginary part:

$$\text{Im}\{\delta E_n^{(4)}\} = \pi |z_{kn}^{(2)}(\omega)|^2 E^4 \delta(\omega_{kn} - 2\omega) \quad . \tag{7.27}$$

Multiplied by two, this quantity determines, in accordance with (2.14), the total two-photon ionization probability for the state n [$z_{kn}^{(2)}$ is defined in (2.13)]. For $|E_n^{(0)}| < \omega$ the imaginary part of the dynamic hyper-polarizability is of no interest, since the one-photon ionization channel is open and the imaginary part of the dynamic hyper-polarizability is small compared to the imaginary part of the dynamic polarizability.

If the polarizability and the hyper-polarizability are compared, the criterion for a small hyper-polarizability is expressed in an analytical form in terms of the ratio of the magnitude of the perturbation to the characteristic differences in atomic energies. However, on the basis of this criterion one cannot give a sufficiently general estimate of the upper limit for the electric field strength, because both the energy differences between bound electronic states and the dipole matrix elements that connect these states greatly depend on the principal quantum number and the particular type of atomic spectrum. Calculations done for the ground states of alkali atoms show that in the nonresonance case the E^4-order terms already become comparable with the E^2-order terms when $E \approx 10^6 V/cm \approx 10^{-3} E_{at}$ [7.8]. For such fields, all of the terms in the power series in E are of the same order, i.e., the series is divergent.

In reality, the hyper-polarizability will correctly determine the shift of an energy level only in regions where the dynamic polarizability is anomalously small, i.e., either in the intervals between resonances or when $\omega_{kn} \pm 2\omega \approx 0$ (in this case the hyper-polarizability increases, unlike the ordinary polarizability, due to resonances). In the latter case, for the above equations to be valid; ω must not exactly coincide with the resonance frequency.

7.2.3 Criteria for Applying Perturbation Theory

To find a quantitative criterion for the equations for the dynamic polarizability to be valid, let us first consider low-lying excited states. For a sufficiently strong field, this problem is equivalent to that of an atomic multiplet in a strong field (Sect.2.5). If Sect.2.5 is recalled, it is seen that the equations for the dynamic polarizability break down for fields that are not very strong, i.e., this happens when

$$\delta E_n^{(2)} \ll \omega_{kn} \quad \text{or} \quad \omega, \tag{7.28}$$

see (2.34) and (2.42). When this is not so, transitions between the states k and n become possible. In Ψ_n, the expansion coefficient a_k becomes comparable to a_n. Hence, there appears a nonresonant mixing of the states k and n similar to that found for an atomic multiplet in a field, i.e., $V_{kn}^{(2)} \gtrsim \omega_{kn}$ (Sect.2.5).

The conditions are less stringent for perturbation theory to be applicable to highly excited states. States with large quantum numbers (e.g., the highly excited states of the hydrogen atom or the ground and the first few excited states of atoms with large Z's) can be examined in the framework of the quasi-classical approximation (Chap.4). In this case it is not advisable to introduce a fixed level and require the inequality (7.28) for such a level, since there are many levels of this kind, each of which provides a small contribution to the dynamic polarizability. Instead, one can substitute an integral for the sum with respect to m in (7.20). The shift in the level n is determined by the real part of the dynamic polarizability, i.e., the principal part of the integral. Levels m which are close to n but on different sides of it make contributions to the integral that are equal in magnitude but of opposite signs. Hence, the smallness of ω_{nm} does not affect the polarizability, and instead of (7.28),

$$\delta E_n^{(2)} \ll |E_n^{(0)}| \quad . \tag{7.29}$$

The equations for the dynamic polarizability also break down close to resonances between the given level n and any of the levels m. The role of resonances will be examined in Chap.8.

7.2.4 The Dynamic Polarizability as a Function of the Perturbing Field

To calculate the dynamic polarizability of different states for different atoms, one must obviously know the matrix elements of the dipole interaction of the given state with all other bound states and with the states in the continuum. To calculate the real part one must sum up the virtual transitions over an infinite number of discrete states and integrate over the continuum. The imaginary part is determined by only one matrix element, i.e., the photoionization probability. The static polarizabilities have been calculated very thoroughly [7.6]. The dynamic polarizability of states has been calculated by applying semi-phenomenological methods, e.g., the quantum defect method and the pseudopotential method [7.9] (Sect.2.6). The phenomenological element in these methods is the use of data from atomic spectra. The calculations were done for alkali atoms and the atoms of inert gases [7.10]. Preciser calculations, which use the Hartree-Fock approximation and take into account interelectron correlations in the atom, have led to results that differ substantially from the results of semi-phenomenological methods (for atoms with two or more valence electrons) [7.11]. A typical curve of the dynamic polarizability of a ground state as a function of the frequency of a linearly polarized field is shown in Fig.7.1. It is evident that the dynamic polarizability differs from the Stark effect in a constant field. The static polarizability of a given state is a constant, while the dynamic polarizability is determined by the frequency of the field. It can vary from zero (in the intervals between resonances) to infinity (at resonance). The infinite value is a result of the perturbation-theory method.

In reality, in passing through resonance, the shift $\delta E_n^{(2)}$ reaches a finite value and then changes sign smoothly rather than abruptly (Chap.8). The behavior of $\delta E_n^{(2)}(E,\omega)$ when it passes through resonance is determined by the finite width of the resonance states or by the degree of the monochromaticity of the field (if it is greater than the former width). The width of the resonance states is determined by the degree of spontaneous relaxation, of ionization, or of resonance interaction between these states (depending on the binding energy, frequency, or strength of the field). In each case, the calculations of dynamic polarizability according to (7.25) are not valid in the neighborhood of a resonance (Chap.8).

The frequency dependence of α is practically the same for all ground states, i.e., the static ($\omega = 0$) value for α is greater than zero, there are regions of anomalous dispersion [where $d\alpha(\omega)/d\omega \gg 1$] when ω passes through resonances, and α vanishes in each interval between resonances (Fig.7.1).

For an excited state, the frequency dependence is different because as the frequency is increased resonances with higher (in relation to state n), $E_m^{(0)} > E_n^{(0)}$, and lower states, $E_m^{(0)} < E_n^{(0)}$, may take place. In particular, the polarizability does not necessarily have to vanish in the intervals between resonances, and regions of negative dispersion, $d\alpha(\omega)/d\omega < 0$, may appear (Fig. 7.2). The limiting values (with $\omega = 0$ and with $\omega \gg |E_n^{(0)}|$) are of similar form as in the perturbation of ground states.

7.2.5 Experimental Data

The nature of the nonresonant perturbation of isolated atomic states in a linearly or circularly polarized field is quite obvious, so that the main goal of most experiments is to obtain some data on the dynamic polarizability. This is where several problems arise. The first is that one can observe only those changes in the energy of a specific transition that are specified by the polarizabilities of the internal and final states. One cannot a priori ignore the perturbation of the ground state since the dynamic polarizability is strongly dependent on the frequency. At a given frequency the excited

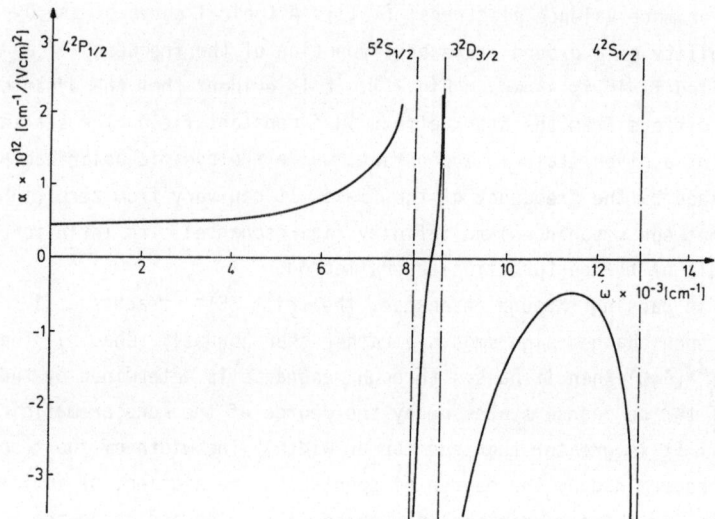

Fig.7.2. Dynamic polarizability α of the excited $4P_{\frac{1}{2}}$ state of potassium as a function of frequency of the external field, ω

state may exhibit a small midresonance polarizability (Figs.7.1,2). The second difficulty is the result of a variety of factors that cause a broadening of the transition line, e.g., the Doppler effect, the non-monochromaticity of the perturbing radiation, and the nonuniformity of the spatial and temporal distribution of the radiation over the target (Sect.6.4).

Summarizing the experimental results cited in [7.3], one can say that

1) in all of the cases in which the criterion for the nonresonant nature of the perturbation was met, a shift in the transition energy was observed which was proportional to the square of the field strength;

2) the experimental values of the real part of the dynamic polarizability agree fairly well with calculated values;

3) the data on the dynamic polarizability was obtained only for selected values of ω and not for a sufficiently wide range of frequencies;

4) there is no data on the imaginary part of the dynamic polarizability when $\omega < E_n^{(0)}$, only when $\omega > E_n^{(0)}$, i.e., when the imaginary part is determined by the one-photon ionization probability [7.12];

5) there is no data on the hyper-polarizability.

The results of typical experiments concerned with the measurement of the level shift in an electromagnetic field make it possible to estimate the magnitude of this effect and the accuracy to which α_n can be measured.

178

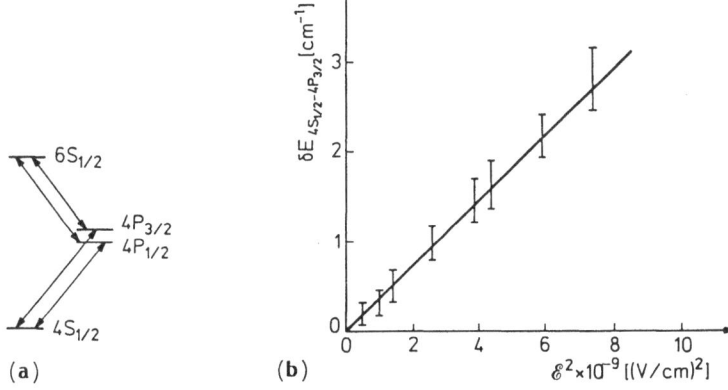

Fig.7.3a,b. Variation in the energy, δE, of the $4S_{1/2} \rightarrow 4P_{3/2}$ transition in potassium as a function of the intensity of the external field E: (a) the level diagram, and (b) experimental results

Two experiments shall now be considered which allow the measurement of the variation of the energy of a transition from the ground state of a potassium atom, $4S_{1/2}$, to the excited state $4P_{1/2,3/2}$, stimulated by a ruby laser (Fig.7.3). The method in which the light is absorbed from an auxiliary source was used. With a field strength of about 5×10^5V/cm $\approx 10^{-4}E_{at}$, $\alpha_{1/2,1/2} = (0.2 \pm 0.1) \times 10^{-11}$cm^{-1}(V/cm)$^{-2}$ and $\alpha_{1/2,3/2} = (1.0 \pm 0.5) \times 10^{-11}$cm^{-1} (V/cm)$^{-2}$, and a shift in the transition energy was observed which was proportional to the square of E [7.13] [the dynamic polarizability can be measured in several different units: 1.0 cm^{-1}(V/cm)$^{-2} = 1.15 \times 10^{14}$a.u. $= 0.14 \times 10^{14}$(Å)^3].

The measurements given in [7.14] were made with a sodium atom (the transition from the ground state, $3S_{1/2}$, to the excited state $4D_{5/2}$) using the two-photon method with counter-propagating beams. A higher degree of accuracy could be obtained and certain conclusions concerning the contributions of the different states could be made. The absorption of light from two lasers with wavelengths 589 and 569 nm resulted in a two-photon resonance between the states. The resonance could be observed because of the fluorescence of the transition (330 nm) $4P \rightarrow 3S$, which follows the spontaneous relaxation of the 4D state. The intensity of the 589-nm radiation (10^3V/cm) was about one order of magnitude higher than that of the 569-nm radiation. The detuning between the transition $3S \rightarrow 3P$ and a single photon of the intense light was relatively small. Both factors are responsible for the much stronger perturbation of the ground state compared to the 4D state. The results are presented in Fig.7.4. The polarizability of the ground state, α_{3S}, is 10^{-8}cm^{-1} (V/cm)$^{-2}$.

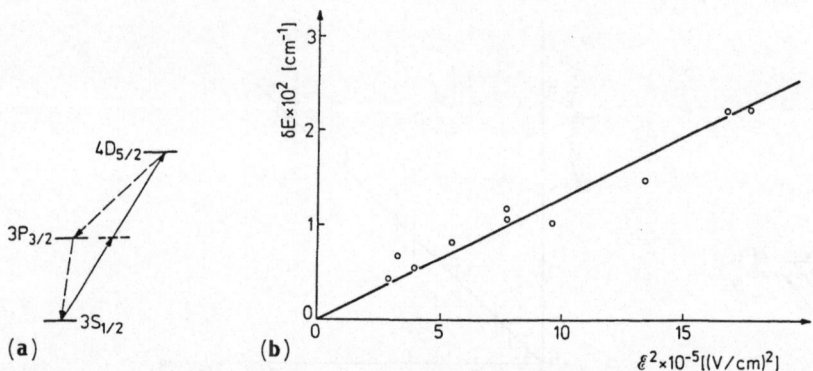

Fig.7.4a,b. Variation in the energy, δE, of the sodium atom in the ground state, 3^2S, as a function of the intensity of the external field E: (a) the level diagram, and (b) experimental results

The data also show that the polarizability depends on the detuning from resonance [7.14]. Since the frequency changes at the same time as the detuning from the resonance with the intermediate state $3P_{3/2}$ changes, it was possible to determine the dependence of the shift of the 3S state on the detuning from the intermediate resonance (Fig.7.5). The experimental data correspond to large detunings from resonances when the magnitude of the shift is propor-

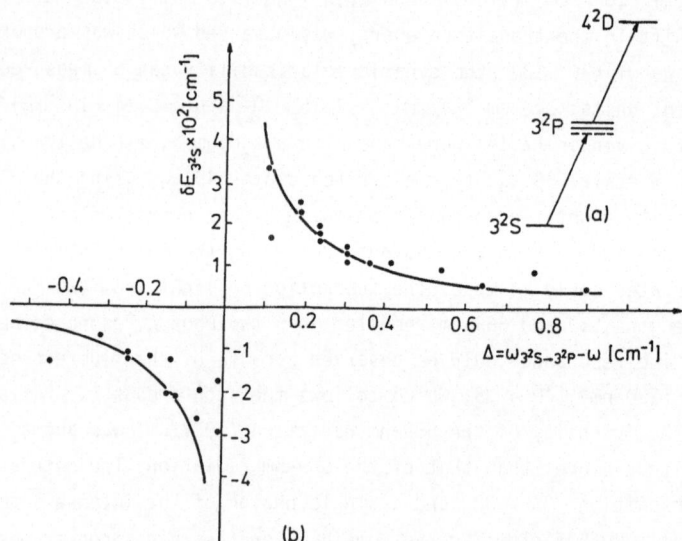

Fig.7.5a,b. Variation in the energy, δE, of the sodium atom in the ground state, 3^2S, as a function of the detuning from resonance, Δ, between the frequency of the external field, ω, and the energy of the $3^2S \rightarrow 3^2P$ transition: (a) the level diagram, and (b) experimental results

tional to the square of the field strength. Resonance phenomena themselves are observed at smaller detunings. Experiments in which such phenomena are observed are discussed in Chap.8.

As an example of an experiment in which the transition energy between two excited states is measured the one described in [7.15] will be discussed. The excited state relaxation method was used to measure the energy of the $7S_1 \rightarrow 6P_2$ and $7S_1 \rightarrow 6P_1$ transitions of mercury (Sect.6.4). The $7S_1$ state was excited by a high-frequency discharge, and lines with $\lambda = 5460$ and 4358 Å were observed. The mercury vapor was irradiated by a pulsed neodymium glass laser ($\lambda = 10\,600$ Å). The observed shift in the two wavelengths was the same. Hence, it could be assumed that the $7S_1$ state experiences the main perturbation. This conclusion is in agreement with the results obtained from theory, i.e., in general the dynamic polarizability is a rapidly growing function of the quantum numbers of the states, especially the principal quantum number.

The polarizability, α_{7S_1}, measured in fields ranging from 10^5V/cm to 5×10^5V/cm was found to be 10^{-12}cm^{-1}(V/cm)$^{-2}$. This coincides, to within 40%, with the results of calculations [7.10], which also show that $\alpha_{7S_1} \gg \alpha_{6P_{1,2}}$.

An example of a shift in the energy of a transition to a highly excited state is the 6S $\rightarrow$ 6F transition of cesium. The resonance multi-photon ionization method and a neodymium glass laser beam were used (Sect.7.5). Figure 7.6 shows the results of this experiment [7.16]; $\delta\alpha_{6S \rightarrow 6F} = (2.4 \pm 0.8) \times 10^{-11}$ cm^{-1}(V/cm)$^{-2}$. The calculated value is in good agreement with this result [7.10]. It is somewhat peculiar that $\alpha_{6S} \sim \alpha_{6F}$.

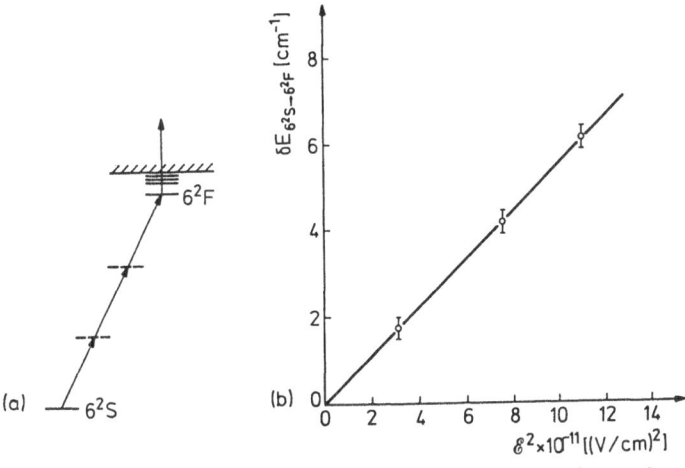

Fig.7.6a,b. Variation in the energy, δE, of the $6^2S \rightarrow 6^2F$ transition in cesium as a function of the intensity of the external field E: (a) the level diagram, and (b) experimental results

7.3 Perturbation of Degenerate States

In a constant field, degeneracy plays a significant role in the shifts of
the atomic levels produced by a perturbation. For example, it is well known
that in the hydrogen atom the degeneracy of the states with respect to the
orbital angular momentum is responsible for the linear rather than quadratic dependence of the shifts on the field. It is then to be expected
that in a variable field the degenerate states will greatly affect the dynamic polarizability. An appropriate form of perturbation theory will be
applied to describe the perturbation of degenerate states (Sect.2.5).

7.3.1 Two Adjacent Levels

Before discussing the general case (an arbitrary number of adjacent levels),
let us first of all consider two adjacent levels, n and k (and any number of
levels that are far from these two). It will·be shown that the results have
a simple appearance. The system of equations (2.35) for two adjacent levels
can be written as follows:

$$i\dot{a}_n = \delta E_{nk}^{(2)} a_k \exp(i\omega_{nk}t) + \delta E_n^{(2)} a_n \quad ,$$

$$i\dot{a}_k = \delta E_{kn}^{(2)} a_n \exp(i\omega_{kn}t) + \delta E_k^{(2)} a_k \quad ,$$

(7.30)

where the $\delta E_{n,k}^{(2)}$ have been defined in (7.17) and are simply the shifts of the
levels n and k when these levels are not adjacent. Furthermore, in (7.30),

$$\delta E_{kn}^{(2)} = - [V_{kn}^{(2)}(\omega) + V_{kn}^{(2)}(-\omega)] = - \frac{1}{4} \sum_{ij} \chi_{ij}^{(1)nk} E_i E_j \quad ,$$

(7.31)

where $\chi_{ij}^{(1)nk}$ is an off-diagonal element of the linear susceptibility; $\delta E_{nk}^{(2)}$
is defined similarly. In general, $\delta E_{nk}^{(2)}$ and $\delta E_{kn}^{(2)}$ are not connected by simple
relationships. Only in the case of exact degeneracy, when $E_k^{(0)} = E_n^{(0)}$, is
$\delta E_{nk}^{(2)} = [\delta E_{kn}^{(2)}]^*$.

Using the substitution

$$a_{k,n}(t) = A_{k,n} \exp[i(\Omega + E_{k,n}^{(0)})t] \quad ,$$

(7.32)

(7.30) can be reduced to a system of two linear equations. In solving the new
sytem, one finds the coefficients $A_{k,n}$ and the two values of Ω. The phases,
Ω, determine the positions of the quasi-energies under the influence of the
perturbation. The result is

$$\Omega_{1,2} = -\frac{1}{2}(E_n + E_k) \pm \left[\frac{1}{4}(E_n - E_k)^2 + \delta E_{kn}^{(2)} \delta E_{nk}^{(2)}\right]^{\frac{1}{2}} , \qquad (7.33)$$

where $E_{n,k} \equiv E_{n,k}^{(0)} + \delta E_{n,k}^{(2)}$ and it has been assumed that $E_k^{(0)} > E_n^{(0)}$.

If $|\delta E_{nk}^{(2)}| \ll \omega_{kn}$, which reflects the fact that the field only slightly perturbs the states n and k, then one can use (7.33) to obtain the usual (i.e., quadratic in the field) shifts. The fact that the levels are adjacent was not taken into account, i.e., $\delta E_{n,k} \to \delta E_{n,k}^{(2)}$, due to the weak mixing of n and k.

If $|\delta E_{kn}^{(2)}| \gg \omega_{kn}$ (strong perturbation), then the fact that n and k are adjacent levels is reflected in their strong mixing. It will be assumed that $\delta E_{kn}^{(2)} = [\delta E_{nk}^{(2)}]^*$, so that using (7.33) yields

$$\Omega_{1,2} = -\frac{1}{2}\left[\delta E_n^{(2)} + \delta E_k^{(2)}\right] \pm \left[\frac{1}{4}\left(\delta E_n^{(2)} - \delta E_k^{(2)}\right)^2 + |\delta E_{nk}^{(2)}|^2\right]^{\frac{1}{2}} . \qquad (7.34)$$

It is seen that the splitting of the levels n and k differs greatly from the shifts, $\delta E_{n,k}^{(2)}$, which did not take into account that the levels are adjacent. The interaction between these states due to the field is of the same order of magnitude as the interaction of each of these states with the field.

Thus, for both a small and a large amplitude of the nonresonant perturbation, the shift of the energy levels proves to be quadratic in the perturbation. However, the dynamic polarizability, which is determined by the proportionality factor of the square of the external field vector, will be different depending on the relationship between the perturbation and the separation of the two levels. For the intermediate case, $\delta E_{kn}^{(2)} \sim \omega_{kn}$. Each level is split into two quasi-levels.

From (7.34) one can see that even if the interaction of the levels n and k is strong, the quasi-levels cannot intersect (the analog of the Wigner-von Neumann theorem for a constant field [Ref.7.17, §79]).

Indeed, two adjacent quasi-levels will only intersect if $\omega_{kn} \sim \delta E_{kn}^{(2)}$ and if

$$\left[\frac{1}{4}(E_n - E_k)^2 + |\delta E_{nk}^{(2)}|^2\right]^{\frac{1}{2}} = 0 . \qquad (7.35)$$

Since ω_{kn} is small, it may be assumed that $\delta E_{nk}^{(2)} = [\delta E_{kn}^{(2)}]^*$, i.e., such an intersection only takes place if $\delta E_{nk}^{(2)} = 0$. For a dipole interaction between electromagnetic radiation and an atom, the two-photon matrix element vanishes either when the states n and k have different parities or, if they have the same parities, when their angular momentum quantum numbers differ by more than two. In this case nonresonant mixing of the levels n and k will not take place.

7.3.2 The General Case

For the general case of N adjacent levels, (7.30) transforms into

$$i\dot{a}_n = \sum_{k \neq n} \delta E_{nk}^{(2)} a_k \exp(i\omega_{nk}t) + \delta E_n^{(2)} a_n \quad , \tag{7.36}$$

where the summation is carried out over all of the adjacent levels. The system of equations (7.36) can be solved in the same way as (7.30). The splitting of the atomic levels can be found by solving an equation of the Nth degree, which can only be done using numerical methods.

All of the above results are valid if the frequency of the external perturbation is comparable to the differences in the frequencies of certain atomic transitions, ω_{ns} (s being any level not adjacent to n), i.e., $\omega > \omega_{ns}$. If $\omega \ll \omega_{ns}$, then the $\cos\omega t$ in the initial equations cannot be considered to be oscillating rapidly. The problem then becomes more complex because one not only has to take into account the two-photon matrix element of the transition between the states n and k but also the one-photon matrix element. Such a one-photon matrix element causes a strong mixing of the states n and k, provided that $z_{nk}E/\omega > 1$ (Sect.2.5). In practice this, in combination with the condition $\omega \ll \omega_{ns}$ and the general requirement that $z_{nk}E \ll \omega_{ns}$, is difficult to achieve [7.18]. This problem will not be considered here.

When is the resonant mixing of adjacent levels in a variable field important? The most obvious example is the hydrogen atom, in which the states that do mix have like parities and belong to the same spectral term. Another example is the multiplet in an elliptically polarized external field of states that are degenerate with respect to the magnetic quantum number M. When the field is linearly polarized, $\Delta M = 0$ and hence the term $\delta E_{nk}^{(2)}$ in (7.36), that determines the degree of mixing, vanishes when $k \neq n$. In a circularly polarized field ΔM is also equal to zero, because the projection of the angular momentum (on any axis) increases or decreases (or vice versa, depending on the direction of the circular polarization) by one unit when one photon is absorbed or emitted. Thus, in linearly or circularly polarized fields there is no mixing of the states of a multiplet.

Finally, when strong fields come into play, the levels of a multiplet of an atom with large Z can be quite close together. If the multiplet is sufficiently isolated in the atomic spectrum, i.e., if $\omega_{ns} \gg \omega_{nk}$ (s is any level of another multiplet), then numerical methods can be used to find the quasi-stationary states [7.19]. With the usual LS coupling, the quasi-stationary states in a weak, linearly polarized field are described by the total angular momentum and its projection in the direction of the field. However, in

a strong field the coupling breaks down and the states are characterized by
the projections of the orbital and spin angular momenta in the direction of
the field (Sect.1.3).

7.3.3 A Perturbation in an Elliptically Polarized Field

A level with a total angular momentum J splits into $2J + 1$ components in an
elliptically polarized field. To find the wave functions of the new states
one must solve equations similar to (7.36). This implies that the final result includes states with different projections of angular momentum. An elliptically polarized field can be thought of as a linear combination of clock-
wise- and counter-clockwise-polarized light. Hence, the states that mix are
those for which $\Delta M = \pm 2$ [in accordance with the selection rules for $\delta E_{nk}^{(2)}$;
see (2.13)]. Unfortunately, it is impossible to formulate any sufficiently
general equations or laws that govern the splitting. Numerical calculations
were made for the 4F level of potassium in the field of a neodymium glass
laser [7.20].

The field completely disrupts the spin-orbit coupling and mixes the
states with $-3 \leqslant M \leqslant 3$, where M is the projection of the orbital angular momentum of the 4F state equal to three. Due to the selection rules ($\Delta M = 0$,
± 2), the system of equations (7.36) splits into two independent subsystems
with $M = \pm 3$, ± 1, and $M = \pm 2$, 0.

Figure 7.7 shows the calculated real parts of the polarizability of the
sublevels of the 4F state of a potassium atom. The values of θ equal to 0
and $\pi/2$ correspond to clockwise and counter-clockwise circular polarizations,
respectively; the value $\pi/4$ corresponds to linear polarization. In all

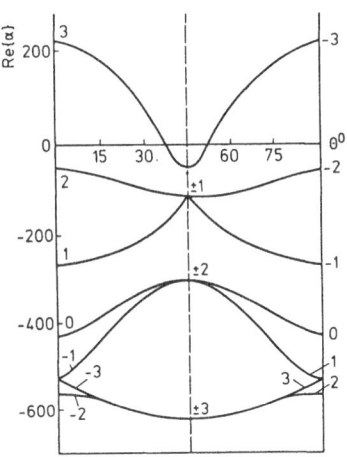

Fig.7.7. Dynamic polarizability α of the
sublevels of the 4F level of the potassium
atom in an elliptically polarized field;
a linearly polarized field corresponds to
$\theta = 45°$, and a circularly polarized field
corresponds to $\theta = 0$ or $90°$

185

three cases the magnitude of the projection of the orbital angular momentum in the direction of the field or of the polarization (from 0 to 3) is given. All of the states except the one with $M = 0$ ($\theta = \pi/4$) are doubly degenerate. This statement is a generalization of Kramer's theorem for a constant electric field to the case of a variable field [7.21].

For all other values of θ (elliptically polarized fields) the states cannot be characterized by M because there is a mixing of "magnetic" substates. Figure 7.7 illustrates how the sign and amplitude of the perturbation and their dependence on the ellipticity of the field are entirely different for different substates, which are described by a new quantum number different from M.

Summarizing all that has been said about the perturbation of a degenerate bound electronic state, one can say that: 1) the initial states are split into components; 2) new quantum numbers are needed to classify the components; and 3) the energy of these components with respect to the energy of the initial state changes.

A study of the effect of an elliptically polarized field on a cesium atom in the ground state (6S) was made using resonance multi-photon spectroscopy (Sect.6.4) [7.22]. The frequency of the laser radiation was chosen in such a way that an intermediate three-photon resonance with the 6F state appeared. When the frequency was varied in the neighborhood of the resonance, a maximum was observed for a linearly polarized field, and several maxima were observed for an elliptically polarized field. Thus, the results of this experiment qualitatively confirm the theoretical predictions. A quantitative comparison is not possible on account of two factors: 1) in an elliptically polarized field not only the resonance state (6F) but also the initial state (6S) is split; hence, 2) one must take into account the selection rules for three-photon transitions from the components of the ground state to the components of the resonance state.

7.3.4 Perturbation of the Hydrogen Atom Spectrum

The spectrum of atomic hydrogen perturbed by a strong, variable monochromatic field exhibits certain peculiarities due to the degeneracy of the spectral terms with respect to the orbital quantum number ℓ, i.e., due to the presence of a constant dipole moment, d. This is the reason why atomic hydrogen shows a linear Stark effect when placed in a constant electric field, whereas other atoms show a quadratic effect. Is this difference also observable in a variable field? A theory that adequately describes the perturbation of the atomic hydrogen spectrum in a variable field will now be developed. The results of this theory will also be compared with experimental data to give an answer to the above question.

If all of the dipole matrix elements that connect the various substates with the same value of n (the principal quantum number) but with different values of ℓ (the orbital quantum number) were identical, then the problem would be reduced to that of a constant electric dipole in a variable monochromatic field (Chap.2). However, in reality this is not the case: with hydrogen the dipole matrix elements strongly depend on ℓ.

According to Sect.2.5, the amplitudes a_ℓ of states with definite values of ℓ satisfy the following system of equations:

$$i\dot{a}_\ell = -E \cos\omega t \sum_{\ell'} \langle n\ell | z | n\ell' \rangle a_{\ell'}$$

$$+ \frac{1}{2} E^2 \cos\omega t \sum_{\ell'NL} \langle n\ell | z | NL \rangle \langle NL | z | n\ell' \rangle$$

$$\times a_{\ell'} \left(\frac{e^{i\omega t}}{\omega_{nN} - \omega} + \frac{e^{-i\omega t}}{\omega_{nN} + \omega} \right) . \qquad (7.37)$$

For the sake of simplicity, it has been assumed that the external field is linearly polarized, so that the magnetic quantum number M is conserved for virtual transitions. To keep the number of equations in (7.37) down to a reasonable number, M has been excluded. In reality a separate system would have to be written for each value of M. The first term in (7.37) corresponds to the virtual transitions from state (n,ℓ) to state (n,ℓ'). In the second term the labels N and L denote the principal and the orbital quantum numbers, respectively, of the shells that are not the nth shell. This term describes the virtual transition $(n,\ell) \to (n,\ell')$, not directly as the first term does, but through the intermediate states (N,L). In the diagrammatic language the matrix elements that correspond to the above transitions can be represented as follows:

If these diagrams are expressed analytically according to the rules of the diagrammatic technique for degenerate states (the diagrams are time-dependent; see Sect.2.5), one arrives at (7.37).

In (7.37) diagrams with orders higher than the second were neglected. This is equivalent to assuming that

$$z_{nN}E \ll \omega_{nN} \quad , \tag{7.38}$$

where ω_{nN} is the frequency of the transition from the nth shell to any other shell, including the adjacent shell N. For the hydrogen atom this inequality implies that $E \ll E_{at}^{(n)}$, where $E_{at}^{(n)}$ is the atomic field for the nth shell. The need for the last inequality is obvious.

The system of equations in (7.37) has a very complicated form. Thus, it is of interest to study the various limiting cases.

Let us first of all assume that the frequency of the external field is sufficiently high, so that

$$\omega \gtrsim \omega_{nN} \quad . \tag{7.39}$$

Neglecting the rapidly oscillating terms in (7.37) yields

$$i\dot{a}_\ell = \frac{1}{4} E^2 \sum_{\ell'} \left[z_{\ell\ell'}^{(2)}(\omega) + z_{\ell\ell'}^{(2)}(-\omega) \right] a_{\ell'} \quad , \tag{7.40}$$

where the two-photon dipole matrix element, $z_{\ell\ell'}^{(2)}$, was defined in (2.13).

The solutions to (7.40) were discussed in Sects.2.5 and 7.3.2. It follows that the nth spectral term is split into steady-state levels that are characterized by a new set of quantum numbers (their number is n, just as the number ℓ of old quantum numbers). The shifts of the levels with respect to the unperturbed value, $E_n^{(0)}$, are proportional to the square of the amplitude E of the external field.

In view of the fact that the states with quantum numbers ℓ and ℓ' have like parities, (7.40) breaks up into two independent subsystems with even and odd values of ℓ. They can therefore be solved separately. Hence, when the condition in (7.39) is met, the state of a hydrogen atom in an external field is specified by its parity.

Let these conclusions be illustrated with specific examples. For example, in the shell with n = 2 the 2s and 2p states do not mix and the orbital quantum number ℓ is still a valid quantum number. In this case the term with n = 2 splits into the s and p sublevels with shifts that are proportional to the field strength. For the same reason, the 3p state does not mix with the s and d states.

Now it is easy to determine what happens in the limiting case of high frequencies, i.e., when

$$\omega \gg \omega_{nN} \quad , \tag{7.41}$$

a condition that is much stricter than the one given in (7.39). In this case the diagonal matrix elements $z_{\ell\ell}^{(2)}$ are proportional to ω^{-2}, and the off-diagonal matrix elements $z_{\ell\ell}^{(2)}$ are proportional to ω^{-4}. This follows from the dipole sum rule, see (7.21). As a result, the off-diagonal elements in (7.40) can be neglected. Also, for frequencies much higher than ω_{nN} the degenerate states no longer mix, i.e., the orbital quantum number ℓ continues to be a valid quantum number. The initial degenerate spectral term, n, splits into states with definite values of ℓ that only undergo shifts which are proportional to the square of the field strength. In this case a purely quadratic shift is small since $\omega \gg \omega_{nN}$.

Now the limiting case of low frequencies shall be discussed, i.e., when

$$\omega \ll \omega_{nN} \quad . \tag{7.41}$$

Here the oscillations of the time-dependent factors in (7.37) become slow and, generally speaking, one cannot neglect them. However, to simplify matters, one can neglect ω compared with ω_{nN} in the denominators of (7.37). Hence,

$$
\begin{aligned}
i\dot{a}_\ell = &-E \cos\omega t \sum_{\ell'} <n\ell|z|n\ell'>a_{\ell'} \\
&+ E^2 \cos^2\omega t \sum_{\ell'NL} <n\ell|z|NL><NL|z|n\ell'> \frac{a_{\ell'}}{\omega_{nN}} \quad .
\end{aligned}
\tag{7.42}
$$

This system of equations can only be solved using numerical methods [7.23].

To analyze this case qualitatively, it will be assumed, for the time being, that all of the dipole matrix elements are equal (the correctness of this assumption was discussed earlier). All of the amplitudes a_ℓ are also equal at any moment of time t if it is assumed that the separate sublevels are all initially equally populated (Sect.2.5). Under these assumptions (7.42) acquires a simpler form:

$$i\dot{a} = \left(-dE \cos\omega t + \frac{1}{2} \alpha E^2 \cos^2\omega t \right) a \quad , \tag{7.43}$$

where

$$
\begin{aligned}
d &= \sum_{\ell'} <n\ell|z|n\ell'> \quad , \qquad a \equiv a_\ell \quad , \\
\alpha &= 2 \sum_{\ell'NL} \frac{1}{\omega_{nN}} <n\ell|z|NL><NL|z|n\ell'> \quad .
\end{aligned}
\tag{7.44}
$$

Note that (7.43) is valid for a dipolar molecule placed in a low-frequency monochromatic field, where d is the dipole moment and α is the static polarizability.

The solution to (7.43) has a simple form:

$$a(t) = \exp\left[i\ \frac{dE}{\omega}\ \sin\omega t + i\ \frac{\alpha E^2}{4}\left(t + \frac{\sin 2\omega t}{2\omega}\right)\right]\ . \tag{7.45}$$

If a(t) is expanded in a Fourier series, then

$$a(t) = \exp\left(i\ \frac{\alpha E^2}{4}\ t\right)\ \sum_{m=-\infty}^{\infty} A_m\ \exp(-im\omega t)\ , \tag{7.46}$$

where

$$A_m = \sum_{s=-\infty}^{\infty} (-1)^m J_s\left(\frac{\alpha E^2}{8\omega}\right) J_{m+2s}\left(\frac{dE}{\omega}\right)\ . \tag{7.47}$$

In contrast to the limiting case of high frequencies a large number of quasi-energy harmonics appear in the atomic spectrum in the low-frequency case. This, together with a term splitting in ℓ, leads to a highly complex system of new stationary states for hydrogen.

However, one can draw some qualitative conclusions if the various limiting cases of (7.46,47) are considered for a fixed frequency that is much lower than ω_{nN} and for a field strength E that is variable.

If the field is so weak that

$$\frac{\alpha E^2}{\omega} \ll 1 \qquad \text{and} \qquad \frac{dE}{\omega} \ll 1\ , \tag{7.48}$$

then the terms in (7.46,47) that do not vanish are only those for which $s = m = 0$. From (7.46) one can see that the shift in a level of a dipolar molecule is equal to $\alpha E^2/4$. If one turns from a dipolar molecule to a real hydrogen atom, it will be discovered that, if the conditions in (7.48) are met, one arrives at a system of equations similar to the one in (7.40), in which one must set ω equal to zero:

$$i\dot{a}_\ell = \frac{1}{2}\ E^2\ \sum_{\ell'}\ z_{\ell\ell'}^{(2)}(0)a_{\ell'}\ . \tag{7.49}$$

As the strength of the field grows, there comes a time when

$$\frac{\alpha E^2}{\omega} \ll 1 \qquad \text{and} \qquad \frac{dE}{\omega} \gtrsim 1\ . \tag{7.50}$$

In this case one can set $s = 0$ in (7.47) and from (7.46) obtain

$$a(t) = \sum_{m=-\infty}^{\infty} (-1)^m J_m\left(\frac{dE}{\omega}\right)\ \exp(-im\omega t)\ . \tag{7.51}$$

This implies that the term is split into quasi-energy levels with quasi-energies equal to

$$E_{nm} = E_n^{(0)} + m\omega \quad . \tag{7.52}$$

Each quasi-energy harmonic, m, has a weight $J_m(dE/\omega)$. In this case the quadratic shift, $-\alpha E^2/4$, can be neglected since it is small compared to $m\omega$.

For hydrogen atoms one can neglect terms that are quadratic in E in (7.37) provided that the conditions in (7.50) are met. As a result this system of equations acquires the following form:

$$i\dot{a}_\ell = -E \cos\omega t \sum_{\ell'} <n\ell|z|n\ell'>a_{\ell'} \quad . \tag{7.53}$$

A spherical system of coordinates is not practical for the description of the unperturbed basis in (7.53) because states with different values of ℓ mix. It is more convenient to describe the basis in terms of a parabolic coordinate system. Thus, the unperturbed states are specified by parabolic quantum numbers (n_1,n_2) instead of the usual quantum numbers (n,ℓ,m). The dipole matrix element for states in one principal shell is diagonal in n_1 and n_2 and can easily be estimated:

$$<nn_1n_2|z|nn_1'n_2'> = \frac{3}{2} n(n_1 - n_2)\delta_{n_1'n_1}\delta_{n_2'n_2} \quad . \tag{7.54}$$

Since the parabolic basis states $|nn_1n_2>$ do not mix, one arrives at a system of independent equations:

$$i\dot{a}_{n_1n_2} = -E \cos\omega t <nn_1n_2|z|nn_1n_2>a_{n_1n_2} \quad . \tag{7.55}$$

The trivial solution to (7.55) is

$$a_{n_1n_2}(t) = A_{n_1n_2} \exp\left(i \frac{3n(n_1 - n_2)E}{2\omega} \sin\omega t \right) . \tag{7.56}$$

It can obviously be expanded in a Fourier series of the form of (7.51) with $d = 3n(n_1 - n_2)/2$.

If the strength of the field is increased still further, so that

$$\frac{\alpha E^2}{\omega} \ll 1 \qquad \text{and} \qquad \frac{dE}{\omega} \gg 1 \quad , \tag{7.57}$$

then the properties of Bessel functions imply that only those terms in (7.37) with numbers $|m|$ equal to or close to dE/ω (the distribution is very narrow) differ significantly from zero. As a result only quasi-energy levels with energies

$$E_n = E_n^{(0)} \pm dE \tag{7.58}$$

are populated to a noticeable degree (7.52). Thus, one is dealing with a splitting of the initial state that is linear in the strength of the variable low-frequency field.

If these results are applied to the hydrogen atom and one substitutes the above value for d, one obtains

$$E_n = E_n^{(0)} \pm \frac{3}{2} n(n_1 - n_2)E \; , \qquad (7.59)$$

which is the well-known expression for the linear Stark effect that is observed for a hydrogen atom in a constant electric field [7.21].

In increasing the field strength still further, one approaches a region in which

$$\frac{\alpha E^2}{\omega} \gtrsim 1 \qquad \text{and} \qquad \frac{dE}{\omega} \gg 1 \; . \qquad (7.60)$$

Since $E \ll E_{at}$ (a condition that is presumed throughout this section), $\alpha E^2 \ll dE$. Hence, the linear effect predominates over the quadratic, so that (7.43) is still valid.

In reality, the degeneracy of the states of the hydrogen atom is not exact, due to spin-orbit coupling. This is the reason why all of the above results are only valid when the interaction between the atom and the external electromagnetic field is stronger than the spin-orbit coupling. For the shell with $n = 2$ the corresponding field strength must be higher than 10^5V/cm. For greater values of n this limit decreases as $1/n^3$.

So far it has been assumed that the radiation was linearly polarized. For the general case of an elliptically polarized field additional mixing occurs with respect to the magnetic quantum number. This increases the rank of the matrix to be diagonalized.

Let us now turn to the numerical calculations of the shifts of the levels of hydrogen. The simplest calculations are for the 1s ground state. Obviously, there is no mixing of substates (since degeneracy is absent), so that one is dealing with the usual (quadratic in the field strength) dynamic polarizability described by (7.17). Assuming that the field is linearly polarized and using (7.24) yields the following expression for the dynamic polarizability:

$$\alpha_1 = 2 \sum_m \frac{(E_m^{(0)} + 1/2)|\langle 1s|z|m\rangle|^2}{(E_m^0 + 1/2)^2 - \omega^2} \; . \qquad (7.61)$$

The sum with respect to m includes all of the discrete states and those of the continuum. It can be evaluated exactly by applying the well-known expression for Green's function for the hydrogen atom (Sect.2.6) [7.24]. The result

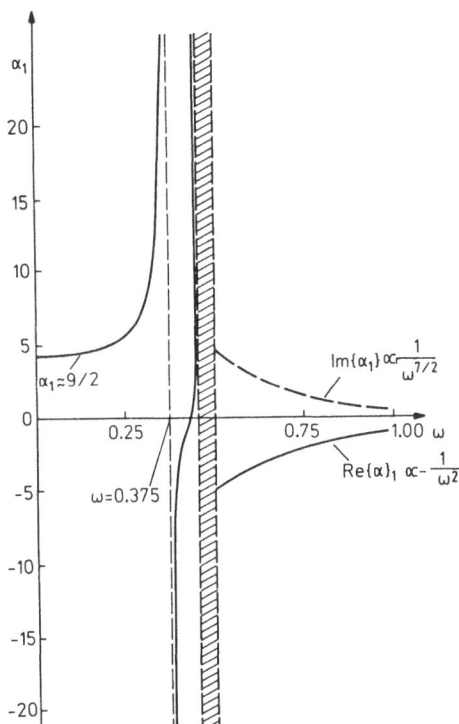

Fig.7.8. Dynamic polarizability α of the hydrogen atom as a function of the frequency of the external field, ω

is depicted in Fig.7.8. At low frequencies, $\omega \to 0$, one approaches the static limit, i.e., $\alpha_1 = +9/2$. For the resonance values of the frequency, i.e., for $\omega = 1/2 - 1/2N_m^2$, where N_m is the principal quantum number of the level m, the shift in the ground state becomes infinite and changes sign in the process, as can be seen from (7.61). As $\omega \to 1/2$, the resonances move closer together. When $\omega > 1/2$, the real part of α_1 is a slowly decaying function, which, when $\omega \gg 1/2$, behaves like $-\omega^{-2}$, see (7.21). In addition, when $\omega > 1/2$ one must substitute $E_m^{(0)} - i\lambda(\lambda = +0)$ for $E_m^{(0)}$ in (7.61). As a result α_1 acquires an imaginary part, which corresponds to the probability of the ionization of a hydrogen atom by one photon.

In Fig.7.8, $\text{Im}\{\alpha_1\}$ is represented by a dashed line. When $\omega \gg 1/2$ the imaginary part of α_1 decreases much faster than the real part (namely, $\text{Im}\{\alpha_1\}$ is proportional to $\omega^{-7/2}$) [7.25].

Experimental studies of the effect of a perturbation on the levels of the hydrogen atom have yet to be fully carried out. Only one experiment is known in which this was done by observing the relaxation of excited states (Sect.6.4) [7.26]. A hydrogen-gas plasma was irradiated with infra-red light

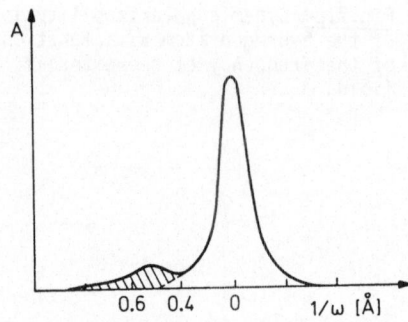

Fig.7.9. The H_δ line of the unperturbed hydrogen atom (not hatched section) and of the hydrogen atom in an external field with $\omega \ll \omega_\delta$ (A is the line intensity).

from a carbon-dioxide laser ($\hbar\omega \approx 0.1$ eV). The transition $n = 6 \rightarrow n = 2$ (the H_δ line) was observed. In the absence of a field the energy of this transition is $\hbar\omega_{6 \rightarrow 2} \approx 3$ eV. The results of this experiment are presented in Fig. 7.9. The main portion of the observed light is the result of the relaxation of an excited atom in a region of the plasma not irradiated by the laser radiation (the principal peak in Fig.7.9). When the laser radiation is focussed on a small volume of the plasma, a weak satellite appears (the shaded section). The change in the maximum of the satellite relative to the principal maximum constitutes the resulting Stark shift of the $n = 6$ level (this shift is large compared to the Stark shift of the $n = 2$ level). However, the separate components of the multiplet are not resolved because the substates are broadened. Under the conditions of this experiment $\omega \sim E_{n=6}^{(0)}$, so that the shift must be quadratic in the field strength, E. The shift in the maximum of the satellite relative to the principal maximum must be given by (7.17) averaged over the values of ℓ for the $n = 6$ level. The experiment was conducted at a fixed intensity, so that the fact that the shift is proportional to the square of the field strength was not directly registered. The maximum value of the shift calculated using (7.37) is in good agreement with the observed value.

Summarizing, one can give the following answer to the question posed at the beginning of this section: the perturbation of the levels of a hydrogen atom by a variable field is similar to the perturbation caused by a constant field, i.e., the shift in the levels is proportional to the field strength, but only for a low-frequency field ($\omega \ll \omega_{nN}$) whose strength is great enough for dE/ω to be greater than unity. In the limiting case of high frequencies ($\omega \gtrsim \omega_{nN}$) the shift is proportional to the square of the field strength.

7.4 Nonlinear Scattering of Light

It was mentioned at the beginning of this chapter that the nonresonant scattering of light is a well-known process (Rayleigh and Raman scattering) [7.27].
This process was already described in Sect.7.1 and will not be discussed
again from the viewpoint of nonlinear scattering in strong fields. In principle, however, other nonlinear processes may be associated with the electromagnetic field of the vacuum, e.g., the spontaneous emission of two or
more photons. Hence, the problem consists of establishing a relationship between the nonlinear scattering processes associated with the field in vacuum and those associated with an external field. In studying the nonlinear
processes caused by the presence of an external field, one must first be
able to establish when one has to take into account the nonlinear processes
associated with the field in vacuum and when one can neglect such processes.

7.4.1 The Interrelation of Nonlinear Scattering Processes Due
to an External Field

Let us first consider those processes in which two or more photons are spontaneously emitted. The probability of such a process is much lower than of
a one-photon emission. Indeed, let us consider the Feynman diagram for the
amplitude of the nonlinearly scattered light. If one adds a branch corresponding to the spontaneous emission of one more photon, the scattering probability will acquire the statistical weight of this additional photon,

$$\frac{d^3k}{(2\pi)^3} = \frac{1}{(2\pi)^3} k^2 dk \, do \quad , \tag{7.62}$$

where do is the solid angle for the emission of this photon, $k = \nu/c$ is the
magnitude of the wave vector, and ν is the frequency. Equation (7.62) leads
to an additional factor $1/c^3$ a.u. in the scattering probability. Since
$c = 137$ a.u., the probability of the emission of one more photon is about
10^6 times less than the probability of a process without such an emission.
It is for this reason that such processes only have meaning when the lower-
order process is forbidden by a selection rule. An example of such a case
is given in Sect.7.4.2.

We shall now turn to the nonlinear scattering of external-field quanta.
If one adds to the Feynman diagram a branch that corresponds to the emission
of a single external-field quantum, the scattering probability is determined
by the factor

$$\left|\frac{z_{kn}E}{\omega_{kn}}\right|^2 \sim \left(\frac{E}{E_{n,at}}\right)^2 \quad , \tag{7.63}$$

where $E_{n,at}$ is the atomic field strength characteristic of the level with the principal quantum number n. Obviously, this factor is also always small compared to unity. Note that it depends on the detuning from resonance and is valid for large detuning, i.e., $\Delta \sim \omega_{kn}$.

Also of interest are those conditions under which the changes in the probability of the emission of a single external-field quantum and of the emission of a vacuum-field quantum are the same, i.e., when

$$\left(\frac{E}{E_{n,at}}\right)^2 \sim \frac{1}{c^3} \quad . \tag{7.64}$$

This corresponds to an external field with an intensity

$$E \sim \frac{E_{n,at}}{c^{3/2}} \sim 10^{-3} E_{n,at} \quad . \tag{7.65}$$

Hence, only if

$$E \gg 10^{-3} E_{n,at} \tag{7.66}$$

can one neglect the nonlinear processes associated with the vacuum field and take into consideration Feynman diagrams with a branch corresponding to the spontaneous emission of a photon. Such diagrams were considered in Sect. 7.1.2. Since, on the whole, perturbation theory is being considered, the field strength cannot be too strong. Thus, the limits of the field strengths are

$$E_{n,at} \gg E \gg 10^{-3} E_{n,at} \quad . \tag{7.67}$$

One can see that the interval is quite narrow. If one were interested in the nonresonant processes which take place at very large values of the detuning, the analysis of the scattering processes would in many cases be considerably more complex because one has to take into account the higher-order effects of the spontaneous emission.

Processes are also of interest for which the detuning from resonance is extremely small. Therefore, the denominators of the matrix element given by (2.21) must be small, and the scattering probability must be high. This refers to cases where

$$\omega_{kn} \gg \Delta \gg \gamma \quad , \tag{7.68}$$

where γ is the resonance width and Δ is the detuning. This type of nonresonant scattering shall be called quasi-resonance scattering. Note that quasi-resonance scattering is still essentially a nonresonant process. It only proceeds with a considerably higher probability. Indeed, only when $\Delta \lesssim \gamma$ do the resonance states mix (Chap.3), which is a strong sign of a resonant process.

Let an estimate be made of the interval, similar to the one in (7.67), for quasi-resonant processes by assuming that $\Delta \sim \gamma$, where γ is the spontaneous width $(\gamma \sim 1/c^3)$. Instead of (7.65) one obtains

$$\left| \frac{z_{kn}E}{\gamma} \right|^2 \sim \left(\frac{E}{E_{n,at}} \right)^2 \left(\frac{\omega_{kn}}{\gamma} \right)^2 \sim \frac{1}{c^3} \quad , \tag{7.69}$$

i.e.,

$$E \sim \frac{1}{c^{9/2}} E_{n,at} \sim 10^{-9} E_{n,at} \quad . \tag{7.70}$$

This yields the following criterion for the strength of the field:

$$E_{n,at} \gg E \gg 10^{-9} E_{n,at} \quad . \tag{7.71}$$

When this condition is met, one only need take into account the spontaneous emission of one photon. The interval of permissible fields is quite wide and covers all of the cases of practical interest.

In the following section quasi-resonant processes will be discussed since they take place with a high probability. The probabilities are taken to be per unit time if the perturbation time T is sufficiently small. In describing nonresonant, nonlinear processes one can limit oneself to small perturbation times because for $E \ll E_{n,at}$, $w \sim (E/E_{n,at})^{2K}$. Also, the following condition is always met at high degrees of the nonlinearity constant K: $wT \ll 1$.

7.4.2 Spontaneous Two-Photon Scattering

In third-order perturbation theory, the linear susceptibility is replaced by the nonlinear susceptibility $\chi^{(2)}$ (Sect.7.1.2). This nonlinear susceptibility consists of a number of terms, to which there correspond various Feynman diagrams with three branches each. The diversity of the diagrams is due to the various possibilities for the absorption or emission of photons of the external field and the spontaneous emission of other photons. The related physical phenomena are also numerous.

What are the selection rules for $\chi^{(2)}$? If the atomic states are nondegenerate, then, as noted in Sect.7.1.2, the nonlinear susceptibility can only have off-diagonal matrix elements because of the conservation of parity, i.e., the final state k (after scattering) cannot coincide with the initial state n (before scattering). However, in the presence of degeneracy (e.g., in the case of the hydrogen atom) the state k can have the same energy but not the same parity as the state n.

In Sect.7.1.2 only diagrams with one spontaneous quantum and two external field quanta were considered. The corresponding terms in the polarization vector are quadratic in E, as shown in (7.9). Also, in the expression for $\chi^{(2)}$ there are terms which take into consideration two spontaneously emitted quanta and one external field quantum. This part of the polarization vector is linear in E. It follows from the results of Sect.7.4.1 that when there are no resonances and the field strength E is much smaller than $10^{-3}E_{n,at}$, the greatest contribution to the polarization will be from the first-order diagrams in E and the second-order diagrams in the field of the electromagnetic vacuum.

Here is one such diagram [for $\chi^{(2)}_{ij\ell}(\nu_2;\ \omega,\ -\nu_1)$ or $\chi^{(2)}_{ij\ell}(\nu_1;\ \omega,\ -\nu_2)$]:

This diagram describes spontaneous nonresonant emission of two photons. The atom spontaneously emits a photon of frequency ν_1, then absorbs a quantum ω of the external field and, finally, spontaneously emits a photon of frequency ν_2. As a result the atom passes from the initial state n to the final state k. The frequencies of the two spontaneously emitted photons are connected by a relationship that follows from the conservation of energy: $\nu_1+\nu_2=\omega-\omega_{kn}$. The frequency of one of the emitted photons can vary continuously from 0 to $\omega-\omega_{kn}$. The frequency of the second photon is determined by the conservation of energy.

Note that if $\omega_{kn}=0$ (this is realized when degeneracy is present), one is dealing with spontaneous two-photon emission, for which $\nu_1+\nu_2=\omega$ (the analog of Rayleigh scattering). If $\omega_{kn}\neq0$, one has Raman scattering.

The contribution to the corresponding polarization vector of the above diagram is:

$$\sum_{sp}\frac{r^i_{np}r^j_{ps}r^\ell_{sk}}{(\omega_{pn}+\nu_1)(\omega_{sn}+\nu_1-\omega)}e^1_ie_je^2_\ell E \quad,$$

where e^1_i and e^2_ℓ are the polarizations of the spontaneously emitted photons, ν_1 and ν_2, and e_j is the fixed polarization of the incident electromagnetic wave. The cross section of the process, σ_2, can be found by 1) adding to-

gether three amplitudes of the type considered above but differing from it
by the orders of the absorption and emission of quanta (all diagrams are
first-order in E), 2) squaring the modulus of the sum, 3) multiplying the result according to the Golden Rule by

$$\int \delta(\omega - \omega_{kn} - \nu_1 - \nu_2) \frac{d^3k_1 d^3k_2}{(2\pi)^6} = \frac{\nu_1^2 \nu_2^2}{(2\pi)^6} d\nu_1 do_1 do_2 \quad , \tag{7.72}$$

where k_1 and k_2 are the wave vectors of the spontaneously emitted photons
($k_{1,2} = \nu_{1,2}/c$), the delta function expresses the conservation of energy, and
do_1 and do_2 are the solid angles into which the photons are emitted, and, finally, 4) dividing the result by the incident flux of photons, i.e., a quantity proportional to cE^2. Thus, as one can see, the cross section of the process
does not depend on the amplitude, E, of the forcing field.

The cross section of spontaneous two-photon emission is $(\hbar c/e^2)^3 \approx 10^6$
times smaller than that of Raman scattering. It can be observed (when degeneracy is not present) when the parity of the final state k differs from that
of the initial state n. Raman scattering is then forbidden by the conservation of parity. The cross section of spontaneous two-photon emission is of
the following order of magnitude:

$$\sigma_2 \sim \frac{1}{c^7} r_0^2 \sim 10^{-14} r_0^2 \quad , \tag{7.73}$$

where r_0 is the Bohr radius. For the hydrogen atom in the ground state,
$\sigma_2 \sim 10^{-30} cm^2$, which is an extremely small quantity. One may recall that the
cross section for the one-photon (Raman or Rayleigh) scattering is of the
order of

$$\sigma_1 \sim \frac{1}{c^4} r_0^2 \sim 10^{-24} cm^2 \quad . \tag{7.74}$$

If an intermediate state p is in "quasi-resonance" with the perturbing
field, i.e., $\omega_{pn} \approx \omega$ (a case considered earlier), then one of the diagrams
for the amplitude of spontaneous two-photon emission yields anomalously
"large" terms:

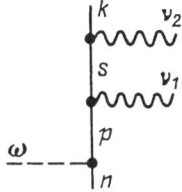

Indeed, this diagram contributes the following to the polarization:

$$\frac{e_i r^i_{np} E}{\omega_{pn} - \omega} \sum_s \frac{r^j_{ps} r^\ell_{sk}}{\omega_{sp} + \nu_1} \; e_j^1 e_\ell^2 \; .$$

Here, from the sum over states p, only the quasi-resonant term is left. It is evident that the probability of this process can be expressed as a product of the population probability of the level p, which, according to first-order perturbation theory, is equal to

$$\left| \frac{e_i r^i_{np} E}{\omega_{pn} - \omega} \right|^2 \; ,$$

and the probability of a spontaneous two-photon transition from the initial state, p, to the final state, k. In this case the nonlinear susceptibility, $\chi^{(2)}_{ij\ell}$, is reduced to a linear susceptibility, $\chi^{(1)}_{j\ell}(\nu_1, -\nu_2)$.

The spontaneous two-photon emission process determines the decay of the 2S state of the hydrogen atom ($p \equiv 2S$), since the one-photon decay process 2S → 1S is forbidden by conservation of the parity ($k \equiv 1S$). This was observed experimentally [7.28], and the results are in good agreement with the theoretical predictions for $\chi^{(1)}$.

7.4.3 Stimulated Two-Photon Emission

According to the results of Sect.7.4.1, when the condition in (7.67) is met for the nonresonant case or the condition in (7.71) is met for the resonant case, stimulated two-photon emission becomes more significant than two-photon spontaneous emission. Stimulated two-photon emission can be described by diagrams similar to those used in Sect.7.4.2 in which one of the spontaneously emitted photons is replaced by a photon from the external field. A typical diagram for the nonlinear susceptibility of this process is:

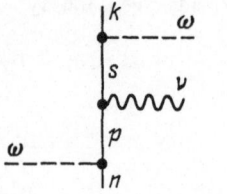

One photon ω of the external field is absorbed, and then one photon ν and one external field photon ω are spontaneously emitted. According to the conservation of energy $\nu = \omega_{nk}$, i.e., one is dealing with Raman scattering.

As in all such cases, bound intermediate "quasi-resonant" p ($\omega_{pn} \approx \omega$) can increase the probability of stimulated two-photon emission by many orders of magnitude. For this resonant case there is one more diagram:

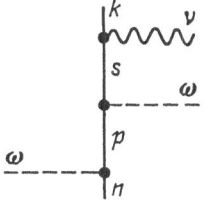

One can see that the probability of such a process can again be written as a product of the population probability of the level p and the probability of Raman scattering, p → k, with the spontaneous emission of a photon of frequency $\nu = \omega_{nk} = \omega_{pk} - \omega$. The Raman scattering is characterized by the suscepti-bility $\chi_{\ell j}^{(1)}$ (ν, $-\omega$) where

$$\chi_{\ell j}^{(1)}(\nu, -\omega) = \sum_s \left(\frac{e_j e_\ell^1 r_{ps}^j r_{sk}^\ell}{\omega_{sp} + \omega} + \frac{e_j^1 e_\ell r_{ps}^j r_{sk}^\ell}{\omega_{sk} - \omega} \right) . \tag{7.75}$$

An experiment designed to observe a stimulated two-photon emission from the deuterium atom was described in [7.29]. A beam of deuterium atoms in the metastable p ≡ 2S state was intersected by a pulsed neodymium glass laser beam. The final state was k ≡ 1S. For laser radiation of energy $\omega = 1.17$ eV and an energy difference $\omega_{pk} = 10.19$ eV, the spontaneously emitted photon has a fre-quency $\nu = \omega_{pk} - \omega = 9$ eV (in the ultraviolet). The number of such photons is proportional to the product $F|\chi_{\ell j}^{(1)}(\nu, -\omega)|^2$, where $\chi^{(1)}$ is given by (7.75), and $F(\propto E^2)$ is the intensity of the incident light. The experimental data is in good agreement with the calculated values. A similar process in the pre-sence of a quasi-resonance was observed for potassium [7.30].

7.4.4 Hyper-Raman Scattering

The next process, characterized by the absorption of two photons, ω, of the external field and the emission of one photon of frequency $\nu = 2\omega - \omega_{kn}$, po-larizability $\chi^{(2)}$, is described by the following Feynman diagrams:

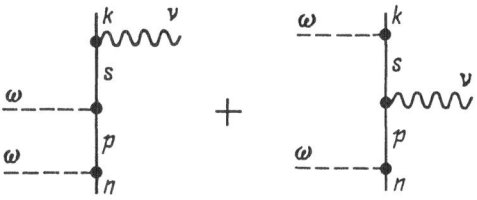

When $\omega_{kn} \neq 0$ the process is called hyper-Raman scattering (Sect.7.1.2). In systems with degeneracy where the condition $\omega_{kn} = 0$ can be fulfilled (e.g., in the hydrogen atom), $\nu = 2\omega$, i.e., the second harmonic is generated. This process is best known in crystals [7.31]. In atoms other than hydrogen, the conservation of parity forbids such a process.

When quasi-resonance sets in, i.e., when $\omega_{pn} \approx \omega$, the number of spontaneously emitted photons is proportional to the square of the susceptibility $\chi_{j\ell}^{(1)}(\nu, \omega)$. The above diagrams can be expressed analytically in the following form:

$$\chi_{j\ell}^{(1)}(\nu,\omega) = \sum_{s} \left(\frac{e_j e_\ell^1 r_{ps}^j r_{sk}^\ell}{\omega_{sp} - \omega} + \frac{e_j^1 e_\ell r_{ps}^j r_{sk}^\ell}{\omega_{sk} + \omega} \right) . \tag{7.76}$$

In this case the frequency of the Raman photon is $\nu = \omega_{pk} + \omega$. Equation (7.76) differs from (7.75) in the sign of the frequency, i.e., $\omega \rightarrow -\omega$.

The above process was observed in an experiment described in [7.32]. As in [7.30], a beam of deuterium atoms in the excited $p \equiv 2S$ state was irradiated by the light from a neodymium glass laser with an energy $\omega = 1.17$ eV. For the $k \equiv 1S$ state, $\nu = \omega_{pk} + \omega = 11.36$ eV (in the far ultraviolet). It was this ultraviolet radiation that was observed in the experiment. Since $\nu > \omega_{pk}$, one is dealing with anti-Stokes Raman scattering.

Anti-Stokes scattering is a process that competes with stimulated two-photon emission. Since the probability for both processes is proportional to ν^2 and the frequency of stimulated two-photon emission, $\nu = \omega_{pk} - \omega$, is lower than the frequency of anti-Stokes scattering, $\nu = \omega_{pk} + \omega$, anti-Stokes scattering dominates. If one takes into account the difference between the susceptibilities $\chi_{\ell j}^{(1)}(\nu, -\omega)$ and $\chi_{j\ell}^{(1)}(\nu,\omega)$ one can see that the cross section of anti-Stokes scattering is six times greater than the cross section of stimulated two-photon emission. In this case theoretical and experimental results agree qualitatively.

7.4.5 Generation of the Third Harmonic

What happens when three photons are absorbed? In order to answer this question one needs to calculate the next term in the expansion of the polarization vector, $\mathbf{P}$, in powers of E, i.e., the nonlinear susceptibility, $\chi_{ij\ell o}^{(3)}$ (defined in Sect.7.1.2). It will be assumed that the conditions in (7.67,71) are fulfilled in the nonresonant and quasi-resonant cases, respectively, i.e., the unlikely process of spontaneous multi-photon emission will not be considered.

The nonlinear susceptibility for the general case was introduced in Sect. 7.1.2, where its properties were also described. In this section the specific physical phenomena that are characterized by $\chi_{ij\ell o}^{(3)}$ will be considered. Because of the extremely unwieldly formulas, the discussion will generally be restricted to that of the Feynman diagrams for the appropriate processes. The first case to be discussed is one in which the initial and final states coincide, i.e., $k = n$ (Rayleigh scattering).

The first of such processes was mentioned in Sect. 7.1.2 — the generation of the third harmonic. In this process three photons of the external field are absorbed and one photon with a frequency $\nu = 3\omega$ is spontaneously emitted. This effect is widely known because of its practical importance in creating high-powered sources of ultraviolet radiation. There is no doubt that this process can only play a noticeable role for intense external fields.

The nonlinear susceptibility $\chi_{ij\ell o}^{(3)}(3\omega,\omega,\omega,\omega)$, which is proportional to the amplitude of the third harmonic generation process, can be expressed as a sum of four Feynman diagrams:

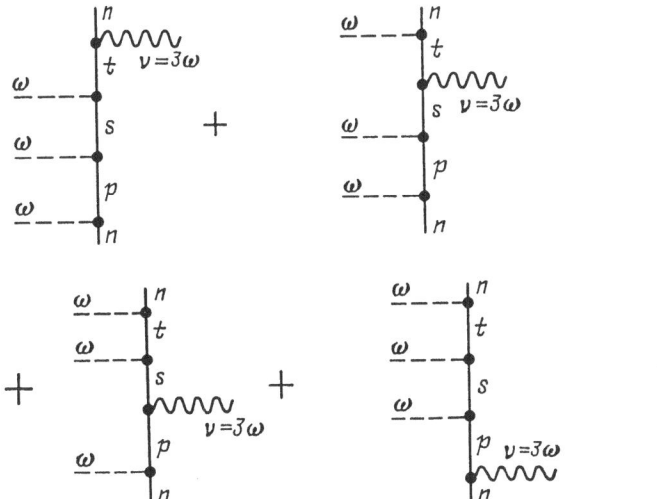

For example, in the second diagram, two photons, ω, of the external field are absorbed, one photon of frequency, $\nu = 3\omega$, is spontaneously emitted, and, finally, a third photon ω is absorbed.

Many studies have been devoted to calculating $\chi^{(3)}$. In [7.33] this was done for alkali atoms. Figure 7.10 depicts $\chi_{zzzz}^{(3)}$ as a function of the wavelength, $\lambda = 2\pi/\omega$, for the ground state of sodium. The radial dipole matrix elements were calculated using the quantum defect method (Chap. 2). Figure

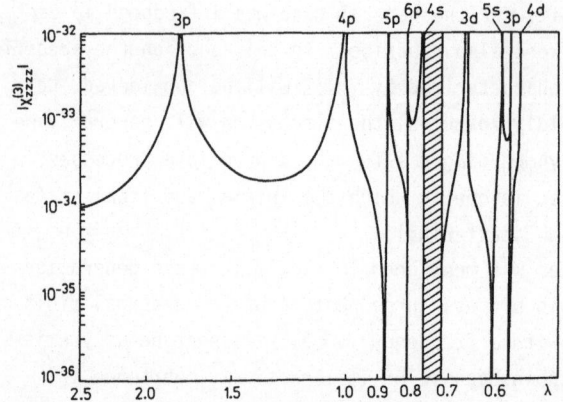

Fig.7.10. Polarizability of the sodium atom in the ground state, $\chi_{zzzz}^{(3)}$, as a function of the wavelength λ of the exciting light

7.10 shows that $\chi^{(3)}$ has numerous resonances, which correspond to the conditions $\omega_{pn} = \omega$, 2ω, etc. (resonances of the type $\omega_{pn} = 3\omega$ are not shown because they correspond to very long wavelengths). In the intervals between resonances, $\chi^{(3)}$ may vanish. In [7.34] $\chi^{(3)}$ was determined more accurately by allowing for the spin-orbit splitting of the unperturbed energy levels of alkali atoms and by substituting exacter values for the dipole matrix elements.

Experimental set-ups for studying the generation of the third harmonic are relatively simple (Fig.6.6). A target consisting of an atomic gas is irradiated by a high-power laser and a suitable absorber is used to measure the light with the frequency 3ω. The absorber must absorb practically all of the exciting light and transmit all of the light with the frequency 3ω. Since the intensity of the third harmonic is proportional to the cube of the intensity of the exciting light, the requirements for the intensity and divergence of the laser radiation are quite high. Experimental data has been primarily gathered for the inert gases [7.35] and for the alkali metals [7.33]. The above-mentioned calculations agree satisfactorily with the experimental results.

It goes without saying that an intermediate resonance for the exciting light at $\omega_{pn} = \omega$, 2ω, or 3ω increases the probability of the generation of a third harmonic (compared to the nonresonant case with the same external field). The effect produced by intermediate resonances can be demonstrated in the generation of the third harmonic in cesium by ruby-laser radiation (Fig.7.11) [7.36]. The laser frequency ω was selected in such a way that a two-photon resonance, $\omega_{pn} = 2\omega$, takes place between the ground state, $n \equiv 6S_{1/2}$, and the excited state, $p \equiv 9D_{3/2}$ (the resonances $\omega_{pn} = \omega$ or 3ω are not observed because the incident or generated light in the neighborhood of the resonances is strongly absorbed by the atomic medium). A resonance sharply increases the probability for the generation of a third har-

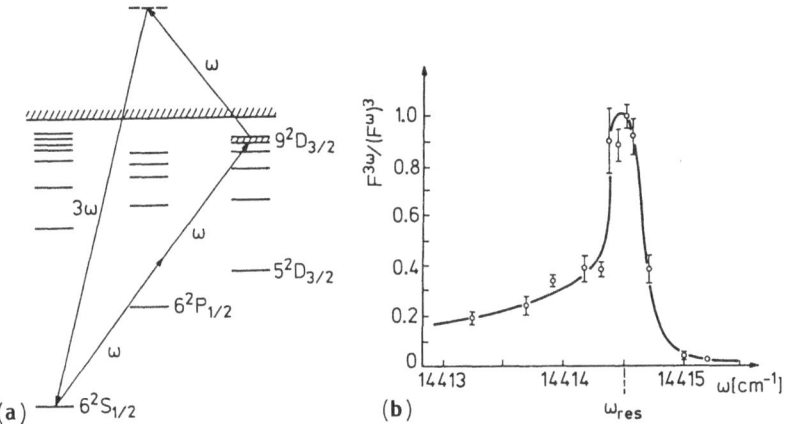

Fig.7.11a,b. The coefficient of the transformation of the light frequency in-to the third harmonic in cesium as a function of the tuning to the interme-diate two-photon resonance: (a) the level diagram, and (b) experimental re-sults

monic. The above theory cannot be used in the immediate vicinity of a reson-ance peak. This case is studied in detail in Chap.8.

The practicability of this generation of the third harmonic is, however, re-stricted by two factors. For a gaseous target the Doppler absorption-line broadening sharply decreases the effectiveness of populating an intermediate resonance state (both the natural width of the resonance state and the width of the exciting field are considerably smaller than the Doppler broadening). In addition, the magnitude of the exciting field is restricted from above by the possibility of ionization (which brings about a broadening of the reson-ance level) and by effects associated with resonant mixing (which is accom-panied by a splitting of the levels into quasi-energy levels). The second factor is of a resonant nature and will be studied in Chap.8.

7.4.6 Other Processes Determined by $\chi^{(3)}$

The above example illustrates the diversity of the physical phenomena de-scribed by the nonlinear susceptibility $\chi^{(3)}$. A number of processes other than the generation of the third harmonic discussed in Sect.7.4.5 will be enumerated but not discussed in detail. Note that the external fields that perturb an atom may differ in frequency and intensity.

The polarization vector $P_i(\omega_a)$ at the frequency ω_a is

$$P_i(\omega_a) = \sum_{j\ell o} \chi_{ij\ell o}^{(3)}(\omega_a;\omega_1,\omega_2,\omega_3)E_1^j E_2^\ell E_3^o \quad . \tag{7.77}$$

The conservation of energy is satisfied:

$$\omega_a = \omega_1 + \omega_2 + \omega_3 - \omega_{kn} \quad , \tag{7.78}$$

where n is the initial state and k the final state. Let us consider a number of particular cases of (7.77).

1) Nonlinear elastic scattering was briefly discussed in Sect.7.1.2. One of the diagrams describing such a Rayleigh process is

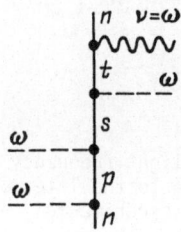

In this process two photons of the external field are absorbed, one is emitted, and another photon of frequency $\nu = \omega$ is spontaneously emitted.

2) The process described by the diagram

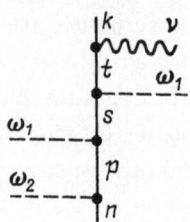

and similar diagrams (in which two photons of frequencies ω_1 and ω_2 are absorbed, one photon of frequency ν is spontaneously emitted, and one photon of frequency ω_1 or ω_2 is emitted) generalizes the hyper-Raman scattering of light because the above diagram is described by the next order of perturbation theory with two different fields. The conservation of energy implies that $\nu = \omega_2 + \omega_{nk}$. If $n = k$, $\nu = \omega_2$, i.e., one has a nonlinear elastic scattering of light of frequency ω_2. Indeed, the frequency of the spontaneously emitted photon, ν, is equal to the frequency of one of the external electromagnetic fields.

If, in addition, the quasi-resonance condition $\omega_{sn} \approx \omega_1 + \omega_2$ is fulfilled, such a process is called two-photon absorption, since the levels are only populated after two photons, ω_1 and ω_2, are absorbed. The nonlinear elastic scattering of light, examined above for one field, is a particular case of absorption.

3) Diagrams of the following type,

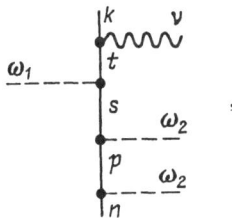

also describe hyper-Raman scattering. One photon is absorbed and three are emitted (two of frequency ω_2 and one of frequency ν). The conservation of energy implies that $\nu = \omega_1 + \omega_{nk} - 2\omega_2$. If $\omega_1 = \omega_2$, then $\nu = \omega_{nk} - \omega$.

4) Diagrams such as

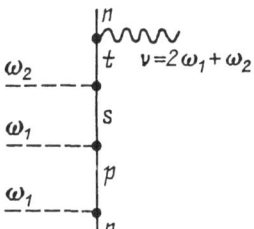

describe what is called the parametric generation of the sum frequency $\nu = 2\omega_1 + \omega_2$. In this process two photons of frequency ω_1 and one of frequency ω_2 are absorbed. If $\omega_1 = \omega_2$, one is again dealing with the third harmonic generation considered in Sect.7.4.5. A resonance gain is attained if $\omega_{sn} \approx 2\omega_1$. An additional gain is attained if $\omega_{ts} \approx \omega_2$. In this way the value of the frequency, ω_2, was transformed from the infrared to the ultraviolet range [7.36]. In this experiment, a two-photon absorption of ω_1 populates the $s \equiv 4P$ state of sodium (from the initial $n \equiv 3S$ state). Then, under the impact of an infrared photon, ω_2, the atom reaches the $t \equiv 5P$ state. A fluorescence of frequency $\nu = 2\omega_1 + \omega_2$ is observed when the atom returns to the initial $n \equiv 3S$ state. The above diagram illustrates this process. The parametric generation of the sum frequency is accompanied by a two-photon absorption of frequency ω_1.

In [7.37] $\chi^{(3)}$ was measured for the generation of the second harmonic. The atom absorbs two photons of the monochromatic field, $E \cos\omega t$, and one photon of the constant field, E'. Afterwards it returns to the ground state. A typical diagram that describes such a process is similar to the one shown above, where $\omega_2 \equiv 0$. The conservation of energy implies that $\nu = 2\omega_1$.

5) The parametric generation of the difference frequency, $\nu = 2\omega_1 - \omega_2$, is described by the following diagram,

$$
\begin{array}{l}
n \\
\text{\Large\char`\~\char`\~\char`\~} \nu = 2\omega_1 - \omega_2 \\
t \\
\text{-----} \omega_2 \\
\omega_1 \quad s \\
\text{-----} \\
\omega_1 \quad p \\
\text{-----} \\
\quad n
\end{array}
$$

,

and similar diagrams. Two photons of frequency ω_1 are absorbed, and one photon of frequency ω_2 and one of frequency $\nu = 2\omega_1 - \omega_2$ are emitted. This type of generation was observed in experiments with rubidium vapor [7.38].

If $\omega_{sn} \approx 2\omega_1$, then a two-photon absorption of ω_1 occurs in the process. If, in addition, $\omega_{pn} \approx \omega_1$, quasi-resonant Raman scattering is also observed.

Concluding this section, it shall be noted that higher-order perturbation theory yields an even more diversified picture of the physical processes. The realization of such processes depends on the availability of high-powered sources of radiation in the required frequency ranges and on the possibility of varying the operating frequencies within a wide range. This makes resonant transitions through intermediate atomic states possible. Such transitions greatly increase the probability of the processes.

7.5 Nonresonant, Nonlinear Ionization

The one-photon ionization (photoionization) of atoms, whose probability is related to the condition that $\omega > E_n^{(0)}$, has been observed at different frequencies ω, and for various bound states n has been studied experimentally, and has been described with great accuracy quantitatively by quantum mechanics ([Ref.7.21, Sect.9.2] and [7.39-42]). The above condition implies that the photoionization of the ground state, n, of an atom can only occur when ultraviolet radiation impinges on the atom. Quantum mechanics has also shown that an electron can reach the continuum by absorbing several photons, i.e., when $K\omega > E_n^{(0)}$ [7.43]. Such multi-photon transitions can lead to the ionization of atoms in their ground states when visible light is directed at them.

It is known that a quite different ionization process can occur in a constant electric field. This is because the electron can tunnel through the potential barrier set up by the field [7.44]. From the qualitative point of view a varying field acts on an electron in the same way as a constant field, provided that the frequency of the field is kept sufficiently low (Sect. 4.5.3). Hence, it was expected that ionization through tunneling could also take place in an electromagnetic field. But how are the multi-photon and tunneling ionization processes related in a field of arbitrary frequency? Do these processes compete or does each have its own criterion for realization?

7.5.1 The Mechanisms of Nonlinear Ionization

The above questions have been answered in Sect.4.5: the ionization probability per unit time, w, is a function of the adiabaticity parameter γ which, in turn, is a combination of three parameters that characterize the potential well (the depth of the well $E_n^{(0)}$) and the variable field (frequency ω and field strength E), (4.86) [7.45]:

$$\gamma = \frac{\omega(2E_n^{(0)})^{\frac{1}{2}}}{E} \; . \tag{7.79}$$

One can specify three typical regions. For $\gamma \gg 1$,

$$w \propto E^{2K} \propto F^K \; , \tag{7.80}$$

where K (= $<E_n^{(0)}/\omega + 1>$) is the number of absorbed photons, and $F(\propto E^2)$ is the intensity of the radiation. Equation (7.80) is typical for multi-photon processes (Chap.2). For $\gamma \ll 1$,

$$w \propto \exp\left(-\frac{4\sqrt{2}|E_n^{(0)}|^{3/2}}{3E}\right) \; , \tag{7.81}$$

where the exponential function is the same one as for tunneling ionization in a constant field E [7.17].

The ionization process is complicated and has no simple analogy for the intermediated case, i.e., for $\gamma \approx 1$. To within exponential accuracy the probability is given by (4.86). Note that for $\gamma \ll 1$, (4.86) leads to (7.81) with a correction factor $(1 - \gamma^2/10)$ in the exponent. Hence, when $\gamma \approx 1$ the process is closer to the tunneling effect than to multi-photon ionization.

7.5.2 Numerical Estimates

Let us now study the behavior of an atom in a field of the frequency range of visible light. The frequencies in this range and the ionization potentials differ by a moderate factor ($E_{n,min}^{(0)} \approx 4$ eV for Cs and $E_{n,max}^{(0)} \approx 24$ eV for He), and γ is proportional to $(E_n^{(0)})^{\frac{1}{2}}$. Hence, according to (7.79) the nature of the ionization process depends primarily on the strength of the field.

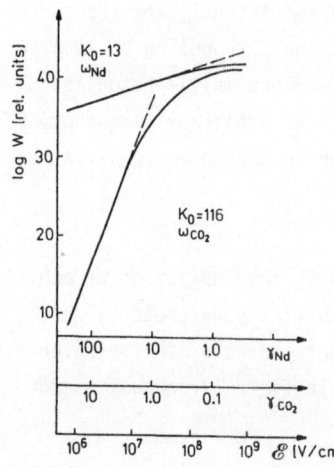

Fig.7.12. Ionization probability for hydrogen as a function of the intensity of the variable external field E (theory); ω_{Nd} ($\sim$ 1 eV) is the operating frequency of a neodymium glass laser, and ω_{CO_2} ($\sim$ 0.1 eV) is the operating frequency of a carbon-dioxide laser

Figure 7.12 illustrates the dependence of w on γ calculated using (4.84) for two typical frequencies and for $E_n^{(0)}$ = 10 eV. One can see that in the frequency range of visible light (in Fig.7.12 this corresponds to the radiation of a neodymium laser) $\gamma \ll 1$ only when $E \approx E_{at}$, i.e., when the general theory of ionization presented in Sect.4.5 cannot be applied [7.46]. Obviously, there are two cases where the tunneling effect can be realized even if $E < E_{at}$: 1) ionization from ground states in the infrared frequency range (in Fig. 7.12 this corresponds to the radiation of a carbon-dioxide laser); 2) ionization from highly excited states in the frequency range of visible light (Sect.9.2).

For a description of the ionization process with $\gamma \gg 1$, one must use perturbation theory in its standard form (Chap.2). However, if $\gamma \approx 1$ or $\gamma \ll 1$, the adiabatic (quasi-classical) approximation must be used (Sect.4.5).

It should be noted that the above reasoning is based on the assumption that intermediate resonances do not play a role. However, one can neglect the intermediate bound electronic states only when the detuning from the resonance between the photons of frequencies K'ω and the bound states, n and m, of energies $E_{n,m}(E)$ (taking into account their perturbation by the field) is sufficiently great, i.e., when

$$\Delta(E) = |E_n(E) - E_m(E) - K'\omega| \gg \Gamma_{n,m}(E) \quad , \tag{7.82}$$

where $\Gamma_{n,m}(E)$ is the width of the bound states. For the particular case of weak fields, i.e., when $\Gamma_{n,m}(E) = \gamma_{n,m}$ (where $\gamma_{n,m}$ is the width of the level n or m), the energies of the states n and m that enter this equation are those of the unperturbed atom. The ionization process in the presence of an intermediate resonance is discussed in Chap.8.

To begin a more detailed study of the possible mechanisms of nonlinear ionization, the ionization of an anion, whose potential well is short-ranged, will be considered. This problem is the simplest for theoretical analysis because of the absence of intermediate bound states and also because the field created by the atom can be neglected for practically all values of the coordinate r of the electron, except for r = 0. However, this case is interesting only from the methodological point of view since no experimental data on the nature of the ionization of anions in a variable field is available.

Next, the nonlinear ionization of an atom will be studied. The electron to be extracted lies in the long-range Coulomb field of the atomic core. From the theoretical point of view this problem is far more complex than the ionization from a short-range potential well because of the presence of numerous intermediate bound electronic states and because the atomic potential can no longer be neglected in a sizeable portion of the coordinate space of the electron.

7.6 Ionization in a Short-Range Potential

The short-range potential is of special interest due to the relative simplicity of the theoretical description of nonlinear ionization. The main simplifying feature is the assumption that an electron in the final state is free and, therefore, we may take into account only the action of the light field on the free electron. When the potential is long-range (e.g., a Coulomb potential), we must allow not only for the action of the light wave on the electron in the final state but for the effect of the tail of the potential as well, which obviously complicates the theoretical analysis of the ionization process [7.46].

The discussion of nonlinear ionization in a short-range potential that we present below is, strictly speaking, valid only for a model potential, namely, the zero-range potential. Even in the case of a narrow well with finite width and with only one level, all the results have an approximate nature, and the wider the well the greater the error. Finally, when the width is such that there appears a second level, there may be resonance between the two and all the results become invalid. As an example we note that the probability of one-photon ionization in a Coulomb field at the ionization threshold is finite rather than zero (which is the case with a delta-like potential).

Finally, we note that nonlinear photodetachment of an electron from a negatively charged ion cannot, strictly speaking, be described as ionization out of a short-range potential (Sect.7.6.5).

7.6.1 Tunneling Ionization

Even this problem is too complex for a general study. Therefore the simple limiting case in which γ is much smaller than unity will be considered first. The exponential nature of the probability of tunneling ionization was discussed earlier, see (7.81). The dependence of the probability on the pre-exponential factor is different for a linearly or circularly polarized field.

The well-known equation for the probability of tunneling ionization from a short-range well in a constant field is [Ref.7.17, §77, Problem 2]

$$w\Big|_{E=\text{const}} = \frac{E}{2(2E_0)^{\frac{1}{2}}} \exp\left(-\frac{2(2E_0)^{3/2}}{3E}\right) \quad , \tag{7.83}$$

where E_0 is the energy of a bound electron in a well of zero width. The equation is even valid for a circularly polarized field because the field does not vary with time. It is a different situation for a linearly polarized field. In this case one must substitute $E \sin\omega t$ for E in (7.83) and average the new expression over the perturbation period T ($2\pi/\omega$). As a result,

$$w = \frac{E^{3/2}(3\pi)^{1/2}}{2(2E_0)^{5/4}} \exp\left(-\frac{2(2E_0)^{3/2}}{3E}\right) \quad . \tag{7.84}$$

It can be seen that even the dependence of the pre-exponential factor on the strength E of the field has changed. It is also easy to see that if $\gamma \ll 1$, the probability ω is smaller than $\omega_{E=\text{const}}$. The reason is evident: the ionization probability in a variable field, $E \sin\omega t$, is obviously lower than that in a constant field of the same amplitude .

7.6.2 Ionization by a Circularly Polarized Field

The problem becomes simpler when the field is circularly polarized [7.47,48]. In this case not only can the values of γ be arbitrary but the amplitude of the perturbation, E, can be of the order of the atomic field, E_{at}.

The operator that represents the interaction between the electron and the field has the following form:

$$V(t) = E(x \cos\omega t + y \sin\omega t) \quad . \tag{7.85}$$

The Schrödinger equation,

$$i = \frac{\partial\Psi}{\partial t} = [H_0 + V(t)]\Psi \quad , \tag{7.86}$$

where H_0 is the Hamiltonian in the absence of an electromagnetic field, can be reduced to the Schrödinger equation for the stationary-state problem (Sect.1.3) by introducing a new function

$$\phi = \exp(i\omega t \hat{L}_z)\Psi \quad , \tag{7.87}$$

where the z axis is in the direction of the wave vector of the circularly polarized electromagnetic wave, and $\hat{L}_z$ is the operator corresponding to the projection of the angular momentum on this axis. The stationary-state Schrödinger equation for $\phi \equiv \phi_E(r)$ is:

$$(H_0 - \omega \hat{L}_z + Ex)\phi_E(r) = E\phi_E(r) \quad . \tag{7.88}$$

In the zero-radius potential model, the boundary condition that determines the behavior of ϕ_E as $r \to 0$ is [7.49]:

$$\phi_E \xrightarrow[r \to 0]{} \frac{1}{4\pi}\left[\frac{1}{r} - (2E_0)^{\frac{1}{2}}\right] \quad , \tag{7.89}$$

where E_0 is the energy of the unperturbed bound state from which the ionization occurs. In a potential well of zero width this is the only bound state.

A solution to (7.88) that satisfies (7.89) defines the real and imaginary parts of the perturbed energy E, which, in turn, determines the Stark shift and the width of the quasi-stationary state in the field of the electromagnetic wave. The shift and the width can be determined from the same equation, whereas in perturbation theory the ionization probability and the shifts of the levels are determined independently of each other.

The wave function ϕ_E has the same singularity at $r \to 0$ as the scalar Coulomb potential. Therefore, using Poisson's differential equation one can write

$$\nabla^2 \phi_E(r) \xrightarrow[r \to 0]{} - \delta(r) \quad . \tag{7.90}$$

Combining (7.88) and (7.90) yields

$$\left(-\frac{1}{2}\nabla^2 - \omega\hat{L}_z + Ex - E\right)\phi_E(r) = \frac{1}{2}\delta(r) \quad . \tag{7.91}$$

Let us consider the solution to this nonhomogeneous equation. By substituting the retarded Green's function [7.49,50]

$$\left(i\frac{\partial}{\partial t} + \frac{1}{2}\nabla^2 + \omega\hat{L}_z - Ex\right)G(r,t;r',t') = i\delta(r - r')\delta(t - t') \tag{7.92}$$

into (7.91), one finds that

$$\phi_E(r) = \frac{i}{2}\int_0^\infty dt \, \exp(iEt)G(r,t;0,0) \quad . \tag{7.93}$$

The problem is thus reduced to finding a solution to (7.92). This can be done by expanding the Green's function, $G(\mathbf{r},t;0,0)$, in a series involving the eigenfunctions of a free electron in a circularly polarized field, see (4.88). The variables of the eigenfunctions can then be transformed relative to a rotating reference frame by applying the operator $\exp(i\omega t\hat{L}_z)$. Denoting these functions, as was done earlier, see (4.88), by $\Psi_p(\mathbf{r},t)$, one obtains the following expression for the retarded Green's function:

$$G(\mathbf{r},t;\mathbf{r}',t') = \int d^3p\,\Psi_p(\mathbf{r},t)\Psi_p^*(\mathbf{r}',t') \quad . \tag{7.94}$$

After evaluating the simple integral in (7.61) and substituting (7.94) into (7.93), one arrives at the following result:

$$\phi_E(\mathbf{r}) = \frac{1}{4\pi} - \frac{1}{(2\pi i)^{\frac{1}{2}}} \int_0^\infty \frac{dt}{t^{3/2}} \exp\left\{i\left[Et + \frac{r^2}{2t}\right.\right.$$

$$+ \frac{Ey}{\omega}\left(\frac{\sin\omega t}{\omega t} - 1\right) - \frac{2Ex}{\omega^2 t}\sin^2\frac{\omega t}{2}$$

$$\left.\left. - \frac{E^2}{2\omega^2}\left(t - \frac{4}{\omega^2 t}\sin^2\frac{\omega t}{2}\right)\right]\right\} \quad . \tag{7.95}$$

If the term in (7.95) that diverges as $r \to 0$ is isolated and substituted into (7.89), the result is

$$(-2E)^{-\frac{1}{2}} = (-2E_0)^{-\frac{1}{2}} - \frac{1}{(2\pi i)^{\frac{1}{2}}} \int_0^\infty \frac{dt}{t^{3/2}} \exp(iEt)$$

$$\times \left\{1 - \exp\left[-\frac{iE^2 t}{2\omega^2}\left(1 - \frac{4}{\omega^2 t^2}\sin^2\frac{\omega t}{2}\right)\right]\right\} \quad . \tag{7.96}$$

This equation implicitly determines the complex energy, E, in terms of E_0, E, and ω. If the adiabaticity parameter, $\gamma = \omega(2|E_0|)^{\frac{1}{2}}/E$, and a dimensionless variable, $u = \omega t$, is introduced into (7.96) and the number of photons needed to ionize an atom (in the framework of perturbation theory) is given by $K = <E_0/\omega + 1>$, then

$$(-E/\omega)^{-\frac{1}{2}} = (-E_0/\omega)^{-\frac{1}{2}} - \frac{1}{2(2\pi i)^{\frac{1}{2}}} \int_0^\infty \frac{du}{u^{3/2}}$$

$$\times \exp\left(i\,\frac{Eu}{\omega}\right)\left\{1 - \exp\left[-\frac{iKu}{\gamma^2} \times \left(1 - \frac{4}{u^2}\sin^2 u\right)\right]\right\} \quad . \tag{7.97}$$

It can be seen that E/ω is a function of only two dimensionless parameters, γ^2 and K.

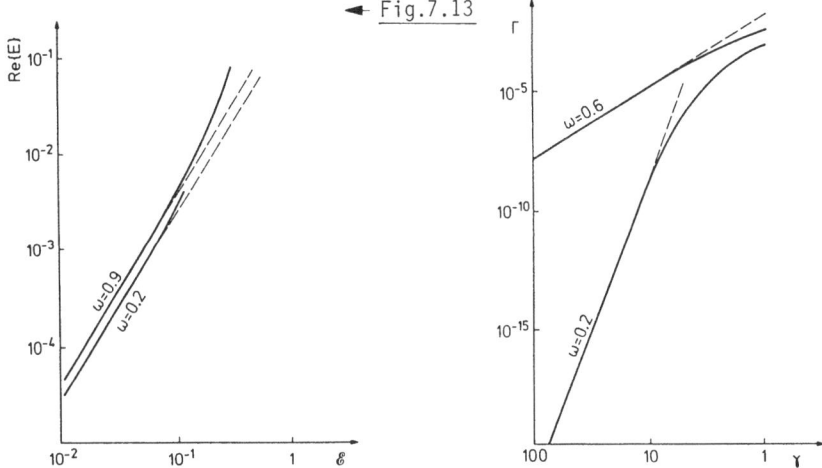

Fig.7.13. Shift in the energy Re {E} of an electron in a zero-radius potential well as a function of the strength of the external field E; the dashed lines represent extrapolations of the quadratic law into the regions of strong fields (all quantities are given in atomic units)

Fig.7.14. Width Γ of the energy level of an electron in a zero-radius potential well as a function of the adiabaticity parameter γ (all quantities are given in atomic units)

Equation (7.97), which implicitly determines E/ω, can only be solved exactly using numerical methods. Figures 7.13 and 7.14 illustrate the results of such calculations for Re{E} and Im{E} = -w/2 and two values of ω (E_0 is assumed to be unity) [7.47]. Figure 7.13 shows that the shift of Re{E} is quadratic in E for weak fields and that it increases faster in strong fields than was to be expected from an extrapolation of this function (the dashed line in Fig.7.13).

The width of the level $w/2 = -\text{Im}\{E\}$ (Fig.7.14) is equal to the ionization probability per unit time. In weak fields this probability is proportional to E^{2K} in agreement with the results of perturbation theory. In strong fields, the probability gradually decreases as the strength of the field increases, contradictory to the predictions of perturbation theory.

If $\gamma \ll 1$, the integral in (7.97) can be evaluated using the saddle-point method. The results $\gamma \ll 1$ and $\gamma \gg 1$ will be examined for fields that are weak compared to atomic fields.

1) With $\gamma \ll 1$ (tunneling ionization) (7.97) yields

$$\text{Re}\{E\} = E_0 - \frac{1}{32|E_0|^2} E^2 \quad , \tag{7.98}$$

$$w = \frac{E}{2(2|E_0|)^{\frac{1}{2}}} \exp\left[-\frac{2(2E_0)^{3/2}}{3E}\left(1 - \frac{\gamma^2}{15}\right)\right] . \tag{7.99}$$

As it should, (7.99) agrees with (7.81). It confirms the statement made in Sect.7.5.1 that when γ is close to unity, the situation comes closer to tunneling than to multi-photon ionization. The factor $(1 - \gamma^2/15)$, which is a result of the varying nature of the field, increases the ionization pro- bability as compared with the ionization probability in a constant field with the same amplitude E, see (7.53).

Equation (7.98) coincides with the expected energy shift of an electron in a short-range potential well resulting from a constant field [7.51].

2) With $\gamma \gg 1$ (multi-photon ionization, i.e., $K \gg 1$) the results established in Sects.4.5 and 7.5.1 show that one can use perturbation theory. Using time-dependent perturbation theory (Chap.2) or the expansion of the solution to (7.97) in a power series in E^2, one arrives at the following expression for the probability of K-photon ionization (when $E \ll E_{at}$):

$$-\text{Im}\{E\} = \frac{1}{2} w_K$$

$$= \frac{e^{\frac{1}{2}}E_0}{(2\pi)^{\frac{1}{2}}(2K + 1)^{3/2}} \left(\frac{eE}{\omega(2E_0)^{\frac{1}{2}}}\right)^{2K} \times \left(\frac{K\omega}{E_0} - 1 - \frac{1}{\gamma^2}\right)^{K+\frac{1}{2}} , \tag{7.100}$$

where e is the base of natural logarithms. This equation is valid for fre- quencies in the range

$$\frac{1}{K - 1} > \frac{\omega}{E_0} > \frac{1}{K} , \tag{7.101}$$

for which the K-photon ionization channel is open and all other channels in- volving fewer photons are closed.

From (7.100) one can see that w_K does not just depend on the power law E^{2K} but also on E_0 and

$$w_K \propto \left(\frac{1}{\omega(2E_0)^{\frac{1}{2}}}\right)^{2K} , \tag{7.102}$$

which, as expected, coincides with (4.92).

The last factor in (7.100) represents a branch point at the threshold of K-photon ionization. Its high order is due to the circular nature of the field, i.e., when K photons from the circularly polarized field are absorbed, the initial S state obtains an orbital angular momentum equal to K, so that the probability of small values of E_{kin} above the threshold is suppressed by a strong centrifugal barrier. The energy of the ejected photoelectron, E_{kin}, is equal to

$$E_{kin} = K\omega - |E_0| - \frac{|E_0|}{\gamma^2} \quad . \qquad (7.103)$$

According to the well-known properties of cross sections near a reaction threshold [Ref.7.49, Chap.9] w_K is proportional to $E_{kin}^{K+\frac{1}{2}}$, which agrees with (7.100).

Finally, the origin of the last term, $1/\gamma^2$, in the threshold factor [or the term $|E_0|/\gamma^2$ in (7.103)] should be explained. If one rewrites the last term in (7.103) as $E^2/2\omega^2$, it becomes evident that this term represents the vibrational energy of the free electron in the circularly polarized field. Just as with (4.81) for a linearly polarized field, this result follows from simple classical considerations. Generally speaking, (7.103) should include the difference in the Stark shifts for the initial (E_0) and the final states in the continuum (i.e., the vibrational energy). However, the Stark shift of the initial bound state (E_0), equal to $E^2/32|E_0|^2$ according to (7.98), is small compared to the vibrational energy, $E^2/2\omega^2$, because $\omega \ll E_0$. For this reason it can be neglected in (7.100,103).

It has been assumed that the Stark shift is the same whether $\gamma \gg 1$ or $\gamma \ll 1$, i.e., it is given by (7.98). This follows from the assumption that $\omega \ll E_0$, whereby in both limiting cases the shift of the level E_0 may be considered to be static. It also remains unchanged when $\gamma \approx 1$.

Thus, each factor in (7.100) has been qualitatively explained.

An estimate of the term of perturbation theory that follows the lowest order determined by (7.100) shows that perturbation theory holds when $\gamma^2 \gg 1$. This coincides with the criterion in Sect.4.5. Strictly speaking, the probabilities change rapidly with small variations in ω, and the above conclusion only holds for frequencies that lie near the middle of the frequency range $|E_0|/K$ to $|E_0|/(K-1)$.

If E_{kin} is sufficiently small, i.e., near a threshold, it may be that the term that follows the lowest-order term in the perturbation theory expansion is of the same order or greater order of magnitude and that the higher-order terms are negligible. This is due to the fact that the absence of a small threshold factor is of greater significance than a small factor $1/\gamma^2$. The criterion which determines whether the next-to-the-lowest-order term is of significance is

$$\gamma^2 (E_{kin}/E_0)^{K+\frac{1}{2}} \lesssim 1 \quad . \qquad (7.104)$$

Of course, this result is of a general nature and does not only refer to the ionization in the short-range potential anion.

Analytical formulas only exist for weak fields, for which perturbation theory is valid. If $\omega > |E_0|$, the one-photon ionization channel is open and E acquires an imaginary part which determines the probability of one-photon ionization. The equation for the shift in the atomic-state energy is:

$$\delta E = - \frac{1}{4} \alpha \, (\omega) E^2 \quad , \tag{7.105}$$

where the dynamic polarizability is equal to

$$\alpha \, (\omega) = - \frac{2}{\omega^2} - \frac{16E_0^2}{3\omega^4}$$

$$+ \frac{8E_0^2}{3\omega^4} \left[\left(1 + \frac{\omega}{|E_0|} \right)^{3/2} + \left(1 - \frac{\omega}{|E_0|} \right)^{3/2} \right] \quad . \tag{7.106}$$

This expression can be found with the help of (7.96) and by applying the standard methods of perturbation theory. The threshold singularity of the type $E_{kin}^{3/2}$ agrees with the general $E_{kin}^{K+\frac{1}{2}}$ law. For $\omega > |E_0|$ the following formula for the probability of one-photon ionization involving the ejection of an electron from a delta-like well is obtained:

$$w = -2 \, \text{Im}\{E\} = \frac{4E_0^4}{3\omega^4} \left(-1 + \frac{\omega}{|E_0|} \right)^{3/2} E^2 \quad . \tag{7.107}$$

At the threshold w vanishes.

7.6.3 Ionization by an Elliptically Polarized Field

In conclusion, the ionization of an anion in an elliptically polarized field will be considered briefly [7.52]. In this case, instead of the closed equation (7.97), an equation that contains an infinite number of terms that depend on the amplitude of the field is obtained for the complex energy, E. This corresponds to a Schrödinger equation in which the temporal and spatial variables are not separable. Such an equation can only be solved numerically. This includes the particular case of a linearly polarized field.

The case for which γ is much smaller than unity was examined at the beginning of this section. What happens when $\gamma \gg 1$ (multi-photon ionization)? The main difference lies in the change in the threshold factor. According to the selection rules, when the S state is ionized by an even number of photons, the final state is again the S state. The threshold behavior is therefore determined by the factor $E_{kin}^{\frac{1}{2}}$. Obviously, when the S state is ionized by an odd number of photons, the final state is the P state, so that the threshold singularity in the ionization probability is determined by the factor $E_{kin}^{3/2}$.

The dependence of the ionization probability on E, E_0, and ω is determined by the same factor,

$$\left(\frac{E}{\omega(2|E_0|)^{\frac{1}{2}}}\right)^{2K} \quad , \tag{7.108}$$

as in (7.100) and (4.92). This expression differs from (7.100) only in numerical factors. The Stark shift of the level E_0 has the same structure, to within a numerical factor, as (7.98) for a circularly polarized field. For an elliptically polarized field the numerical factors include the degree of ellipticity of the electromagnetic field; the dependence on E, E_0, and ω does not change. Finally, in the expression for E_{kin} the shift is half as large as the Stark shift $|E_0|/\gamma^2$ in a linearly polarized field, since the classical vibrational energy of a free electron in a linearly polarized field is $E^2/4\omega^2 = |E_0|/2\gamma^2$.

7.6.4 The Intermediate Case $(\gamma^2 \sim 1)$

Let us turn to the problem of calculating the pre-exponential factor in the expression for the ionization probability for an anion when γ is of the order of unity. In the adiabatic approximation, the exponential factor is determined by (4.51,52), where it was assumed that $E \ll E_{at}$ and $\omega \ll E_0$. The use of these approximations in (7.97) is difficult and only suitable for a circularly polarized field.

The main problem lies in calculating the matrix element that describes the ionization process:

$$S_{fi} = \int_{-\infty}^{\infty} <\Psi_f|V|\Psi_i^{(0)}> \, dt \quad , \tag{7.109}$$

where V is the amplitude of the perturbation by the electromagnetic field, $\Psi_i^{(0)}$ is the unperturbed, (initial) stationary-state wave function of the electron in the field of the anion, and Ψ_f is the exact wave function that takes into account both the atomic potential and the electromagnetic field.

Instead of (7.109) it is convenient to use a different expression for the ionization matrix element, based on the general formulas of quantum mechanical scattering theory:

$$S_{fi} = \int_{-\infty}^{\infty} <\Psi_{Ef}|U|\Psi_i> \, dt \quad , \tag{7.110}$$

where $U(\mathbf{r})$ is the atomic potential, i.e., the potential of the short-range field of the neutral atomic core, Ψ_{Ef} is the wave function of the electron in the electromagnetic field without the atomic core [it is given by (4.88)], and Ψ_i is the exact wave function, which is simply $\Psi_i^{(0)}$ when the field is switched off.

The approximation $\Psi_i \to \Psi_i^{(0)}$ rests on the fact that because $U(\mathbf{r})$ is short-ranged and the electromagnetic field is weak compared to $U(\mathbf{r})$ (this is equivalent to saying that $E \ll E_{at}$), the exact wave function, Ψ_i, differs little from $\Psi_i^{(0)}$ inside the region where $U(\mathbf{r})$ is of significance. Hence, one can substitute $U_0\delta(\mathbf{r})\Psi_i^{(0)}$ for $U(\mathbf{r})\Psi_i$. After this only the definite integral in (7.70) needs to be evaluated, which generally must be done numerically. If $\omega \ll E_0$, the saddle-point method can be used to obtain an analytical expression.

Using the fact that

$$\Psi_i^{(0)} \propto \exp(-iE_0 t) \quad , \tag{7.111}$$

and the expression for Ψ_{Ef}, (4.55), the saddle points in (7.70) are found to be those points for which

$$\frac{1}{2}\left(\mathbf{p} - \frac{1}{\omega}E \cos\omega t\right)^2 = E_0 \quad , \tag{7.112}$$

where $\mathbf{p}$ is the momentum of the ejected photoelectron.

Such calculations were carried out in [7.53-55] and made it possible to find the pre-exponential factors in (4.51,52) for the ionization probability in linearly, circularly, and elliptically polarized fields. These expressions will not be given here because they are cumbersome.

Since the saddle points in (7.110) are situated in the plane of complex "time", the above method is also known as the "imaginary time method" (Sect. 4.5). From the physical point of view the ionization process consists of an electron tunneling through a slowly varying potential barrier.

Finally, it should be noted that, strictly speaking, the problem considered is only exact for a zero-radius potential. Even for a potential well with a small but finite width and only one discrete level the above solution is only approximate, and the wider the well the worse the approximation. For a sufficiently wide well a second level emerges and resonances with this level as an intermediate level may be observed. In this case the solution is not even qualitatively applicable. For example, the probability of one-photon ionization in a Coulomb field at the threshold ($E_{kin} = 0$) does not vanish, as it does with a delta-like potential.

7.6.5 Nonlinear Detachment of an Electron from a Negatively Charged Ion

In conclusion we shall briefly discuss the process of nonlinear detachment of an electron from a negatively charged ion. Only one experiment is known [7.46] in which a two-photon detachment of an electron from an I⁻ ion was observed. It was found that the dependence of the electron yield on the intensity of laser radiation can be expressed by a quadratic law quite accurately. It was also found that the two-photon cross section of photodetachment α^2 is $(3.5 \pm 1.5) \times 10^{-49} \, cm^4 s$. The results of this experiment clearly show that the model of a short-range potential used in calculating the photodetachment cross section in [7.56] gives poorer agreement with the experiment than the more complex model [7.57], which assumes that the electron consitutes an effective potential that accounts, for one, for the polarization of the atomic core. Detailed theoretical studies on the role of the polarizability of a negatively changed ion [7.58] have shown that allowance for polarizability changes not only the absolute values of the cross sections but leads to a qualitatively different dependence of the cross sections on the frequency of the radiation. The absence of a sufficient body of experimental data makes it impossible at present to give a complete description of this interesting phenomenon.

7.7 The Ionization of Atoms

In real atoms the potential of the atomic core cannot be assumed to be short-ranged. While for the hydrogen atom, it is a Coulomb potential, for alkali atoms it is approximately a Coulomb potential. The long-range nature of the potentials of atomic cores is also inherent in other atoms. Hence, an electron extracted from an atom is affected by the atomic as well as the electromagnetic field. In general the solution of the corresponding Schrödinger equation is very complex because the spatial and temporal coordinates are not separable.

Let us consider multi-photon ionization when ω is much less than $E_n^{(0)}$. When $\gamma \gg 1$, the process is known as direct multi-photon ionization. Of course it is assumed that there are no intermediate resonances. Qualitatively this process is described by (4.92), if one is not interested in pre-exponential factors.

A direct multi-photon transfer of an electron from a bound state to the continuum consists of a sum of virtual transitions, in each of which the conservation of energy is realized to within the limits of the time-energy uncertainty principle. When

$$\omega \ll E_n^{(0)} \qquad \text{and} \qquad \gamma \gg 1 \; , \tag{7.113}$$

the probability of a direct transition is related to the intensity of the light by a power law.

7.7.1 The Power-Law Dependence of the Ionization Probability on the Field Strength

Equation (4.92) was experimentally verified under the conditions given in (7.113) for a large variety of atoms, for light of various frequencies and different degrees of polarization, and for degrees of nonlinearity that varied from 2 to 22 [7.46,59]. In principle, the upper limit is not of importance. It is stipulated by the degree to which the multi-photon ionization process is effective in placing restrictions on the strength of the external field and thereby on the conditions given in (7.113).

Figure 7.15 illustrates the power-law dependence of the ionization probability on the intensity of the radiation, see (4.57), using as an example the three-photon ionization of the metastable helium atom [7.60].

The absolute value of the probability per unit time, w, is related to the absolute value of the radiation intensity, F, by a factor $\alpha^{(K)}$, called the multi-photon cross section:

$$\alpha^{(K)} = w/F^K \; . \tag{7.114}$$

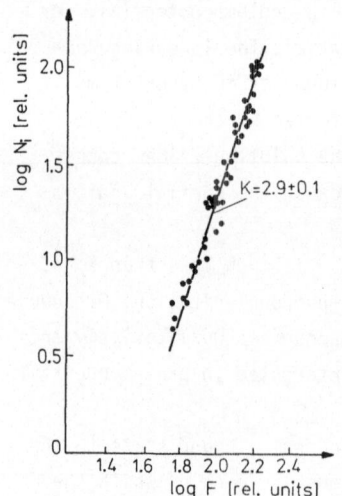

Fig.7.15. Ion yield N_i as a function of the radiation intensity F in relative units for the three-photon ionization of the metastable 2^1S state of helium

Although the dimensions of $\alpha^{(K)}$ for $K > 1$ are not cm^2 and depend on the value of K (if the dimensions of w and F are s^{-1} and photons/cm^2s, respectively, the dimensions of $\alpha^{(K)}$ are $cm^{2K}s^{K-1}$), this definition of multi-photon cross section has a physical meaning. Indeed, this factor does not depend on the radiation intensity and when $K = 1$, i.e., for one-photon processes, $\alpha^{(1)}$ is measured in cm^2, which corresponds to the accepted cross section of a process in which one particle (one photon) is absorbed. Since the dimension of multi-photon cross sections depends on the degree of the nonlinearity of the transition, there is no sense in comparing multi-photon cross sections for processes with different values of K; what must be compared are the probabilities at a fixed intensity. A comparison of multi-photon cross sections is only advisable for processes with the same degree of nonlinearity.

7.7.2 Dependence of the Multi-Photon Cross Section on the Frequency and the Degree of Ellipticity of the Light

The magnitude of the multi-photon cross section at a fixed degree of nonlinearity depends on the frequency ω and the degree of ellipticity ρ of the light and also on the bound states of the particular atom. These parameters and the selection rules for one-photon and multi-photon transitions determine which virtual transitions are allowed and the magnitude of the detuning between $K'\omega$ and the differences in the energies between the bound electronic states.

The conditions in (7.113) suggest that in order to calculate the K-photon cross section one must apply K-order time-dependent perturbation theory. In accordance with Sect.2.4, the matrix element that corresponds to a transition of an electron from a bound state to the continuum is represented by (2.21).

Figure 7.16 illustrates the predicted dependence of the three-photon ionization cross section of the hydrogen atom on the frequency of linearly

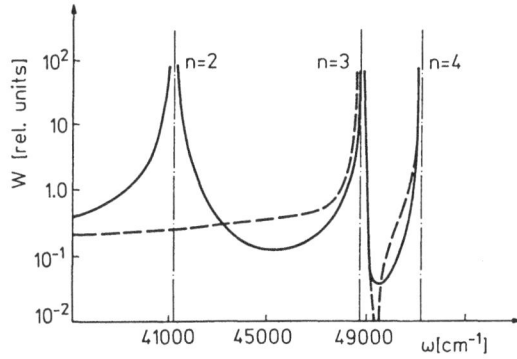

Fig.7.16. Probability of a three-photon ionization of the hydrogen atom as a function of the radiation frequency ω (theory); the solid line represents linearly polarized light, the dashed line circularly polarized light, and n is the principal quantum number of the bound states

223

and circularly polarized light [7.61]. Intermediate two-photon resonances with discrete spectral terms can be clearly seen. Intermediate one-photon resonances with the same terms occur at higher frequencies and are not shown. One can see that the resonance maxima for linear and circular polarizations coincide, except for the term with $n = 2$, in which the two-photon resonance is absent because the selection rules forbid the $1S \to 2S$ transition in a circularly polarized field. The dependence of the cross sections on the frequency and degree of ellipticity is similar for more complex atoms. Obviously, in complex atoms the spectrum is considerably richer. The probability of a nonresonant two-photon ionization of metastable helium by linearly and circularly polarized light has been computed [7.62]. A slight dependence on the degree of the ellipticity of the field in the neighborhood of the one-photon resonances with the intermediate states was noted. The probability of K_0-photon ($K_0 \leqslant 16$) ionization of a hydrogen atom has recently been calculated [7.63].

The function $\alpha^{(K)}(\omega, \rho)$ qualitatively reflects the basic relationships that are valid for all atoms and processes with any degree of nonlinearity. The maxima of $\alpha^{(K)}(\omega, \rho)$ are due to the emergence of one-photon and multi-photon intermediate resonances. The minima depend on the polarization of the light. For circularly polarized light and an initial S state, there is a frequency in each interval between resonances for which the multi-photon cross section vanishes. For linearly polarized light the cross section in the minima is always nonzero. This is due to differences in the transitions that contribute to the compound matrix element given in (2.21).

For circularly polarized light each subsequent transition from the S state takes place with a unit increase in the orbital quantum number. All of the photons exhibit either right-hand or left-hand helicity. The final state has a fixed angular momentum, so that the ionization probability is determined by a single matrix element. Since the sign of the compound matrix element changes when the frequency passes through a resonance with a discrete intermediate level and the signs of the one-photon matrix elements remain the same for transitions in which the orbital quantum number increases by unity, there exists a frequency in each interval between resonances for which the ionization probability vanishes.

On the other hand, for linearly polarized light, transitions can take place with either an increase or decrease in the orbital quantum number. The light consists of photons with opposite helicities. Thus, transitions into states with different angular momenta are possible, so that the total probability is a sum of the squares of the partial amplitudes. Since each of these amplitudes vanishes at different frequencies in the interval between resonances, the total probability is always nonzero. A similar situation occurs when the ionization is brought about by a circularly polarized field and proceeds from an initial state with a nonzero angular momentum. Hence, the final states can exhibit different values of the angular momentum.

The ratio R of the probabilities of ionization by circularly polarized and linearly polarized fields poses an interesting problem. As already noted, due to the ionization of an s-state by a circular field, the orbital angular momentum in the composite matrix element of photon absorption increases. When a linear field is used instead of a circular one, the corresponding term is not the only one, but in the quasi-classical approximation it nevertheless gives the main contribution to the composite matrix element (Chap.9). For this reason, the radial matrix elements and the propagators (in R) cancel out and only the ratios of the angular parts of the matrix elements remain. These can easily be calculated. Thus, we find (a detailed consideration of this problem and references to original works have been given in [7.64]) that

$$R = (2K - 1)!!/K! > 1 \quad . \tag{7.115}$$

This estimate is valid only for small values of K. The reader must bear in mind, however, that the above relationship represents an upper limit and can in some cases be invalid. This is especially true when resonance maxima for ionization by circular and by linear fields do not coincide. In this case $R < 1$.

Reviewing the dependence of the ionization probability on the frequency and the degree of ellipticity of the light, one can draw two important conclusions. First of all, the probability exhibits no regular dependence on the degree of ellipticity of the light. Also, the fact that the ionization probability vanishes under certain conditions is a new feature of nonlinear phenomena.

7.7.3 Criteria for Applying Perturbation Theory

The applicability of perturbation theory in the calculation of multi-photon cross sections is restricted: the degree to which the initial and the final free state of the electron are perturbed by the external field must be small. It is obvious that as the intensity of the external field increases, this criterion will first of all break down for a free electron. The requirement that the perturbation be small means that the vibrational energy of the free electron must be small compared to its kinetic energy:

$$E_{vib} \sim \frac{E^2}{\omega^2} \ll E_{kin} \quad . \tag{7.116}$$

Since

$$E_{kin} = K\omega - E_n^{(0)} - E_{vib} \quad , \tag{7.117}$$

where $E_n^{(0)}$ is the energy of the electron's initial state, one cannot reduce the criterion given in (7.116) to a criterion for E. The frequency still plays a significant role. In principle one can always select a frequency so small that (7.116) is no longer valid for a given field.

In the frequency range of visible light, the upper limit for E can be estimated from (7.116) if it is assumed that the electron is ejected with an energy of the order of a photon:

$$E_{kin} \sim \hbar\omega \sim 0.1 \text{ Rydberg} = 0.05E_{at} \quad (E_{at} \approx 26 \text{ eV}) \quad . \tag{7.118}$$

The inequality in (7.116) is reversed for fields greater that $10^{-2}E_{at}$. Note that the estimate (7.118) gives the same magnitude of the intensity of the light field as does the condition $\gamma \sim 1$. Since the amplitude of the perturbation is proportional to E^K, a decrease in the field strength causes a sharp decrease in the amplitude and thereby widens the lower range of admissible kinetic energies for the emitted electron.

The perturbation of intermediate resonance states also plays an important, although indirect, role. It can lead to a breakdown of the condition that no resonances take place, a criterion established for bound states in the absence of a field. There is no common criterion which can be used to decide when the intensity of a field is sufficient to perturb the intermediate bound states, since in this case the frequency of the light plays a decisive role. Indeed, a perturbation of the bound electronic states leads to a change in the detuning of resonances in the compound matrix element, whereby resonances may even be induced by the field. Obviously, the smaller the initial detuning (which for a given atomic spectrum depends on the frequency of the radiation), the weaker the field need be for a resonance to occur. For example, a perturbation of the helium atom by an electromagnetic field of about $10^{-4}E_{at}$ leads to the appearance of an intermediate resonance at a frequency that differs considerably from the resonance frequency of the unperturbed atom (Fig.7.17, [7.60]). Accordingly, for a fixed intensity of light, the calculations done for the first nonvanishing order of perturbation theory do not predict an ionization process in a definite frequency range about a resonance. Due to a shift in the atomic levels and their broadening because of the external field, this region must be wider than the natural widths of the atomic levels.

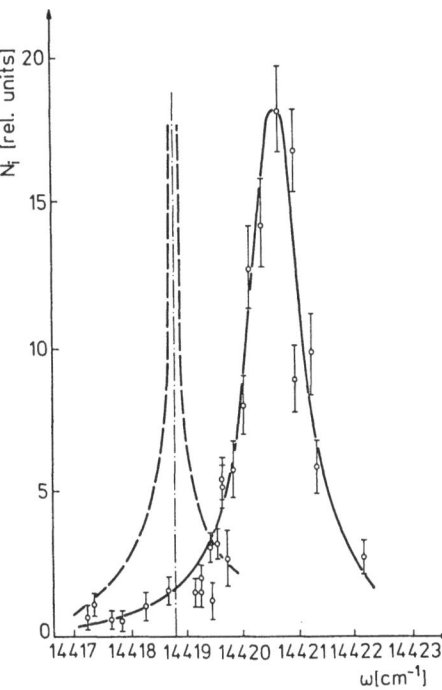

<u>Fig.7.17.</u> Ion yield N_i as a function
of the radiation frequency in the region of an intermediate two-photon
resonance for the three-photon ionization of the metastable 2^1S state of
helium; the solid line represents experimental results, and the dashed
line represents the results of calculations using perturbation theory
in the first nonvanishing order

However, even if one takes into account the restrictions brought about by the
perturbation of electronic states, there still remains a wide range of permissible field strengths and many frequency intervals for which perturbation
theory and (7.113) are valid.

7.7.4 Calculation of Multi-Photon Cross Sections and Their Comparison with Experimental Data

The variable parameter of multi-photon ionization is the frequency of the
field. Thus, the main objective of a theoretical description is to calculate
the frequency dependence of the multi-photon cross section and to compare it
with the experimental data obtained for fixed frequencies or for relatively
narrow frequency intervals.

Two difficulties arise in the calculation of multi-photon cross sections.
The first is related to the presence of an infinite sum in the compound matrix element given in (2.21). The second difficulty is characteristic of
calculations of any effects in complex atoms; it has to do with the choice
of an approximate expression for the wave function and energy terms of the
optical electron. An exact calculation of the sum in (2.21) is possible if
one knows the Green's function whose poles coincide with the bound states

227

(Sect.2.6). Among the approximate methods for dealing with the bound electronic states in a complex atom, the two that have been used with success
are the quantum defect method and the model-potential method (Sect.2.6). The
results of the calculations depend strongly on the type of approximation;
the probabilities can even differ by an order of magnitude. The probability
of K_0-photon ($K_0 \leqslant 16$) ionization of a hydrogen atom has recently been computed [7.63]. Experimental data on multi-photon cross sections was obtained
for a large number of atoms with different electronic configurations in the
valence shell, for linearly and circularly polarized light with wavelengths
ranging from 1 μm to 2.5 μm, and for degrees of nonlinearity, K, that lie
between 2 and 22 [7.46,64,65]. Most of the data is not very accurate because
radiation from a multi-mode laser and an absolute method of measuring the
multi-photon cross sections were used (Sect.6.5). When single-mode laser radiation and a relative method of measuring the multi-photon cross sections
was used, the results were much more accurate (Sect.6.5). For example, cross
sections could be measured with one hundred percent accuracy. Such accuracy
allows one to define the optimum method for calculating a given cross section.
However, experimental results only exist for a few frequencies which correspond to the operating frequencies of high-power solid-state lasers. These frequencies usually lie in the intervals between resonances, far from the characteristic frequencies that correspond to the maxima or minima in the dispersion curve (which represents the frequency dependence of the ionization probability). For a fairly wide range of frequencies, data only exists for processes involving a small number of photons. One such process is the three-photon ionization of the potassium atom using a dye laser (Fig.7.18) [7.66,67].

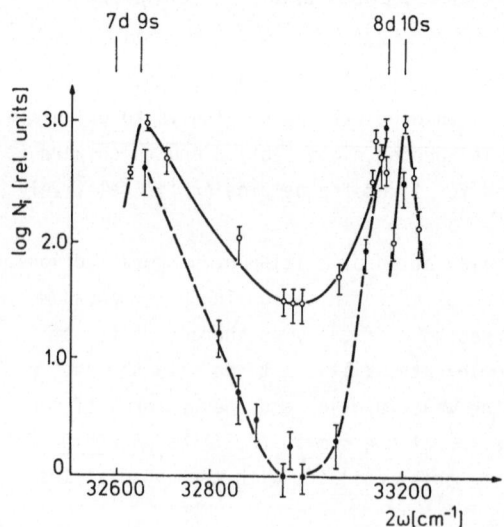

Fig.7.18. Ion yield N_i as a
function of the radiation frequency ω for the three-photon
ionization of the potassium
atom; the results of the calculations made for linearly
and circularly polarized light
are represented by the solid
and dashed curves, respectively

7.7.5 The Angular Distribution of the Emitted Photoelectrons

In a linearly polarized field the angle between the polarization vector,
$\mathbf{e} = E/|E|$, and the momentum $\mathbf{p}$ of the emitted photoelectron shall be labelled.
To find the dependence of the differential probability, $dw^{(K)}/d\Omega$, on θ, one
must first determine the dependence of the one-photon matrix elements $(\mathbf{r} \cdot \mathbf{e})_{mn}$
on θ.

The variable of integration, $\mathbf{r}$, in $(\mathbf{r} \cdot \mathbf{e})_{mn}$ can be written in terms of
spherical coordinates, i.e., $\mathbf{r} = (r,\theta',\phi')$ whereby the polar axis lies in the
direction of the momentum of the electron. Using the addition theorem for
spherical harmonics, one can write [7.44]:

$$(\mathbf{r} \cdot \mathbf{e})_{mn} = \alpha_{mn} \cos\theta + \beta_{mn} \sin\theta \quad , \tag{7.119}$$

where α_{mn} and β_{mn} do not depend on θ.

In multiplying these matrix elements together to form the multi-photon
matrix element given in (2.21) and in raising the modulus of the product to
the second power, the probability, $dw^{(K)}/d\Omega$, is found to be a homogeneous,
2K-degree function in $\cos\theta$ and $\sin\theta$.

Since the initial state is degenerate with respect to the orbital angular momentum ℓ, only the probability averaged over the magnetic quantum
number of the initial state has physical meaning. This averaged probability
does not depend on the choice of the quantization axis, i.e., on the angle
ϕ. Hence, the angular distribution of the averaged probability only depends
on θ. It is obvious that a reverse direction for $\mathbf{e}$ should not affect the
observable quantities. When the sign of $\mathbf{e}$ changes (and the direction of $\mathbf{p}$
remains unchanged), θ changes to $\pi - \theta$, i.e., $\cos\theta$ changes sign. This implies
that the probability is an even function of $\cos\theta$. Since the averaged probability is a homogeneous function of even degree, it is also an even function
of $\cos\theta$. Since the averaged probability is a homogeneous function of an even
degree, it is also an even function of $\sin\theta$. The terms that contain odd powers
of $\cos\theta$ and $\sin\theta$ vanish when averaged over the magnetic quantum number M.
Finally, because even powers of sines can be written in terms of even powers
of cosines, the final expression for the angular distribution of the photoelectrons in the event of a K-photon ionization by a linearly polarized field
is

$$\overline{dw^{(K)}}/d\Omega = E^{2K}[\alpha_0^{(K)} + \alpha_1^{(K)} \cos^2\theta + \alpha_2^{(K)} \cos^4\theta +$$

$$\dots + \alpha_K^{(K)} \cos^{2K}\theta] \quad , \qquad\qquad (7.120)$$

where the bar over the differential probability represents an averaging over the magnetic quantum numbers. The coefficients $\alpha_s^{(K)}$ are determined in terms of atomic parameters and the frequency of the light, ω. Their calculation for complex atoms obviously involves certain models of the unperturbed atomic wave functions. The form of the distribution in (7.120) with respect to θ depends to a great extent on ω due to the different frequency dependences of the $\alpha_s^{(K)}$.

For a one-photon ionization, (7.120) is simpler [7.44]:

$$\overline{dw^{(1)}}/d\Omega = E^2(\alpha_0^{(1)} + \alpha_1^{(1)} \cos^2\theta) \quad . \qquad\qquad (7.121)$$

In the two-photon case one has [7.68]

$$\overline{dw^{(2)}}/d\Omega = E^4(\alpha_0^{(2)} + \alpha_1^{(2)} \cos^2\theta + \alpha_2^{(2)} \cos^4\theta) \quad . \qquad\qquad (7.122)$$

For an arbitrary value of K, (7.120) also describes the ionization of the S state without an averaging (since M is zero) [7.69].

Actual calculations of angular distributions have only been made for cases in which the initial state is an S state. No experimental data of sufficient accuracy has yet been obtained.

What happens if the light is circularly polarized? Simple expressions for the angular distribution are only obtained if the initial state is an S state. According to the selection rules (Sect.1.3), the final state has a nonzero angular momentum with a projection equal to K. The wave function of the final state, $\psi_p^{(0)}(\mathbf{r})$, can be expanded in a series involving all of the Legendre polynomials in the angle θ between $\mathbf{r}$ and $\mathbf{p}$:

$$\psi_p^{(0)}(\mathbf{r}) = \sum_{\ell=0}^{\infty} R_{E\ell}^{(0)}(r) P_\ell(\cos\theta') \quad , \qquad\qquad (7.123)$$

where each of the $R_{E\ell}^{(0)}(r)$ is a solution to the radial Schrödinger equation with an angular momentum quantum number ℓ, an energy $E = p^2/2$, and an atomic potential as the interaction Hamiltonian. Using the addition theorem for Legendre polynomials, one can rewrite (7.123) as follows:

$$\psi_p^{(0)}(\mathbf{r}) = \sum_{\ell=0}^{\infty} \frac{4\pi}{2\ell + 1} \sum_{M=-\ell}^{\ell} R_{E\ell}^{(0)}(r) Y_{\ell M}^*(\phi,\psi) Y_{\ell M}(\beta,\gamma) \quad , \qquad\qquad (7.124)$$

where ϕ and ψ specify the direction of $\mathbf{r}$, and β and γ specify the direction of the momentum p of the ejected electron. Taking into account the already mentioned selection rules, the term in (7.124) that is proportional to $Y_{KK}(\beta,\gamma)$ can be isolated. Thus, the angular distribution of the probability of the ejection of a photoelectron from an S state through ionization by a circularly polarized field is

$$dw^{(K)}/d\Omega = C_K E^{2K} \sin^{2K}\beta \quad , \tag{7.125}$$

where C_K is a constant, and the angle β defines the direction of the ejected photoelectron (specified by the direction of $\mathbf{p}$) relative to that of the electromagnetic wave. Hence, in contrast to the case for a linearly polarized field, the shape of the angular distribution of the photoelectrons does not change when the frequency of the circularly polarized field is varied. Finally, it should be noted that as the intensity of the electromagnetic field increases, the next-to-lowest order of perturbation theory begins to make itself felt and the shape of the angular distribution begins to change compared to that for weak fields. The reason for this is that a higher order of perturbation theory corresponds to the absorption of a larger number of photons, which in turn brings about an increase in K in (7.119). This change in the shape of the angular distribution can be seen in the calculations for the hydrogen atom in a linearly polarized field [7.70].

7.7.6 Tunneling Ionization of Atoms ($\gamma \ll 1$)

An expression for the probability of the tunneling ionization of real atoms in variable fields can be obtained quite simply if the well-known formula for the probability of ionization of the hydrogen atom in a constant field is used as a basis [7.17]:

$$w|_{E=\text{const}} = \frac{4}{E} \exp\left(- \frac{2}{3E}\right) \quad . \tag{7.126}$$

In full analogy with the ionization in the short-range potential (Sect. 7.6.1), the probability of ionization in a variable circularly polarized field is equal to the probability of ionization in a constant field. For a linearly polarized field the ionization probability is found by substituting $E \sin\omega t$ for E and averaging over the period of the field, $T = 2\pi/\omega$:

$$w = 4\left(\frac{3}{\pi E}\right)^{\frac{1}{2}} \exp\left(- \frac{2}{3E}\right) \quad . \tag{7.126}$$

The above equation refers (as does the previous equation for a constant field) to the ionization of the hydrogen atom in the ground state.

Comparing (7.127) with (7.84), one can see that there is a difference (in the pre-exponential factor) and a similarity (in the exponential function) in the probabilities of tunneling through a short-range or a Coulomb potential barrier. This coincidence of exponential functions agrees with the general statement made in Sect.4.5, i.e., the exponential functions do not depend on the type of potential.

If one stays in the frequency range of visible light and deals with real atoms, the condition $\gamma \ll 1$ is equivalent to $E \gtrsim E_{at}$ (Sect.7.5.1). For this reason, when studying the tunneling ionization process in a variable field, one must ensure conditions such that the ionization of atoms in their ground states will take place using infrared radiation. Such experiments have yet to be conducted. Tunneling ionization has only been observed at $\gamma \approx 0.7$ [7.72].

7.7.7 The Intermediate Case ($\gamma^2 \sim 1$)

There are no theoretical findings for the ionization probability in this case. By analogy with the limiting cases, $\gamma^2 \gg 1$ and $\gamma^2 \ll 1$, one can only say that, to within exponential accuracy, the ionization probability for real atoms is described by the equations valid for short-range potentials (Sect. 7.6). In the frequency range of visible light, where the condition that γ be of the order of unity is realized for very strong fields (close to the intensity of atomic fields), the situation becomes more complex because it becomes increasingly difficult to separate the direct and resonance ionization processes due to the strong perturbation of the bound electronic states. Although a large broadening of the resonance greatly decreases their role, it also greatly decreases the frequency range in which the resonances can be neglected. In any case, any theory that claims to describe the ionization process that takes place in the intermediate region must account for the resonances and their shifts when the intensity of the visible light is varied. Resonances were observed in experiments with an adiabaticity parameter γ as small as two [7.71].

For the intermediate case, ionization was only observed in some inert gases at $\gamma \approx 0.7$ [7.59]. The experimentally determined dependence of ion yield on the field strength has a threshold, see (7.127).

In conclusion, it must be noted that the intermediate region encompasses a comparatively narrow range of field strengths. When establishing the boundaries of the intermediate region it must be remembered that the criterion $\gamma^2 \sim 1$ must be fulfilled and is reduced to the usual form, $\gamma \gg 1$ or $\gamma \ll 1$, only in the limiting cases. Therefore, strictly speaking, only the interval $0.7 \lesssim \gamma \lesssim 2$ can be considered to belong to the intermediate case.

8. Resonance Phenomena

Resonance phenomena can be observed if the energy of a photon of the external field (or the energy of several such photons) is equal to the difference in energy between bound electronic states. This may be the case for an unperturbed atom or for an atom perturbed by the field. A necessary condition for resonance is an allowed transition with a given number of photons. In resonance, the lifetime of an excited electronic state is determined by the time of spontaneous or induced transitions from this state, i.e., by a quantity that is much greater than the time required for virtual transitions, which, in turn, is determined by the time-energy uncertainty principle. Hence, the probability of the occurrence of a resonance process is much higher than that of a nonresonance process for a fixed field strength.

To describe most of the resonance phenomena it may seem appropriate to use the two-level approximation in which n is the initial and m the final level. However, in the overwhelming majority of real problems the final state is excited. For this reason it must be borne in mind that the electron may pass over to a state with a greater or smaller energy. In the first case, not only must the external field which is responsible for the resonance transition be considered, but also the field of the electromagnetic vacuum, which is responsible for the spontaneous relaxation of excited states. Naturally, a transfer to higher states can only occur under the influence of an external field. A particular case is a transition to the continuum, i.e., when the field ionizes the atom. Thus, in real situations, one usually has to consider a three-level system, n, m, k, where the third state k may be a discrete state or a state in the continuum. Only with relaxation from the excited state to the ground state is the three-level system reduced to a two-level system.

For a resonance transition to occur, the frequency of the external field must almost be equal to the frequency of a transition between bound electronic states. As usual, it will be assumed that the external field is ideally monochromatic. Then the above criterion for resonance must be satisfied within a resonance width that is determined by the width of the resonance states. The widths of the states depend on the intensity of the external field and on the nature of the transition the electron makes from the resonance state m (it entered this state under the influence of the field). This means that one may speak of spontaneous and field (or ionization) widths.

In any field, vacuum or external, the probability of finding the electron in a given state is a smooth function of the frequency. The Lorentzian line shape for a spontaneous decay is a well-known example of such a function (Sect.8.1.2). Hence, there are no strict limits on the resonance frequency of the external field. As the detuning from resonance $\Delta = \omega_{mn} - \omega$ (where ω is the frequency of the monochromatic external field, and ω_{mn} is the frequency of the transition between the states n and m) increases, the nonresonant interaction of these states with other states, bound or free, starts to compete with the resonant interaction. The effect of a resonance is clearly seen in the dispersion profiles for the dynamic polarizability (Fig.7.1) and the probability of direct multi-photon ionization (Fig.7.15): as the frequency of the external field approaches the transition frequency, the influence of the resonance increases, and leads to an increase in the amplitude of the resonance. Hence, the detuning must not exceed Γ_{mn} (the resonance transition width from n to m) for a resonance to appear. Of course, a resonance can take place when the transition $n \rightarrow m$ is allowed by selection rules. We must note, however, that the smallness of the quadrupole matrix elements may be compensated for by the resonant increase of the probability (Sect.8.4).

Among the various resonance processes one must be able to distinguish the scattering that leads to the appearance of light of another frequency. Indeed, a real target always contains a large number of atoms so that many such photons are produced. In this way the atomic target may be affected by two fields, the main resonance (external) field of frequency $\omega \approx \omega_{nm}$ and the field of frequency ω_{mk} created in the target. As long as the intensity of the scattered light is low, the transition $m \rightarrow k$ is due to the vacuum field (similar to a transition in an isolated atom), e.g., spontaneous Raman scattering. As soon as the intensity of the scattered light is strong enough, the transition is "forced" by the field of frequency ω_{mk}. This is known as stimulated Raman scattering. (If the scattering process refers to an incident photon, it is always a stimulated process.) Therefore, in the framework of the model in which an atom is placed in a monochromatic field whose frequency is equal to one of the transition frequencies, stimulated Raman scattering is characteristic of a number of atoms and not of just one atom. This is also the case for complexer processes which involve the excitation of atoms by fields of various frequencies. In spite of the practical importance of spontaneous Raman scattering (multi-photon parametric phenomena), such phenomena will not be studied since the subject of this book deals with elementary processes. A different situation arises when the field of frequency ω_{mk} is produced by another source. A three-level system in two resonance fields is a model for many elementary processes that are often realized in practice.

The theory of resonance processes is based on the resonance approximation (Chap.3). In fact, the criterion for applying this approximation, see (3.5), is that the detuning from resonance be smaller than the frequency of the resonance field, i.e., $\Delta \ll \omega$. Only such processes that are characterized by diagrams whose "branches" alternate with respect to the "trunk" will be considered (Sect.2.3). All other processes will be neglected. The probability ratio of the second process to the first, equal to Δ/ω, is much less than unity. This is obviously a less strict criterion than that needed for a resonance. Indeed, one can only isolate a two-level system from the entire set of bound electronic states when $\Gamma_{mn} \ll \omega_{mn}$. Hence, strictly speaking, the resonance approximation can only be used to describe resonance processes involving discrete states. Whether or not it can be used when there are transitions into the continuum is not so obvious.

In Chap.3 the resonance approximation was used to describe the behavior of a two-level system in a resonance field. The example examined did not reflect reality, since the vacuum field, which may transfer the electron from an excited state to the ground state or to another excited state, was not taken into account. This will be done in Sects.8.1-3.

In Chap.3 the criterion for the applying perturbation theory $(\mathbf{r}_{mn} \cdot \mathbf{E})/\Delta \ll 1$, was introduced. In a real situation this criterion is no longer valid due to the presence of spontaneous processes. An analysis of the results of this chapter leads to a new criterion:

$$\frac{(\mathbf{r}_{mn} \cdot \mathbf{E})}{(\Delta^2 + \gamma_{m,n}^2)^{\frac{1}{2}}} \ll 1 \quad , \tag{8.1}$$

where $\gamma_{m,n}$ are the spontaneous widths of states m and n, and Δ is the detuning from resonance. It may be noted that the equations of the resonance approximation do not exactly transform to those of perturbation theory. Since $\Delta \ll \omega$, the results of this chapter have not taken into account resonant terms that do not contain Δ in the energy denominators.

This chapter deals with phenomena, in contrast to those considered in Chap.7, that are characterized by their resonant nature. Such phenomena are much more complex than nonresonance phenomena because the dependence of the atomic polarization on the amplitude of the field is essentially nonlinear in an external field. As in the nonresonance case, the spontaneous emission of photons can be considered to be a scattering of the incident light. It is obvious that the scattering can be both elastic (Rayleigh) and inelastic [8.1].

8.1 Spontaneous Emission of Light by an Atom in a Resonance Field

Before proceeding with this problem, it is advisable to recall the solution to the simple problem involving the spontaneous emission of photons when an atom in an excited state relaxes to the ground state in the absence of a perturbing field [8.2].

8.1.1 The Natural Line Width

Let us consider a two-level system consisting of a ground state n and an excited state m. At $t = 0$ the electron is in the upper state. What is the probability of a transition into the lower state per unit time (for small times) if this transition does not depend on time?

Let V' be the operator that represents the electric dipole interaction of the electron with the quantized electromagnetic field in vacuum [8.3]:

$$V' = i \sum_{k\alpha} (2\pi\nu_k)^{\frac{1}{2}} [(r \cdot e_\alpha') b_{k\alpha} \exp(-i\nu_k t)$$

$$+ (r \cdot e_\alpha'^*) b_{k\alpha}^+ \exp(i\nu_k t)] \quad , \tag{8.2}$$

where ν_k, k, and α are the frequency, the wave vector ($k = \nu_k/c$), and the polarization of the photons of the vacuum, respectively; e_α' is their polarization vector, and $b_{k\alpha}^+$ and $b_{k\alpha}$ are the creation and annihilation operators of a photon in the state (k,α). From (8.2) one can see that the matrix element of V', which describes the emission of a photon, ν_k, when the system goes over from state m to state n, has the following form:

$$V'_{nm} = i(2\pi\nu_k)^{\frac{1}{2}} (e_\alpha'^* \cdot r_{nm}) \exp(i\nu_k t) \quad . \tag{8.3}$$

By isolating the time dependence in the same way as was done in Chap.3, one arrives at

$$V'^{(1)}_{nm} = i(2\pi\nu_k)^{\frac{1}{2}} (e_\alpha'^* \cdot r_{nm}) \quad . \tag{8.4}$$

The total probability per unit time, γ_m, of a spontaneous transition from the upper level m to the lower level n (i.e., the spontaneous relaxation probability for a two-level system), summed over the two polarizations α and integrated over the emission angles of the photons, is equal to

$$w_m \equiv \gamma_m = \frac{\pi}{2} \sum_{\alpha=1,2} \int |V'^{(1)}_{nm}|^2 \delta(\omega_{mn} - \nu_k) \frac{d^3k}{(2\pi)^3}$$

$$= \frac{4\omega^3_{mn}}{3c^3} |r_{nm}|^2 \quad . \tag{8.5}$$

The spontaneous transition can be described in terms of the wave function of the electron in the upper level if one substitutes $\exp(-iE_m t)$ for $\exp(-iE^{(0)}_m t)$, where the notation $E_m = E^{(0)}_m - i\gamma_m/2$ has been introduced. Hence, the spontaneous transition leads to a broadening of the mth level. The quantity γ_m is called the spontaneous, or natural, line width. The way in which this broadening was introduced is known as the Breit-Wigner procedure [8.4].

8.1.2 The Lorentzian Line Shape

The equation for the spontaneous emission probability per unit time is valid for small times, i.e., $t \ll 1/\gamma_m$. For large times the dependence of the absolute transition probability on time deviates from a linear one. This section is devoted to the study of such deviations and deals with time intervals of any duration.

Let us again consider a two-level system consisting of a ground state n and an excited state m, in which the electron is in the upper state at $t = 0$. According to Sect.2.1, the system of equations for the probability amplitudes a_{m0} of finding the electron in the state m and the field in the state without photons and for the probability amplitudes $a_{nk\alpha}$ of finding the electron in the state n and the field in the state with one photon having wave vector k and polarization α has the form:

$$i\dot{a}_{m0} = \sum_{k\alpha} \frac{1}{2} <m0|V'^{(1)}|nk\alpha> a_{nk\alpha} \exp[i(\omega_{mn} - \nu_k)t] \quad ,$$

$$i\dot{a}_{nk\alpha} = \frac{1}{2} <nk\alpha|V'^{(1)}|m0> a_{m0} \exp[i(-\omega_{mn} + \nu_k)t] \quad , \tag{8.6}$$

where $V'^{(1)}$ is the amplitude of the dipole interaction between the vacuum field and the electron, see (8.4). This interaction only has one time-dependent exponential factor. The second time-dependent exponential factor does not contribute to the interaction because the corresponding matrix elements vanish. The quantity ν_k is the frequency of the emitted photons.

By a simple substitution it is easy to check that the solution to this system is of the form

$$a_{m0} = \exp(-\gamma_m t/2) \quad ,$$

$$a_{nk\alpha} = - <nk\alpha|V'^{(1)}|m0> \frac{\exp[i(\nu_k - \omega_{mn})t - \gamma_m t/2] - 1}{\nu_k - \omega_{mn} + i\gamma_m/2} \tag{8.7}$$

where γ_m is determined by (8.5). In this solution the level shift due to the vacuum field (the Lamb shift) has not been taken into account.

The above expression for $a_{nk\alpha}$ shows that for times that meet the condition $\gamma_m t \ll 1$ (t being the perturbation time), one can determine the probability per unit time of the spontaneous emission of a photon with frequency $\nu_k = \omega_{mn}$, and finds that the total spontaneous emission probability is equal to $\gamma_m t$.

If t is large, i.e., if $\gamma_m t \gg 1$ (steady state; the atom is in the lower state n), one can neglect the exponential in (8.7) compared to unity and thus obtain a formula for the absolute probability of the emission of a photon of frequency ν_k:

$$|a_{nk\alpha} (t \to \infty)|^2 = |<nk\alpha|V'^{(1)}|m0>|^2 \frac{1}{(\nu_k - \omega_{mn})^2 + \gamma_m^2/4} \quad . \tag{8.8}$$

If this expression is summed up over the various polarizations of the emitted photon and is integrated over the emission angles, the spectral distribution of the emitted photons is obtained:

$$g(\nu) \, d\nu = \sum_{\alpha} \int |a_{nk\alpha}(t \to \infty)|^2 d\Omega$$

$$= \frac{1}{\pi} \frac{\gamma_m/2}{(\nu - \omega_{mn})^2 + \gamma_m^2/4} \, d\nu \quad . \tag{8.9}$$

As expected, the integral of $g(\nu)$ over the various frequencies of the emitted photons is equal to unity. Equation (8.9) represents the natural line profile, or the Lorentzian line shape. From a mathematical standpoint the above result can be seen as a Lorentzian "broadening" of the delta function, $\delta(E_m - E_n^{(0)} - \nu)$,

$$\delta(E_m - E_n^{(0)} - \nu) \to \frac{1}{\pi} \frac{\gamma_m}{2} \frac{1}{(\nu - \omega_{mn})^2 + \gamma_m^2/4} \quad , \tag{8.10}$$

where $E_m = E_m^{(0)} - i\gamma_m/2$. Indeed, both functions have a maximum at $\nu = \omega_{mn}$, and their integrals with respect to ν are equal to unity. In this case the substitution of $E_m^{(0)} - i\gamma_m/2$ for $E_m^{(0)}$ is also equivalent to the well-known Breit-Wigner procedure (see above).

8.1.3 Spontaneous Emission of Photons in a Weak Resonance Field

Let us assume that the two levels m and n have radiative widths γ_m and γ_n, respectively, and that, in addition, the system is placed in a weak field, $E\cos\omega t$, of a frequency $\omega \approx \omega_{mn}$. The term "weak field" is not an absolute characteristic of the field strength. It is a relative characteristic related to the width γ_{mn} and the detuning Δ. What is meant by the word "weak" is that $(\mathbf{r}_{mn} \cdot \mathbf{E})/(\Delta^2 + \gamma_{mn}^2)^{1/2} \ll 1$.

According to Sect.2.1, in first-order perturbation theory the probability amplitudes for the upper and the lower levels are

$$a_n(t) = \exp(-iE_n^{(0)}t) \quad , \qquad \text{and} \tag{8.11}$$

$$a_m(t) = \frac{\mathbf{r}_{mn} \cdot \mathbf{E}}{2\Delta} [1 - \exp(i\Delta t)] \exp(-iE_m^{(0)}t) \quad , \tag{8.12}$$

whereby spontaneous transitions were not taken into account, $\Delta = \omega_{mn} - \omega$, and it has been assumed that the initial conditions at $t = 0$ are $a_n(0) = 1$ and $a_m(0) = 0$, i.e., the electron is in the lower state, n.

In accordance with the Breit-Wigner procedure discussed in Sect.8.1.2, the decay of the states m and n can be accounted for by substituting $E_{n,m}^{(0)} - i\gamma_{n,m}/2$ for $E_{n,m}^{(0)}$. As a result one obtains

$$a_n(t) = \exp(-\gamma_n t/2) \quad , \tag{8.13}$$

$$a_m(t) = \frac{\mathbf{r}_{mn} \cdot \mathbf{E}}{2\Delta - i(\gamma_m - \gamma_n)} \exp(-iE_m^{(0)}t)$$
$$\times \{\exp(-\gamma_m t/2) - \exp[i(\Delta + i\gamma_n/2)t]\} \quad , \tag{8.14}$$

where for the sake of simplicity it has been assumed that $E_n^{(0)} = 0$.

The first exponential term in $\{...\}$ in (8.14) is due to a real intermediate state, since it does not oscillate and decreases at the rate γ_m of the damping of the intermediate state m, i.e., it exhibits features inherent in the evolution of the unperturbed state m. It is also a reflection of the cascade transition from the initial state n through the real intermediate state m to the final state.

The second exponential term in (8.14) oscillates for fairly large values of Δ and decreases at the rate of the damping of the initial state n. It is due to a virtual intermediate state and describes a two-photon transition from the initial state n through the intermediate state m to the final state.

The total probability of such a transition is

$$W_m = \gamma_m \int_0^\infty |a_m(t)|^2 \, dt \quad . \tag{8.15}$$

Substituting (8.14) into (8.15) yields

$$W_m = \gamma_m \frac{(r_{mn} \cdot E)^2}{4\Delta^2 + (\gamma_m - \gamma_n)^2}$$

$$\times \left[\frac{1}{\gamma_m} + \frac{1}{\gamma_n} - 2\text{Re}\left\{ \frac{1}{(\gamma_m + \gamma_n)/2 - i\Delta} \right\} \right] \quad . \tag{8.16}$$

The first two terms in the square brackets of (8.16) can be considered to be equal to the probability of finding the atom in the "real" and the "virtual" intermediate state m. They are the result of taking the square of the modulus of the first and second terms in curly brackets of (8.14).

The third term is due to the interference between the first and second terms in the curly brackets of (8.14). It describes the interference between cascade and two-photon transitions [8.5]. The interference term does not play a role if

$$|\Delta| \gg \gamma_{m,n} \quad . \tag{8.17}$$

When this condition is met, the cascade and two-photon transitions can be considered to be independent processes because the total transition probability is equal to the sum of the probabilities for each process. However, when $\Delta \lesssim \gamma_{m,n}$, i.e., in the resonance case, the cascade and two-photon transitions cannot be separated, and one can only speak of an indivisible transition process.

If the condition in (8.17) is met and the decay rates γ_m and γ_n differ significantly, the greatest contribution to the total probability W_m will be made by the term which corresponds to the smaller decay rate. For example, if the intermediate state m has the longest lifetime, i.e., $\gamma_m \ll \gamma_n$, the process can be assumed to be a two-stage process. Otherwise, when $\gamma_m \gg \gamma_n$, the process is of a two-photon nature.

8.1.4 Spontaneous Emission of Photons in a Strong Resonance Field

In Sect.8.1.3 first-order perturbation theory was used, i.e., it was assumed that the field strength was low. Now let it be assumed that the field is so strong that perturbation theory cannot be applied, but one can apply an adiabatic perturbation (Chap.3). In the following discussion it will be assumed that a spontaneous broadening of the level m does not lead to transitions to the state n and that all of the spontaneous transitions proceed along channels that do not include the states n and m and that are independent of each other. One can then use the solution to Rabi's problem, i.e., (3.6), with the same initial conditions, $a_n(0) = 1$ and $a_m(0) = 0$, as in Sect.8.1.3. Substituting $E_{mn}^{(0)} - i\gamma_n/2$ for $E_{mn}^{(0)}$ in (3.6) yields

$$a_m(t) = -i \frac{\mathbf{r}_{mn} \cdot \mathbf{E}}{2\tilde{\Omega}} \sin\tilde{\Omega}t \times \exp[- iE_m^{(0)}t + i\Delta t/2 - (\gamma_m + \gamma_n)t/4] \quad , \qquad (8.18)$$

where the notation

$$\tilde{\Omega} = \frac{1}{2} \left\{ \left[\Delta + \frac{i}{2} (\gamma_n - \gamma_m) \right]^2 + (\mathbf{r}_{mn} \cdot \mathbf{E})^2 \right\}^{\frac{1}{2}} \qquad (8.19)$$

has been introduced. Since the expression in the square brackets of (8.18) is complex-valued, (8.18) is very cumbersome in such a general form. However, one thing is clear: one cannot separate cascade and multi-photon processes in a strong field. To simplify matters, only the case in which the upper and lower levels have the same widths will be considered. The Rabi frequency is thus real-valued and coincides with that given in (3.4). As a result (8.18) becomes simpler. If $\gamma_m = \gamma_n = \gamma$, then

$$a_m(t) = -i \frac{\mathbf{r}_{mn} \cdot \mathbf{E}}{2\Omega} \sin\Omega t \times \exp(- iE_m^{(0)}t + i\Delta t/2 - \gamma t/2) \quad , \qquad (8.20)$$

where the Rabi frequency Ω is defined in (3.4), i.e.,

$$\Omega = \frac{1}{2} [\Delta^2 + (\mathbf{r}_{mn} \cdot \mathbf{E})^2]^{\frac{1}{2}} \quad . \qquad (8.21)$$

After substitution of (8.20) into (8.15), the following expression is obtained for the total transition probability:

$$W_m = 2\gamma \frac{(\mathbf{r}_{mn} \cdot \mathbf{E})^2}{16\Omega^2} \left(\frac{1}{\gamma} - \text{Re}\left\{ \frac{1}{\gamma - 2i\Omega} \right\} \right) \quad . \qquad (8.22)$$

It can be seen that as the intensity of the external field increases, the interference between the "real" and "virtual" states, so important in weak fields, vanishes. At the same time, however, the distinctions between cascade and two-photon transitions disappear.

Several remarks on terminology are now in order. It is reasonable to retain the names "cascade" and "two-photon" when the two transitions can be distinguished. If the interference between the "real" and "virtual" states is important or if the intermediate and initial states mix due to a strong resonance field, the most commonly used name to describe such a transition is "resonance two-photon transition".

8.1.5 Introduction of an External Field into the Resonance Excitation

In describing the resonance approximation in Chap.3, an idealized two-level system in an external field was considered while the presence of the vacuum field was ignored. Due to this field another parameter appears in addition to the detuning Δ and the build-up time δT for the perturbation. This third parameter is the spontaneous width γ of the levels. The role of γ is quite obvious: the detuning from resonance cannot be less than γ and, as a result, cannot be equal to zero. Hence, for small values of Δ and, in particular, for an exact resonance, the criterion for the sudden introduction of a perturbation is that $\gamma \delta T \ll 1$. Reversing the inequality sign yields the criterion for the adiabatic introduction of a perturbation. Naturally, when $\Delta > \gamma$ the important factor is Δ and the criteria are of the same form as those obtained for the idealized system discussed in Chap.3.

In the "giant pulse" operation of a high-power laser, the duration of the leading edge of a pulse, δT, is of the order of or less than $10^{-9}-10^{-8}$s. The lifetime of excited states, $1/\gamma$, is usually greater than 10^{-8}s. Hence, to attain exact resonance, a perturbation must be introduced suddenly (in contrast to the nonresonance case).

In a real, two-level system, the field strength does not determine the introduction of a perturbation but the Rabi frequency, i.e., the rate at which the amplitudes of the resonating states oscillate. The possibility of a mixing of resonating states is determined by the competing spontaneous decay of the upper state. It is obvious, however, that one can always use an external field strong enough to be able to neglect the effect produced by the spontaneous decay. For an exact one-photon resonance the field strength is not at all high. Indeed, if the dipole moment of the atom, $d_{mn} \sim er_0$, is of the order of one Debye and the time required for a spontaneous decay is of the order of 10^{-8}s, the Rabi frequency becomes comparable to the spontaneous width of the upper level for field strengths between 10 and 100 V/cm, i.e., for relatively weak fields. If all possible transitions (including multi-photon transitions) and all of the allowed values of Δ are taken into account, the above estimate of the critical field will be too low. In all other cases it will be higher. For induced transitions the lifetime may be several orders of magnitude shorter, and hence the critical field strength will be several orders of magnitude greater.

8.2 Resonance Fluorescence

Fluorescence is usually called resonance fluorescence if the frequency of the atomic transition coincides with that of the external field [8.6]. A resonance absorption leads to the excitation of the atom. It is known that fluorescence phenomena are the result of the interaction of an excited atom with the vacuum field. The vacuum field causes spontaneous transitions to occur from an excited level to lower-lying levels and, finally, to the ground state. The nature of the resonance fluorescence depends to a great extent on the intensity of the exciting external field. The field can be considered to be weak if the width of the excited state is determined by the vacuum field (the natural width). It is considered to be strong if the width of the excited state (the induced width) is determined by the external field itself. Resonance fluorescence in weak fields has been widely studied [8.2]. For this reason this section will mainly deal with strong fields. However, to simplify matters, a weak field will be discussed first. The very essence of resonance fluorescence makes it possible to study a two-level system using the resonance approximation. But, contrary to the procedure applied in Sect.3.1, the interaction of the vacuum field with the system must be taken into account.

8.2.1 A Weak External Field

Let us consider a two-level system in a field E of frequency ω, which is almost equal to the energy difference, $\omega_{mn} = E_m^{(0)} - E_n^{(0)}$, of the resonant levels (it is assumed that these levels are nondegenerate): i.e., $\Delta = |\omega_{mn} - \omega| \ll \omega$. At $t = 0$ it is assumed that the system is in the lower state n. In absorbing a photon of the external field the electron is excited into the upper state m by the weak perturbation, $V = r \cdot E \cos\omega t$. The state m is then spontaneously de-excited and the electron goes back to the state n.

Let us consider resonance fluorescence in a weak field. This is a two-photon process in which one photon, ω, is absorbed (perturbation V) and another, ν_k, is emitted (perturbation V'). According to (2.13), the corresponding two-photon resonance matrix element has the following form:

$$V_{nn}^{(2)}(\omega) = \quad\quad = \frac{V_{nm}^{\prime\,(1)}(r_{mn} \cdot E)}{4(\omega_{mn} - \omega)} \quad , \tag{8.23}$$

where $V_{nm}^{\prime\,(1)}$ is defined in (8.4).

Equation (8.23) is valid when ω is not close to the energy difference ω_{mn}. At the same time ω must not differ considerably from ω_{mn}, so that one can neglect the nonresonant part of the fluorescence, which corresponds to the emission of a photon of frequency ν_k and the absorption of a photon of frequency ω (the two-photon matrix element $V_{nn}^{(2)}(-\omega)$ corresponds to the nonresonant part). If ω is almost equal to ω_{mn}, then (8.23) is valid, provided that the energy of the resonating state m is given by the complex-valued energy $E_m^{(0)} - i\gamma_m/2$. This involves nothing more than the familiar Breit-Wigner procedure studied in Sect.8.1.2. It enables one to correctly determine the matrix element of a two-photon process near a resonance in a weak field.

From (8.23) and (2.16) the well-known expression for the probability of resonance fluorescence per unit time is obtained [Ref.8.3, §64]:

$$dw_{nn} = \frac{\omega^3}{8\pi c^3} \frac{|(r_{mn} \cdot e)(r_{mn} \cdot e_\alpha'^*)|^2}{(\omega_{mn} - \omega)^2 + \gamma_m^2/4} E^2 do\delta(\nu - \omega) \, d\nu \quad , \tag{8.24}$$

where do is the solid angle into which the photon is ejected, e is the polarization vector of the incident photon, and e_α' is the polarization vector of the emitted photon. From (8.24) one can see that according to the conservation of energy the frequency of the emitted photon, ν, is strictly equal to the frequency of the absorbed photon, ω (elastic scattering). In other words, the spectral distribution of the emitted photons is that of a delta-function, $\delta(\nu - \omega)$. Equation (8.24) is the well-known Breit-Wigner equation for elastic resonance scattering by a quasi-discrete level [Ref.8.4, §134].

The incident light is never strictly monochromatic but exhibits a certain finite spectral width $\Delta\omega$. This finite width leads to a similar broadening of the spectral line of the emitted light. If $\Delta\omega \ll \gamma_m$, the fluorescence line is much narrower than the line due to the spontaneously emitted radiation. Therefore, in this case, absorption and emission cannot be considered as being cascade processes, i.e., as being two successive and independent transitions, because then the atom would not "remember" which photon it absorbed first and as a result would emit radiation with a natural spectral width. In terms of the nomenclature of Sect.8.1, the above process is a two-photon transition.

A different situation arises when the atom is irradiated by light with a large line width, i.e., $\Delta\omega \gg \gamma_m$. Integrating (8.24) with respect to ω from $\omega_{mn} - \Delta\omega$ to $\omega_{mn} + \Delta\omega$ and then carrying the limits to $-\infty$ and $+\infty$ yields

$$dw_{nn} = [(\omega_{mn} - \nu)^2 + \gamma_m^2/4]^{-1} \, d\nu \quad, \tag{8.25}$$

i.e., the width of the scattered light is equal to the natural line width, γ_m, of the decaying state, m. With respect to the shape of the absorbed and emitted lines the resonance fluorescence behaves as if it consists of two independent processes, i.e., absorption and subsequent emission. Hence, in terms of the nomenclature of Sect.8.1, this process is a two-stage, or cascade, process.

Numerous experiments in which resonance fluorescence has been observed were carried out under conditions for which $\Delta\omega \gg \gamma_m$. This is usually the case when the spectral width of the radiation of the exciting source is greater than the natural line width. As a result a fluorescence spectrum of width γ_m is observed. Experiments can also be carried out under conditions for which the above inequality has the opposite sign. An example of such an experiment was given in [8.7]. A barium atom was excited from the ground state into the excited state, $^1P_1$ (which has a natural width $\gamma_m = 5 \times 10^{-4} \, \mathrm{cm}^{-1}$),by laser radiation with an effective spectral width $\Delta\omega$ that is several times smaller than γ_m. The fluorescence spectrum was observed to be approximately half as broad as the natural width of this transition (Fig.8.1).

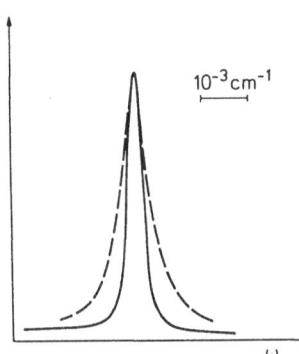

I

$10^{-3} \mathrm{cm}^{-1}$

ω

Fig.8.1. The resonance fluorescence spectrum that results when a narrow band (*solid line*) and a wide-band of weak exciting radiation (*dashed line*) is used. A barium atom was excited from the ground state into the excited state P_1

When the resonant state m is degenerate (e.g., in one projection of the angular momentum), (8.24) must be modified by the following substitution:

$$(\mathbf{r}_{mn} \cdot \mathbf{e})(\mathbf{r}_{nm} \cdot \mathbf{e}_\alpha'^*) \rightarrow \sum_m (\mathbf{r}_{mn} \cdot \mathbf{e})(\mathbf{r}_{nm} \cdot \mathbf{e}_\alpha'^*) \quad, \tag{8.25}$$

where the summation is over all of the degenerate states.

If the cross section dσ is calculated using (8.24), dσ is summed over the final polarizations α and the directions of the emitted photon, and the polarizations, **e**, of the absorbed photon are averaged, the total cross section for resonance fluorescence (elastic scattering) is (e.g., for the case where the initial state n is an S state):

$$\sigma_{el} = \frac{3\pi c^2}{2\omega^2} \frac{\gamma_m^2}{(\omega_{mn} - \omega)^2 - \gamma_m^2/4} .$$ (8.26)

For other values of the angular momenta, J_n and J_m, of the initial and final states, a certain statistical factor g, that can easily be calculated, replaces the factor 3/2 in (8.26) [Ref.8.3, Chap.6].

To obtain (8.26) in a simpler way, one only needs to multiply the probability of finding the electron in the upper level m by the probability per unit time, w_{mn}, of a spontaneous transition, $m \rightarrow n$, divide it by the average flux density of the incident photons, $cE^2/8\pi\omega$, and apply (8.5).

It follows from (8.26) that the resonance width is γ_m. The influence of the vacuum field also leads to a shift in the position of the resonance, $\omega_{mn} - \omega$, i.e., to a Lamb shift in the real part of the energy E_m [8.8]. Due to its smallness it will not be considered any further.

8.2.2 Criterion for Applying Perturbation Theory

What are the highest fields for which the above equations are valid? In a resonance field the atomic levels are split into quasi-energy levels separated by energies of the order of $|r_{mn} \cdot E|$ (Chap.3). This splitting cannot be observed if it is small compared to the natural width γ_m [8.9],

$$|r_{mn} \cdot E| \ll \gamma_m .$$ (8.27)

From (8.27) it follows that the value of E is restricted from above by the critical value

$$E_{cr} = \left(\frac{\gamma_m}{\omega_{mn}}\right) E_{at} ,$$ (8.28)

above which the field cannot be considered to be weak. For example, the critical field is fairly weak for the 1S $\rightarrow$ 2P transition in the hydrogen atom, i.e., $E_{cr} \sim 10^{-8} E_{at} \approx 50$ V/cm. This is in good agreement with an estimate of E_{cr} that was made at the beginning of this chapter.

Comparing the estimates (8.28) and the critical strengths of the classical fields (Chap.1), we can easily see that the strengths equal to or higher than E_{cr} are always classical.

The vast body of experimental data on resonance fluorescence in weak fields is described by the above relationships fairly accurately. The data will not be discussed since it is well known and has already been discussed in [8.2].

246

8.2.3 The Density-Matrix Method

Let us now turn our attention to fields of arbitrary strength. The Breit-Wigner procedure can be used to describe the departure of the electron from the upper state but cannot be used to describe its arrival at the lower state. We therefore conclude that it is impossible to obtain a solution in terms of the customary quantum mechanical amplitudes. This does not contradict what was said in Chap.3; there the idealized problem was studied, in which the vacuum field was ignored.

One way to solve the problem is to introduce the amplitudes $a_k|K>$, where a_k ($k = m,n$) describes the state of the two-level system, and $|K>$ describes the vacuum state with K photons. Since the amplitudes with different numbers of photons are linked to one another in the corresponding Schrödinger equation, an infinite set of equations emerges. These can be solved, but the solution is cumbersome [8.3].

A simpler method is to use the density matrix instead of the wave function $\Psi(x,q,t)$ to describe a system that consists of an atom, a vacuum field, and an external field [Ref.8.4, §14]:

$$\rho(x,x',t) = \int \Psi(x,q,t)\Psi^*(x',q,t) \, dq \quad , \tag{8.29}$$

where q denotes the coordinates of the photons of the vacuum, and x and x' are the atomic coordinates. The density matrix, ρ, also satisfies the following equation:

$$i \frac{\partial \rho}{\partial t} = [H,\rho] \quad , \tag{8.30}$$

where H is the Hamiltonian of the atom-field system. Expanding ρ in a series involving the stationary states of the atom, $\psi_m^{(0)}(x)$ and $\psi_n^{(0)}(x)$, yields

$$\rho(x,x',t) = \sum_{i,j=m,n} \rho_{ij}(t)\psi_j^{(0)*}(x')\psi_i^{(0)}(x) \quad . \tag{8.31}$$

The diagonal matrix elements, ρ_{nn} and ρ_{mm}, denote the probabilities of finding the electron in the states n or m, respectively.

If the vacuum field is ignored, $H = H_0 + V$, where H_0 is the Hamiltonian of the atom, and V describes the interaction between the atom and the external field. Equations (8.30,31) then yield

$$i \frac{\partial \rho_{ij}}{\partial t} = [(H_0 + V)\rho]_{ij} - [\rho(H_0 + V)]_{ij} \quad . \tag{8.32}$$

Suppose only the vacuum is responsible for the interaction V' and the external field is switched off. The equation for ρ_{ij} would then have the following form:

$$\frac{\partial \rho_{ij}}{\partial t} + \frac{1}{2}(\gamma_i + \gamma_j)\rho_{ij} = \delta_{in}\gamma_m\rho_{mm}\delta_{ij} + \omega_{ij}\rho_{ij} \quad . \tag{8.33}$$

The appearance of the attenuation constants, γ_i and γ_j, on the left-hand side of this equation is due to the decay of the states i and j. In the absence of the field E,

$$\rho_{ij}(t) = \rho_{ij}(0)\exp[i(E_i^* - E_i)t] \quad , \tag{8.34}$$

where $E_{i,j} = E_{i,j}^{(0)} - \gamma_{i,j}/2$ and $\gamma_n = 0$.

The right-hand side of (8.33) represents the influx of electrons into the state n from the state m per unit time.

When both fields, E and E', are present, (8.32 and 33) must be combined:

$$i(\partial\rho_{nn}/\partial t) = i\gamma_m\rho_{mm} + V_{nm}\rho_{mn} - \rho_{nm}V_{mn} \quad ,$$

$$i(\partial\rho_{nm}/\partial t) = -i\gamma_m\rho_{nm}/2 + \omega_{nm}\rho_{nm} + V_{nm}(\rho_{mm} - \rho_{nn}) \quad , \tag{8.35}$$

$$i(\partial\rho_{mn}/\partial t) = -i\gamma_m\rho_{mn}/2 + \omega_{mn}\rho_{mn} + V_{mn}(\rho_{nn} - \rho_{mm}) \quad .$$

The equation for ρ_{mm} was omitted because $\rho_{mm} + \rho_{nn} \equiv 1$.

According to Sect.3.1, the time dependence of the resonance approximation on the external field, cos ωt, is only evident in one exponential term:

$$V_{mn} = \frac{1}{2}\mathbf{r}_{mn} \cdot \mathbf{E}\exp(-i\omega t) \quad , \qquad V_{nm} = \frac{1}{2}\mathbf{r}_{nm} \cdot \mathbf{E}\exp(i\omega t) \quad . \tag{8.36}$$

Substituting (8.36) and $\rho_{mm} = 1 - \rho_{nn}$ into (8.35) yields a linear system of three nonhomogeneous differential equations. If the following substitutions are introduced: $\rho_{nm} = \alpha\exp(i\omega t)$ and $\rho_{mn} = \alpha^*\exp(-i\omega t)$, a linear system of three nonhomogeneous differential equations with constant coefficients is obtained [8.10]:

$$i\left(\frac{d\rho_{nn}}{\pm dt}\right) + i\gamma_m\rho_{nn} + \frac{1}{2}\mathbf{r}_{mn} \cdot \mathbf{E}\alpha - \frac{1}{2}\mathbf{r}_{nm} \cdot \mathbf{E}\alpha^* = i\gamma_m\rho_{mm} \quad ,$$

$$i\left(\frac{d\alpha}{dt}\right) + \left(\Delta + \frac{i}{2}\gamma_m\right)\alpha + \mathbf{r}_{nm} \cdot \mathbf{E}\rho_{nn} = \frac{1}{2}\mathbf{r}_{nm} \cdot \mathbf{E} \quad , \tag{8.37}$$

$$i\left(\frac{d\alpha^*}{dt}\right) + \left(-\Delta + \frac{i}{2}\gamma_m\right)\alpha^* - \mathbf{r}_{mn} \cdot \mathbf{E}\rho_{nn} = -\frac{1}{2}\mathbf{r}_{mn} \cdot \mathbf{E} \quad ,$$

where $\Delta = \omega_{mn} - \omega$ is the detuning from resonance.

Hence, with the help of the density matrix method the resonance fluorescence problem is reduced to a system of three ordinary differential equations (8.37).

Regardless of the specific solution to (8.37), one can note the general character of the solutions for different time intervals.

For times $T \ll 1/\gamma$ radiation damping is not important. In this case the elements of the density matrix can be expressed in terms of the products of amplitudes: e.g., $\rho_{ij} = a_i^* a_j$. The equations for these amplitudes are the usual Schrödinger equations with the Hamiltonian $H_0 + V$, and depend on the manner in which the field was introduced.

If $T \sim 1/\gamma$, this transition is characterized by a very complicated form of solution.

Finally, for $T \gg 1/\gamma$ a stationary state occurs, which is characterized by atomic levels with constant populations. A distinguishing feature of this stationary state is that the solutions do not depend on the initial conditions, in particular, on the method of switching on the external field (adiabatically or suddenly).

8.2.4 Elastic (Rayleigh) Scattering in Strong Fields

The solution to (8.37) can be expressed in terms of the general solution to the homogeneous system of equations and a particular solution of the non-homogeneous system. Let us start with the simpler problem of finding a particular solution. The solution consists of a set of constants ρ_{mm} and α:

$$\rho_{mm}(\infty) = \frac{(\mathbf{r}_{mn} \cdot \mathbf{E})^2}{4\Delta^2 + 2(\mathbf{r}_{mn} \cdot \mathbf{E})^2 + \gamma_m^2} \quad ,$$

$$\alpha(\infty) = -\frac{2(\Delta - i\gamma_m/2)\mathbf{r}_{mn} \cdot \mathbf{E}}{4\Delta^2 + 2(\mathbf{r}_{mn} \cdot \mathbf{E})^2 + \gamma_m^2} \quad .$$

$$(8.39)$$

The general solution to a homogeneous set of equations with constant coefficients exhibits an exponential dependence on time. For example, $\rho_{mm} = R_0 \exp(s_0 t) + R_+ \exp(s_+ t) + R_- \exp(s,t)$, where s_0, s_+, and s_- are the roots of the characteristic equation of (8.37). The expressions for α and α^*, not given here, have a similar form. The real parts of these roots can only be negative; otherwise the law of the conservation of the number of particles will not be satisfied. Therefore, as $t \to \infty$, the general solution to the homogeneous system vanishes, i.e., the particular solution to the non-homogeneous system is equal to the general solution.

Equation (8.38) enables one to find the total probability of a spontaneous emission per unit time:

$$w = \gamma_m \rho_{mm}(\infty) \quad . \tag{8.39}$$

This probability is a sum of the probability of elastic scattering ($\nu = \omega$) and the probability of inelastic scattering ($\nu \neq \omega$). The elastic scattering process is due to the dipole moment of the atom and is of a classical nature.

It is this time-dependent moment that emits radiation in accordance with the well-known principles of radiation theory. According to the general rules for finding average quantities, the average dipole moment is [8.11]

$$\mathbf{d}(t) = \int \mathbf{d}\rho(x,x) \; dx \quad . \tag{8.40}$$

Applying (8.31) and bearing in mind that the dipole moment only contains off-diagonal matrix elements yields

$$\mathbf{d}(t) = \rho_{mn}\mathbf{d}_{nm} + \rho_{nm}\mathbf{d}_{mn}$$

$$= \alpha(\infty) \; e^{i\omega t}\mathbf{r}_{mn} + \alpha^{*}(\infty) \; e^{-i\omega t}\mathbf{r}_{nm} \quad . \tag{8.41}$$

It can be seen that the average dipole moment oscillates harmonically with a frequency equal to that of the forcing field, ω. Also, it is a well-known fact that the dipole will emit radiation of the same frequency. Therefore, the particular solution to the nonhomogeneous system in the form of (8.38) corresponds to radiation whose frequency is exactly equal to that of the incident light, ω, (elastic scattering). Hence, the radiation probability is [Ref.8.3, §45]

$$dw = \frac{\omega^3}{2\pi c^3} \; |\mathbf{d} \cdot \mathbf{e}_{\alpha}^{'*}|^2 do' \quad . \tag{8.42}$$

Substituting (8.41) into (8.42) yields

$$dw_{el} = \frac{\omega^3}{2\pi c^3} \; |\alpha(\infty)|^2 |\mathbf{r}_{nm} \cdot \mathbf{e}_{\alpha}^{'*}|^2 do\delta(\nu - \omega) \; d\nu \quad . \tag{8.43}$$

If one then substitutes (8.38) into (8.43), the differential probability of elastic resonance fluorescence (elastic scattering) in a strong field can be found:

$$dw_{el} = \frac{1}{8\pi} \left(\frac{\omega}{c}\right)^3 \; |(\mathbf{r}_{nm} \cdot \mathbf{e}_{\alpha}^{'*})(\mathbf{r}_{mn} \cdot \mathbf{e})|^2 E^2$$

$$\times \frac{[(\omega_{mn} - \omega)^2 + \gamma_m^2/4]do\delta(\nu - \omega) \; d\nu}{[(\omega_{mn} - \omega)^2 + \gamma_m^2/4 + |\mathbf{r}_{mn} \cdot E|^2/2]^2} \quad . \tag{8.44}$$

If the external field is weak, i.e., if $|\mathbf{r}_{mn} \cdot E| \ll \gamma_m$, see (8.15), then (8.44) becomes (8.9). However, if the field is so strong that $|\mathbf{r}_{mn} \cdot E| \gg \gamma_m$, then

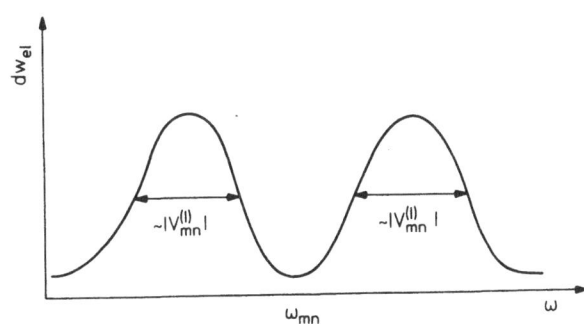

Fig.8.2. The resonance
fluorescence probability
dw_{el} as a function of the
frequency ω of a strong
external field; ω_{mn} is
the transition frequency
in the two-level system

$$dw_{el} = \frac{1}{8\pi} \left(\frac{\omega}{c}\right)^3 |(\mathbf{r}_{nm} \cdot \mathbf{e}'^*_\alpha)(\mathbf{r}_{mn} \cdot \mathbf{e})|^2 E^2$$

$$\times \frac{[(\omega_{mn} - \omega)^2 + \gamma_m^2/4]d\sigma\delta(\nu - \omega) \, d\nu}{[(\omega_{mn} - \omega)^2 + |\mathbf{r}_{mn} \cdot \mathbf{E}|^2/2]^2} . \tag{8.45}$$

As can be seen from (8.45), the probability of observing resonance fluorescence in a very strong field has two maxima with widths of the order of the Rabi resonance frequency (Fig.8.2). This is the main difference between the probability in a very strong field and the probability in a weak field [the latter only exhibits one maximum with a width γ_m, see (8.9)].

The total probability of elastic resonance fluorescence can be determined using (8.44) in the same way as was done for a weak field:

$$w_{el} = \frac{(\omega_{mn} - \omega)^2 + \gamma_m^2/4}{(\omega_{mn} - \omega)^2 + \gamma_m^2/4 + |\mathbf{r}_{mn} \cdot \mathbf{E}|^2/2} \, w , \tag{8.46}$$

where w was defined in (8.39).

Thus, for very long times t, i.e., when $\gamma_m t \gg 1$, the interaction of the two-level system (n,m) with a resonance field results in elastic resonance fluorescence whose cross section can be determined using (8.44-46) and whose frequency is exactly equal to ω.

8.2.5 Inelastic Scattering

In a strong field elastic scattering is described by (8.44). All of the corresponding Feynman diagrams only contain one spontaneously emitted photon. A typical diagram of this type is:

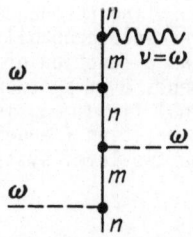

As a consequence of the energy conservation law, the frequency of the emitted photon, ν, is exactly equal to the frequency of the incident light, ω. From (8.46) one can see that w_{el} is always less than w. The probability of elastic scattering is given by $w - w_{el}$. The Feynman diagrams that represent inelastic scattering contain at least two spontaneously emitted photons. A typical diagram is shown below:

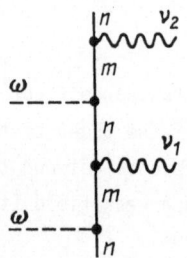

This diagram shows that, according to the energy conservation law, $\nu_1 + \nu_2 = 2\omega$ and that the frequencies of the emitted photons, ν_1 and ν_2, may differ from ω. Inelastic scattering is a result of spontaneous transitions between the quasi-energy levels of a two-level system.

Let us now turn to a mathematical description of inelastic scattering. It follows from what has been said that the spectral distribution of the inelastic scattering should exhibit maxima at frequencies that correspond to the differences in the energy of the quasi-energy levels. In accordance with the pattern of these levels discussed in Chap.3, there are three such maxima: a central maximum corresponding to a scattering at the frequency ω, and two satellites that are shifted with respect to the central maximum by a quantity that is proportional to twice the Rabi frequency. The widths of these maxima are of the order of γ_m. In a weak field, where the Rabi frequency is small compared to the spontaneous width γ_m, all three maxima merge, and the solution coincides with that given by perturbation theory (Sect.8.2.1).

The aim of this section is to quantitatively describe inelastic scattering. According to the general principles of radiation theory, the probability

$dw(\nu)$ is proportional to $<|\hat{d}_\nu|^2>$ in the dipole approximation, where d_ν is the Fourier component of the operator for the dipole moment, $d(t)$, i.e.,

$$\hat{d}_\nu = \int_0^T \hat{d}(t)\, e^{-i\nu t}\, dt \quad . \tag{8.47}$$

The brackets $<...>$ denote a quantum mechanical averaging over the initial state of the system containing one electron, and T is the time during which the field is switched on.

The probability of elastic scattering is determined by the square of the average value of the dipole moment, i.e.,

$$dw_{el}(\nu) \propto |<\hat{d}_\nu>|^2 \quad . \tag{8.48}$$

Consequently, the probability of inelastic scattering is determined by the difference

$$<|\Delta \hat{d}_\nu|^2> = <|\hat{d}_\nu|^2> - |<\hat{d}_\nu>|^2 \quad , \tag{8.49}$$

i.e., by the fluctuation of the dipole moment. While elastic scattering can be described in the framework of classical radiation theory, inelastic scattering is essentially a quantum mechanical effect. For this reason it is necessary to introduce quantum operators for the creation and annihilation of particles: a_i^+, a_j. It is also convenient to express these operators in terms of the Heisenberg representation, i.e., with the variable time [8.12].

To calculate d, the operator for the atomic density matrix must be introduced:

$$\hat{\rho} = \sum_{ij} a_i^+ a_j \rho_{ij} \quad . \tag{8.50}$$

Generally, the operator for the total density matrix of the system must also depend on the operators for the photons of the electromagnetic field. However, if it is assumed that the changes that occur in the atom have little effect on the vacuum states, then the operator for the total density matrix can be expressed as a product of two operators; one that depends on the operators for the atom, and the other that depends on the operators for the photons. This is what is meant by the "Markov", or factorization, approximation. However, it cannot be applied when $\gamma_m \sim \omega_{mn}$. In such a case the time required for the emission of a photon, $1/\gamma_m$, is comparable to the time it takes the system to change from one state to the other, $1/\omega_{mn}$. This follows from the time energy uncertainty principle.

If the operator for the total density matrix is averaged over the states of the photons according to the Markov approximation, one obtains the operator for the atomic density matrix. This operator enables one to predict changes in atomic states without having to analyze the state of the electromagnetic field.

Let us examine what has been said in about Feynman diagrams. The exact approach to solution involves finding the amplitude for the emission of a large number of photons from the quasi-energy states. A typical diagram is shown below:

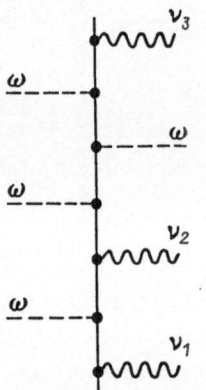

The total amplitude is represented by a sum of such diagrams. A transition to a description of the total amplitude using the atomic density matrix corresponds to an averaging over all of the frequencies the spontaneously emitted photons except the recorded frequency.

The average value I for any operator $\hat{I}$ can be calculated using the following equation:

$$I = <\hat{I}> = Tr\{\hat{I}\hat{\rho}\} \quad . \tag{8.51}$$

The operator for the dipole moment is equal to

$$\hat{d}(t) = \sum_{ij} d_{ij} a_i^+ a_j \quad , \tag{8.52}$$

where the d_{ij} are the matrix elements of the dipole moment. In the expression

$$<|\hat{d}_\nu|^2> = \int_0^T \int_0^T <\hat{d}^*(t)\hat{d}(t')> e^{-i\nu(t-t')} \, dt \, dt' \quad , \tag{8.53}$$

we substitute $\tau = t - t'$, and when $T \to \infty$ [8.13],

$$<|\hat{d}_\nu|^2> = T \int_{-\infty}^{\infty} |d_{mn}|^2 \, \overline{<\hat{\rho}_{nm}(t)\hat{\rho}_{mn}(t + \tau)>} \, e^{-i\nu\tau} d\tau + \text{compl. conj.} \quad , \tag{8.54}$$

where the bar over the expression in brackets denotes an averaging over time t, and $\hat{\rho}_{nm} \equiv a_m^+ a_n$. Equation (8.54) only has meaning for the stationary state when t is large compared to $1/\gamma_m$.

A calculation of d_ν and $\hat{\rho}$ for simultaneous operators is relatively simple. For example,

$$<\hat{\rho}_{mn}> = \sum_{ij} \text{Tr}\{a_n^+ a_m a_i^+ a_j\} \rho_{ij} \tag{8.55}$$

$$= \sum_{ij} \delta_{mi} \delta_{nj} \rho_{ij}$$

$$= \rho_{mn} \quad .$$

In the above derivation the commutative properties of the operators a_m and a_i^+ have been used. Also, the fact that the state over which the averaging was performed only contains one particle nullifies the action of the product of two operators, $a_m a_j$.

Equation (8.54) contains an average of the product of two operators at different instants of time. To evaluate such an average it is necessary to transform the operators to the same time since only then do the Heisenberg operators exhibit the usual commutative properties. To this end one can formally expand $\hat{\rho}_{mn}(t+\tau)$ in terms of a complete set of operators $\hat{\rho}_{ij}$:

$$\hat{\rho}_{mn}(t + \tau) = \sum_{ij} \rho_{mn}^{ij}(t,\tau) \hat{\rho}_{ij}(t) \quad . \tag{8.56}$$

In view of the linear dependence of the operators $\hat{\rho}_{mn}(t+\tau)$ on the new functions $\rho_{mn}^{ij}(t,\tau)$ (considered to be functions of τ), the equations for these new functions formally coincide with those for the density matrix ρ (8.35). Indeed, (8.56) is valid not only for the operators but also for their averages, i.e., for the elements of the density matrix ρ. The only change lies in the new initial conditions. They can be derived from (8.56) if τ is set equal to zero. Hence, $\rho_{k\ell}^{ij}(t,0) = \delta_{ki} \delta_{\ell j}$.

Substituting (8.56) into (8.54) yields

$$|<\hat{d}_\nu|^2> = T|d_{mn}|^2 \sum_{ijk\ell} \text{Tr}\{a_m^+ a_n a_j^+ a_i a_k^+ a_\ell\}$$

$$\times \int_0^\infty \overline{\rho_{k\ell}(t)\rho_{mn}(t + \tau)} \, e^{-i\nu\tau} \, d\tau + \text{compl. conj.} \tag{8.57}$$

In this equation the lower limit of integration is equal to zero, and not $-\infty$, as in (8.54). This is due to the fact that the operator $a(t+\tau)$ can only annihilate a particle if it exists at time t, i.e., for $\tau \geq 0$.

Evaluating the trace of the product of six operators, a^+, a, in the same manner as was done for the product of four operators, and going over from $\langle|\hat{d}_\nu|^2\rangle$ to the probability of the emission of one photon per unit time, $dw(\nu)$, yields

$$dw(\nu) = \gamma_m 2\ \text{Re}\ \int\limits_0^\infty [\rho_{mn}(t)\rho_{mn}^{mn}(t,\tau) + \rho_{nm}(t)\rho_{mn}^{nn}(t,\tau)] \times e^{-i\nu\tau}\ d\tau\ \frac{d\nu}{2\pi}\ .$$
(8.58)

Averaging over t is unnecessary in (8.58) since the stationary state does not depend on time. This can easily be verified if one substitutes the stationary solutions (8.38) into (8.58).

If (8.58) is integrated with respect to ν and the initial conditions for the ρ_{mn}^{kn} are taken into account, the total probability of a spontaneous emission (8.39) is obtained. This coincidence is not accidental. It is predicted by the so-called optical theorem [8.4].

Furthermore, if in (8.58) $\tau \to \infty$, then $\rho_{mn}^{mn} \to 0$, while $\rho_{mn}^{nn} \to \rho_{mn}(t+\tau)$. Substituting these values into (8.58) yields the total probability of elastic scattering (8.46).

The solutions to (8.35) are generally cumbersome due to the Cardano formulas for the characteristic values. A similar situation arises for the ρ_{mn}^{kn}, since the equations for these coincide with the ones given in (8.35). In various limiting cases this problem becomes more manageable (only the case in which the detuning from resonant is equal to zero will be considered). In this case the roots of the characteristic cubic equation of (8.35) have a simple form:

$$s_0 = -\gamma_m/2\ ,\quad s_\pm = -\frac{3}{2}\gamma_m \pm [(\gamma_m/4)^2 - |r_{mn}\cdot E|^2]^{\frac{1}{2}}\ ,$$
(8.59)

Any solution to (8.35) can be represented as follows:

$$\rho = a + be^{s_0 t} + ce^{s_+ t} + de^{s_- t}\ ,$$
(8.60)

where the coefficients a, b, c, d are determined by the initial conditions.

For a weak field, in particular for a field in which $|r_{mn}\cdot E|$ is less than $\gamma_m/4$, all of the roots are real. In addition to coherent (elastic) scattering (determined by the constant a), inelastically scattered light of frequency ν (near ω) which is a superposition of three resonances with the same peak frequency ω position but of different widths, is observed. All of these widths are of the order of γ_m. In this case the inelastic scattering is small compared to the elastic scattering.

If the field is strong, such that $|r_{mn}\cdot E| > \gamma_m/4$, two symmetrically situated satellites with widths equal to $3\gamma_m/2$ appear at a distance

$[(r_{mn} \cdot E)^2 - (\gamma_m/4)^2]^{\frac{1}{2}}$ from a central peak at a frequency ω and with a width γ_m. For a very strong field the satellites are shifted from the central peak by a quantity equal to twice the Rabi frequency $|r_{mn} \cdot E|$.

Calculations using (8.58) lead to the following result (as noted above, the elastic scattering is negligible in a very strong field):

$$dw(\nu) = dw_{inel}(\nu)$$

$$= \frac{1}{2\pi} \left(\frac{1/3}{[4(\nu - \omega + |r_{mn} \cdot E|)/3\gamma_m]^2 + 1} + \frac{1}{[2(\nu - \omega)/\gamma_m]^2 + 1} \right.$$

$$\left. + \frac{1/3}{[4(\nu - \omega - |r_{mn} \cdot E|)/3\gamma_m]^2 + 1} \right) d\nu \quad . \qquad (8.61)$$

Integrating this expression with respect to ν yields $w = \gamma_m/2$, which corresponds to a spontaneous emission from the level m (the probability of finding the electron in level m is equal to 1/2).

Equation (8.61) could also have been obtained using a much simpler method without resorting to (8.58) and the function $\rho_{k\ell}^{ij}$, which depend on two different times. It is clear that transitions between quasi-energy levels can occur as a result of the spontaneous emission of four photons: one of frequency $\omega - |r_{mn} \cdot E|$, one of frequency $\omega + |r_{mn} \cdot E|$, and two of frequency ω. In very strong fields and when $\Delta = 0$, the probability of finding the electron in any of the quasi-levels is the same, so that the probability per unit time for the emission of each of these photons is equal to $\gamma_m/8$. Taking into account the spectral widths of the satellites, $3\gamma_m/2$, and of the central peak, γ_m, (8.61) is obtained directly. Naturally, such a simple approach cannot be used for a finite detuning from resonance or for weaker fields.

8.2.6 Comparison with Experiment

Resonance fluorescence has been observed in strong fields by exciting a transition between the hyperfine levels of a sodium atom: $3S_{1/2}(F' = 2)$ $\rightarrow 3P_{3/2}(F' = 3)$ [8.7,14]. The fluorescence spectrum observed in [8.7] is shown in Fig.8.3 for different values of the power of the exciting radiation. It can be clearly seen that as the intensity of the resonance field increases, satellites appear in addition to the inelastic scattering. This happens for field strengths that satisfy the condition $|r_{mn} \cdot E| > \gamma_m/4$. In a very strong field the height of the central peak is approximately three times greater than the amplitude of the satellites, and the width of the satellites is one and a half times greater that that of the central peak. The frequency of the satellites differs from the inelastic frequency

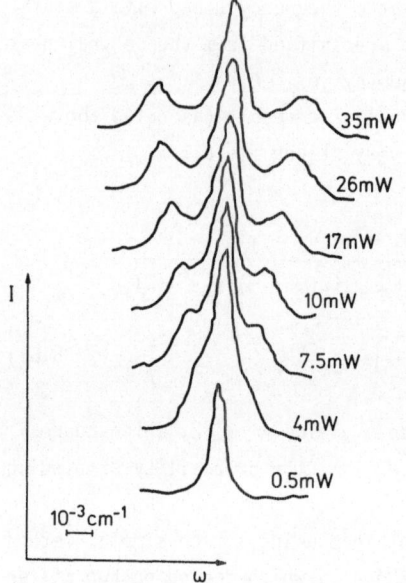

35mW

26mW

17mW

10mW

7.5mW

4mW

0.5mW

10^{-3}cm^{-1}

I

ω

Fig.8.3. The resonance fluorescence
spectrum of a sodium atom as a function
of the power of the exciting field

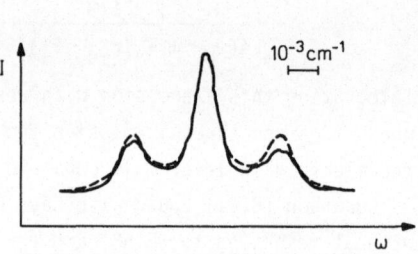

I

10^{-3}cm^{-1}

ω

Fig.8.4. Resonance fluorescence
spectrum of a sodium atom in a
strong field

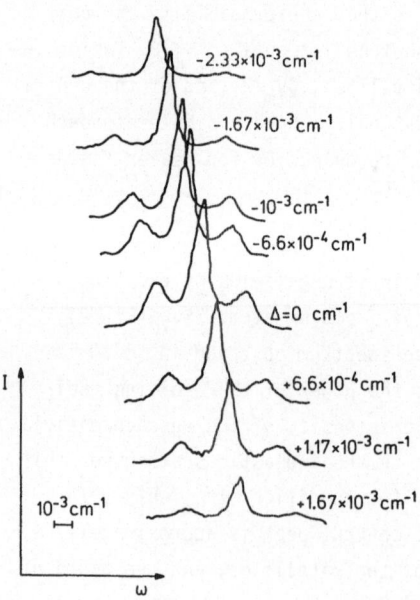

-2.33×10^{-3}cm^{-1}

-1.67×10^{-3}cm^{-1}

-10^{-3}cm^{-1}

-6.6×10^{-4}cm^{-1}

$\Delta = 0$ cm^{-1}

$+6.6 \times 10^{-4}$cm^{-1}

$+1.17 \times 10^{-3}$cm^{-1}

$+1.67 \times 10^{-3}$cm^{-1}

I

10^{-3}cm^{-1}

ω

Fig.8.5. The resonance fluorescence spectrum of a sodium atom as
a function of the detuning from
the resonance between the strong
external field and the transition

by a quantity that is twice the Rabi resonance frequency. Hence, both qualitatively and quantitatively (8.61) is in good agreement with the experimental data. Figure 8.4 illustrates such a comparison on the basis of calculations carried out in [8.15] (dashed line) and the experimental data given in [8.7] (solid line).

The dependence of the fluorescence spectrum on the detuning from resonance is illustrated in Fig.8.5. As expected, the satellites cease to be observed when the detuning is of the order of the width of the central peak. The results obtained in [8.14] are of a similar nature.

In conclusion, one can say that the experimental data on resonance fluorescence in strong fields can be adequately described by theory both qualitatively and quantitatively.

8.2.7 The Bloch-Siegert Shift

The above theory concerning the behavior of a two-level system in a monochromatic field was based on the resonance approximation, see (8.36). One of the results was that the central line in the fluorescence spectrum is not shifted, i.e., $\nu = \omega$. This assertion is, in fact, only approximate. If the terms that did not enter into the resonance approximation are taken into account, it can be seen that the central line is indeed shifted. This is known as the Bloch-Siegert shift [8.16]. In this subsection this shift will be calculated. To do this one can use the diagrammatic technique (Sect. 2.3). It is clear from Sect.3.1.1 that in deriving (3.2) only diagrams of the type

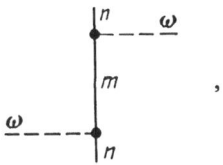

,

and other diagrams obtained by adding dashed lines from above were used. Diagrams of the type

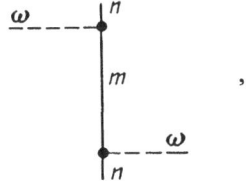

,

were neglected. This was justified by the fact that the denominator of the propagator in this diagram, equal to $\omega_{mn} + \omega$, is not a resonance denominator and in the neighborhood of the resonance, $\omega_{mn} \approx \omega$, is simply equal to $2\omega_{mn}$. For this reason the contribution of this diagram is small. It gives the diagonal matrix element of the state n, i.e., the shift in energy of the state n. Applying the rules of the diagrammatic technique, one finds that

$$\delta E_n = - \frac{|r_{mn} \cdot E|^2}{8\omega_{mn}} \, . \tag{8.62}$$

The state m undergoes a similar shift, only in the opposite direction. The combined effect is a "repulsion" of the two states by a quantity that is equal to

$$\frac{|r_{mn} \cdot E|^2}{4\omega_{mn}} \, .$$

The time-dependent diagrams,

which were not taken into account, contribute nothing to the shift δE_n after being averaged over time. For example, the left diagram exhibits an exponential dependence as $\exp(2i\omega t)$.

Since a real atomic system has levels i other than m and n, the shift in energy is also determined by the positions of the levels i. In view of the nonresonant nature of the energy denominators in this case, one can use lowest-order perturbation-theory diagrams for the diagonal matrix element of the perturbation of the state n:

If the perturbation is a field whose frequency lies in the visible spectral range, these diagrams reflect a shift in the level n equal to the Bloch-Siegert shift (δE_n). Together the two diagrams determine the shift in the level n. The situation is different when the field is in the radiofrequency

range. The distance between the resonating levels n and m is of the order of
the radiofrequency, whereas the distance to the other levels is of the order
of optical frequencies. The Bloch-Siegert shift is large compared to the nonresonant shifts due to other atomic levels.

The Bloch-Siegert shift has been measured over a wide range of field
strengths in a number of experiments [8.17]. The results are in good agreement with the above estimates except for strong fields.

8.2.8 The Test Field

There is another way of observing the satellites, i.e., to apply a weak test
field to the system. The test field should not produce shifts in or cause a
splitting of the atomic levels, or saturate transitions. It should only in-
duce transitions. Such induced transitions between the quasi-levels of the
system may proceed, in contrast to spontaneous transitions, not only from
higher to lower levels but also in the opposite direction, or to an external
level. A theoretical analysis of this situation is simple. In (8.35) there
are terms that represent the test field. A solution is determined following
the lines of ordinary perturbation theory, i.e., the solution that is obtained without the test field is used as the zeroth approximation. The details of such calculations can be found in [8.18].

Progress in experimental techniques has made it possible, in a number of
cases, to obtain sufficiently accurate quantitative results that describe
the resonance splitting in the optical frequency range [8.19-22]. For exam-

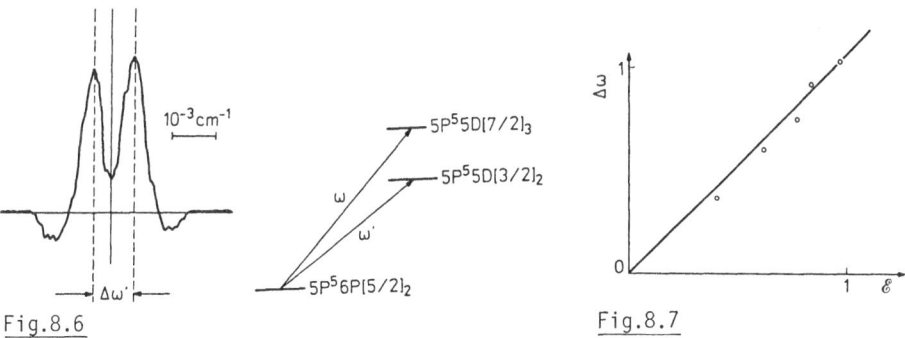

Fig.8.6   Fig.8.7

Fig.8.6. The absorption spectrum of the test radiation; the maxima correspond
to the quasi-energy states of the $5P^56P[5/2]_2$ level of the xenon atom

Fig.8.7. The splitting $\Delta\omega'$ of the quasi-energy states of the $5P^56P[5/2]_2$ level of the xenon atom as a function of the strength of the strong field E

ple, the strong 3.51 μm (infrared) field of a laser was used to perturb the $5P^56P[5/2]_2 \rightarrow 5P^55D[7/2]_3$ transition of the xenon atom [8.19]. A test field of 4.54 μm wavelength was used to measure the population of the $5P^55D[3/2]_2$ state as a function of the detuning between the test field and the initial state $5P^56P[5/2]_2$. Figure 8.6 shows two maxima in the absorption spectrum of the test radiation that correspond to the quasi-energy states of the $5P^56P[5/2]_2$ level. The distance between the maxima is a linear function of the intensity of the strong field. From the proportionality factor (Fig.8.7) one can determine the dipole matrix element of the above resonance transition.

8.2.9 Multi-Photon Resonance

The resonance between a given pair of levels and a strong field may be due to not only one but several photons. All of the above results remain valid if one replaces the one-photon matrix element $\mathbf{r}_{mn}$ by the multi-photon matrix element $\mathbf{r}_{mn}^{(K)}$ (Sect.3.1). At first the multi-photon problem appears to be harder to solve than the one-photon problem because in a multi-photon transition some of the photons from the external field can be replaced by photons from the vacuum field.

In reality one can ignore the possibility of such a substitution. When $|\mathbf{r}_{mn} \cdot \mathbf{E}|^K \gtrsim \gamma_m$, i.e., when the perturbation is strong, $|\mathbf{r}_{mn} \cdot \mathbf{E}| \gg \gamma_m$. In other words, if one substitutes the vacuum field for the external field in the multi-photon matrix element for some intermediate transition, only small correction terms are obtained. Neither do problems arise in the case of a weak external field, when $|\mathbf{r}_{mn} \cdot \mathbf{E}|^K \ll \gamma_m$, since the vacuum field does not contribute to the transitions to higher levels (due to the absence of photons from the vacuum field and also since the transitions to lower levels are accompanied by the spontaneous emission of a single photon with a frequency equal to that of the atomic transition). In this case the equations of Sect.8.2 are modified by simply changing $\mathbf{r}_{mn}$ to $\mathbf{r}_{mn}^{(K)}$ and $E_m - E_n^{(0)} - \omega$ to $E_m - E_n^{(0)} - K\omega$.

New effects are discovered when one wants to determine the spectral widths. Although the new multi-photon matrix elements, corrected with due regard to the spontaneous field, are small compared to $\mathbf{r}_{mn}^{(K)}$, they determine the width, not the shift, of a state (in contrast to $\mathbf{r}_{mn}^{(K)}$). For this reason they broaden the spectral lines of the emitted photons (without changing their position) when the levels m and n are not connected by a dipole transition. Hence, a direct, spontaneous dipole broadening (which is large compared to the widths) is absent.

Multi-photon resonance fluorescence has not been observed experimentally; which is not surprising since it is a rare process. In a weak external field the probability of a cascade decay of an excited state through intermediate bound states is always prevalent. When the external field is strong, other competing processes begin to play a role (Sect.8.3). For this reason, multi-photon resonance fluorescence can only be expected to be observed for a multi-photon transition to the lowest excited state.

8.3 Multi-Photon Excitation and Emission

Among the various phenomena that take place in weak fields the photoexcitation of an atom is defined as an induced transition, allowed by the selection rules for electric dipole transitions, from a lower state n to an upper state m. The atom stays in the state m for the duration of its natural lifetime.

In a strong field the situation is much more complex. In this case there are three channels open for the electron to leave the excited state, i.e., induced bound-bound transitions to both higher- and lower-lying states and induced bound-free transitions to the continuum (ionization). Thus, it is difficult to clearly say what processes excite the atom. Only those processes will be considered in which the fate of the electron after it has transferred to the long-lived excited state m is of no interest. The special features of the interaction of the strong field and the atom in this excitation process are the multi-photon nature of the transition between n and m and the resonance mixing of these states.

Therefore, this section is a logical continuation of the previous section, in which the one-photon resonance process in a two-level system (n,m) was studied.

When studying multi-photon, bound-bound transitions in real atoms one must bear in mind the possibility of a resonance with an intermediate state. The probability of a multi-photon dipole transition may prove to be of the same order of magnitude as the probability of a cascade transition through an intermediate quadrupole-bound state. Thus, one cannot only use the dipole approximation in the general case.

Multi-photon excitation is extremely important in multi-photon spectroscopy, optical pumping, and a wide range of processes involving the selective action of resonance radiation on atoms.

One would think that multi-photon emission, i.e., a multi-photon bound-bound transition, from a higher to a lower state, is the reverse of multi-photon absorption. Actually there is a big difference because some of the virtual transitions from higher to lower states may be spontaneous as well as induced transitions.

8.3.1 Selection Rules for Multi-Photon Transitions

For transitions to the continuum, the final state is infinitely-fold degenerate with respect to the angular momentum, and, hence, there are no selection rules. However, for bound-bound transitions the final and initial states must satisfy certain selection rules.

Let us recall the selection rules considered in Sect.1.2.3. In the dipole approximation, the selection rules for multi-photon transitions are based on the selection rules for the absorption of a single photon of the dipole radiation. Hence, for a transition involving K photons,

$$\Delta J = J_m - J_n = -K, -K + 1, \ldots, K - 1, K \ . \tag{8.63}$$

The selection rules with respect to the magnetic quantum number M depend on the polarization of the applied perturbation. For a linearly polarized perturbation, $M_m - M_n = 0$; when it is circularly polarized, $M_m - M_n = K$ or $-K$ depending on the helicity. If the field is elliptically polarized,

$$\Delta M = M_m - M_n = -K, -K + 1, \ldots, K - 1, K \ . \tag{8.64}$$

It is often the case that the initial state is an S state and the perturbing field is circularly polarized. Then, $M_m = K$ and $J_m = K$, since the angular momentum cannot be smaller than its projection.

The degeneracy with respect to the magnetic quantum number can be ignored when the field is linearly or circulary polarized, since the states of the degenerate level do not mix. If the field is elliptically polarized, each of the final states can be viewed as a linear combination of states with different magnetic quantum numbers. These combinations are determined in the process of solving (2.28). There is no need to solve the secular equation for the intermediate states, which are summed over in the compound matrix element. Because the summation is made over all of the states of the intermediate degenerate level, the choice of the basis states does not influence the final result of the summation. Another case where degeneracy is present is the hydrogen atom, in which the states with different angular momenta mix [8.23]. Such mixing is possible in any type of polarized field. The solution to (2.32) yields new basis states that are no longer

264

characterized by their orbital quantum numbers (it is no longer a "good" quantum number). Therefore, for the hydrogen atom there is no sense in seeking the probability of transitions to excited states with definite values of angular momentum. The atomic multiplets of more complex atoms can be similarly analyzed. A mixing of the components of a multiplet is only possible in sufficiently strong fields. According to Sect.2.5.3, the criterion for such fields is that $r_{nn'}^{(2)}E^2 \gtrsim \omega_{nn'}$, where $\omega_{nn'}$ is the distance between adjacent components in the multiplet.

8.3.2 The Probability of Multi-Photon Bound-Bound Transitions

In this section the problem of determining the probability of a transition between the discrete atomic states n and m will be considered. The diagram that corresponds to the respective matrix element in the lowest-order perturbation theory has the appearance of a "lopsided tree" (Sect.2.4.1):

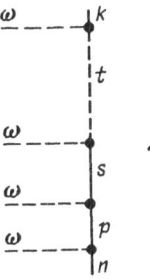

According to (2.22) the probability (per unit time) of a K-photon transition under the influence of a field $E \cos\omega t$ is equal to

$$dw_{nm}^{(K)} = 2\pi |V_{mn}^{(K)}|^2 \delta(\omega_{mn} - K\omega) \quad , \tag{8.65}$$

where the compound matrix element, $V_{mn}^{(K)}$, has the following form:

$$V_{mn}^{(K)} = E^K \sum_{p,s,\dots,t} \frac{1}{2^K} \frac{(\mathbf{e}\cdot\mathbf{r})_{mt} \dots (\mathbf{e}\cdot\mathbf{r})_{sp}(\mathbf{e}\cdot\mathbf{r})_{pn}}{[\omega_{tn} - (K-1)\omega]\dots [\omega_{sn} - 2\omega][\omega_{pn} - \omega]} \quad . \tag{8.66}$$

The magnitude of the compound matrix element of multi-photon bound-bound transitions strongly depends on the frequency of the external field and increases as a resonance with some intermediate bound state is approached. Equation (8.66) contains an infinite sum over the intermediate states t,s,p, ... because of the existence of a continuum for a real atomic system. The problem of choosing the correct electronic wave functions for complex atoms and then summing them up does not in any way differ from the similar problem for bound-free transitions (Sect.8.5).

From (8.66) one can see that the expression for the transition probability, (8.65), is only applicable when none of the energy denominators are small, i.e., when $|\omega_{sn} - K'\omega| \gg \gamma_s$, where γ_s is the spontaneous width of the level s, and K' is the number of photons involved in the transition from the state n to the state s.

The delta-function in (8.65) is, of course, physically unrealistic. In a real situation there is a transition from the state m to some other state (Sect.8.3). The delta-function must be replaced by the shape of the corresponding line. Such reasoning is valid for cascade processes.

On the other hand, a process can only be called a cascade process if the excited state m has a sufficiently long lifetime (Sect.8.1). According to the Breit-Wigner procedure one can substitute $E_s^{(0)} - i\gamma_s/2$ for $E_s^{(0)}$ in (8.66) when the energy denominators are small, i.e., when $|\omega_{sn} - K'\omega| \lesssim \gamma_s$.

The multi-photon matrix elements of bound-free transitions have been measured for many values of K, for many different atoms, and for many frequencies (Sect.7.5). A similar detailed experimental study for bound-bound transitions has not yet been undertaken. However, in a large number of experiments involving multi-photon excitation, the processes could be described with sufficient accuracy by the above equations even when approximate electronic wave functions were used. In the majority of the cases the measurements and calculations involved transitions from the ground state to the lowest-lying excited states of alkali atom. There were either no intermediate states, or very few, so that only a few matrix elements had to be taken into account. This simplified the interpretation of the results of experiments and increased the precision of the calculations. Finally, the transitions in these experiments were usually of a two-photon nature. There is practically no sufficiently reliable data on multi-photon excitation when a large number of intermediate levels are involved.

The results of an experiment in which the obvious and practically important role of an intermediate, long-lived quasi-resonant state, i.e., a state with an extremely small detuning from resonance, was illustrated, are shown in Fig.8.8 [8.24]. Observations were made of the dependence of the probability of the two-photon excitation of the sodium atom from the 3S state to the 4D state on the detuning from resonance with the 3P state. A decrease in the denominator of the expression for the two-photon matrix element can increase the probability of a two-photon excitation process by many orders of magnitude. But due to the anharmonicity of the atomic levels, it is necessary to use two lasers with different operating frequencies for the resonance excitation of the 4D state. A satisfactory description of the behavior

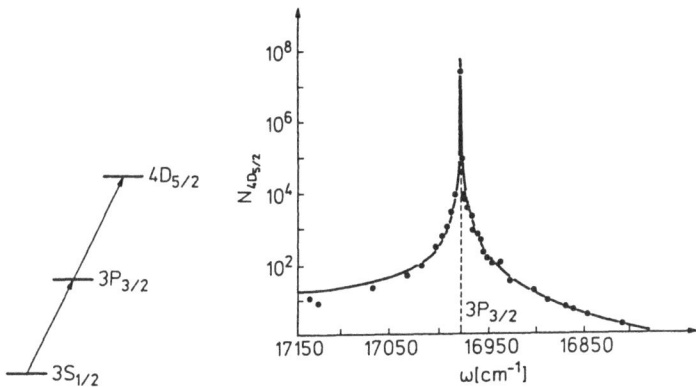

Fig.8.8. The population N of the $4D_{5/2}$ level of a sodium atom as a function of the frequency ω

of the resonance curve is achieved by substituting $E_{3P}^{(0)} - i\gamma_{3P}/2$ for $E_{3P}^{(0)}$ in the denominator of the two-photon matrix element:

$$V_{3S,4D}^{(2)} = EE' \frac{<3S|z|3P><3P|z|4D>}{\omega_{3P,3S} - i\gamma_{3P}/2 - \omega} \quad . \tag{8.67}$$

8.3.3 Quadrupole Transitions

Until now it has been assumed that the interaction between the electromagnetic field and the electron is of a dipolar nature. Generally the electric quadrupole matrix elements are negligible compared to the dipole matrix elements. But, if any of the energy denominators in the multi-photon matrix element $V_{mn}^{(K)}$ of a quadrupole transition proves to be small, such a quadrupole transition may contribute more to the probability of multi-photon excitation than a dipole transition that takes place with a greater detuning from resonance. The relatively small numerator (matrix element) is compensated by the smaller denominator (the small detuning from resonance).

The role of quadrupole transitions was demonstrated in an experiment in which the 3P → 4F transition in the sodium atom was induced by a dye laser [8.25]. Two intersecting beams, an atomic beam and a beam of light, were used. One laser was used to populate the 3P state. The intensity of its radiation was weak enough for there to be no resonance mixing of the 3S and 3P states. Hence, the 3P state had a spontaneous width. These conditions were necessary for simple estimates of the transition probabilities to be made from the observed pulse heights. The operating frequency of the second laser could be varied so that a resonance between the 3P and 4D states

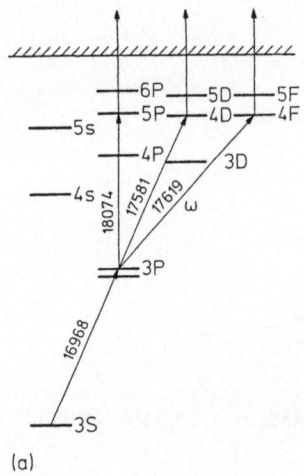

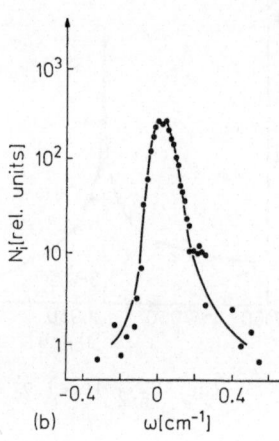

Fig.8.9a,b. The quadrupole transition 3P → 4F of a sodium atom: (a) the level diagram, and (b) the ion yield as a function of the frequency ω of the second laser

(a)　　　　　　　　　　　　　(b)

($\Delta \ell = 1$; dipole transition) or between the 4F and 3P states ($\Delta \ell = 2$; quadrupole transition) could take place (Fig.8.9a). The line width of the second laser was about $10^{-2} cm^{-1}$, so that the fine structure of the doublets could be clearly resolved ($\Delta E \sim 1 cm^{-1}$). The occurrence of resonances at the operating frequency of the second laser was registered using the resonance multi-photon ionization method (Sect.6.4). The number of ions produced during the three-photon ionization of sodium was measured.

The ion signal is a function of the operating frequency of the second laser (Fig.8.9b). The measurement of the ratio between the ion yields for the dipole (3P-4D)- and quadrupole (3P-4F)-resonance made it possible to obtain reliable data on quadrupole matrix elements, with the help, of course, of the well-known data on dipole matrix elements. The experimental value of the matrix element $<3P|r^2|4F>$ was found to be approximately $10^{-15} cm^2$. This is in good agreement with the results presented in [8.26]. This is also true of the ratio between the matrix elements for dipole and quadrupole transitions (which was found to be proportional to n^{-2}, where n is the principal quantum number).

Thus, the above experiment made it possible to measure the magnitudes of the quadrupole matrix elements that enter into the expression for the compound two-photon matrix element.

Finally, for the hydrogen atom and in other cases of degeneracy, dipole transitions can take place at the frequency of the external field in addition to the quadrupole transitions. In this case the contribution of the quadrupole transitions can be neglected.

8.3.4 Multi-Photon Mixing of Resonance States

The various processes that compete with multi-photon excitation were mentioned at the beginning of Sect.8.3. Next, the phenomenon of multi-photon
resonance mixing that takes place in a two-level system (n and m) and leads
to a decrease in the population rate of state m shall be discussed.
(Sometimes the rather unfortunate term "saturation" is used in the literature. The basis for this is the experimentally observed decrease in the
growth rate of the population of state m. However, the term is ambiguous
since the same effect can be observed when $\int_0^T w \, dt \sim 1$, where w is the transition probability per unit time, and T is the observation time.) This mixing phenomenon was already studied in Chap.3. It was discovered that an increase in the strength of the external field and a decrease in the detuning
from resonance (the more so, with the simultaneous variation of both parameters) lead to a mixing of the resonance states. It was also noted that in the
event of multi-photon resonance the mixing rate is determined by the multi-
photon Rabi frequency, which in exact resonance is equal to the multi-photon matrix element, $V_{mn}^{(K)}$, for the transition but is small compared to the
one-photon Rabi frequency. This means that the experimenter must use a fairly strong field to achieve mixing. In reality, however, one must bear in
mind that induced transitions from the state m to other bound or free states
(other than n) and virtual transitions, which determine the nonresonant perturbation of the states n and m, are also possible. A nonresonant perturbation detunes the resonance and, in this way, prevents resonance mixing from
taking place. The relative importance of these competing effects is largely
determined by the degree of the nonlinearity of the transitions $n \rightarrow m$ and
$m \rightarrow E$ (note that the label E defines the energy of the electron in the continuum, so that the latter transition represents an ionization process). For
transitions between bound states or to the continuum, the kind of dependence
of the corresponding matrix elements on the field strength is the decisive
factor. A process of higher order can always be neglected (note that the
probability of a K-photon resonance mixing is proportional to E^K, whereas
the probability of a K-photon ionization is proportional to E^{2K}.

If a mixing of resonance states occurs in multi-photon excitation, two
effects may be observed. First of all, one may observe a deviation from the
power law for the emission cross section of the spontaneous decay of the
state m. In addition, the spectral distribution of the emitted photons may
change due to a splitting of the state m into two quasi-levels (Sect.3.1).

8.3.5 Competing Processes in the Multi-Photon Excitation of Atoms

It is obvious that all of the competing processes that have just been discussed manifest themselves in the same way, i.e., as the intensity of the
exciting radiation increases, the probability of excitation increases more
slowly than it would according to the respective power law. Competing effects have been observed in the two-photon excitation of cesium and thallium [8.27]. With cesium, a two photon-transition to the 9D state was achieved
using a tunable ruby laser. The detector of the fluorescence produced in
the decay process 9D → 6P was placed at right angles to the laser beam.

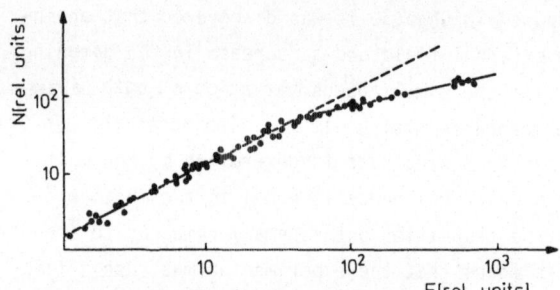

Fig.8.10. Fluorescence
yield N as a function
of the intensity of
the exciting radia-
tion

Figure 8.10 shows the dependence of the fluorescence yield on the intensity of the laser radiation. It can be seen that as the intensity of the laser
radiation increases, the two-photon excitation probability starts to increase
more slowly than it would if the $N \propto F^2$ law were valid (the power law is realized for weak fields). Deviations from the quadratic law were also observed in the excitation of thallium. Regardless of the nature of the competing effects, it is important that such effects are observed and that a
threshold field, $E_{thr} \sim 10^4 V/cm$, exists.

8.3.6 Stimulated Multi-Photon Emission

The probability of spontaneous multi-photon emission is fairly low. A well-
known example of such a process is the two-photon decay of the 2S state of
hydrogen. The lifetime of this decay process is of the order of a tenth of
a second. In the spontaneous two-photon decay process photons of various
energies are emitted [8.28]. The sum of these energies is equal to the energy difference between the two states. The probability of such a decay is
determined by integrating over all possible pairs of photons.

In a strong external field whose frequency ω is less than ω_{mn} (here m and
n are the upper and lower states, respectively), the atom emits a stimulated

photon of frequency ω before spontaneously emitting a photon of frequency $v_k = \omega_{mn} - \omega$ (the sequence may be reversed). If the forcing field is not polarized, the angular distribution of the spontaneously emitted photons has the following form: $1 + \cos^2\theta$, where θ is the angle between the wave vector $\mathbf{k}$ of a spontaneously emitted photon and the wave vector $\mathbf{k}_\alpha$ of a stimulated photon from the external field. However, this process must compete successfully with spontaneous one-photon emission or spontaneous cascade emission for stimulated multi-photon emission to occur. As a result stimulated emission has thus far only been observed under rather unusual conditions, as in the decay of the metastable 2S state of deuterium in the field of a neodymium glass laser [8.29]. When the excitation energy of this state is around 10 eV and the energy of the photons of the external field approximately 1 eV, photons with an energy of about 9 eV were observed. Hence, a two-photon decay process involving the metastable state must have taken place, in which the emission of a photon with an energy of about 1 eV was of a stimulated nature. The experimentally measured cross section of this process is in good agreement with the results of calculations.

8.4 Spontaneous Raman Scattering

Sections 8.2,3 were devoted to processes that take place in a two-level system in an external resonance field. Now processes will be considered that are, in fact, beyond the scope of two-level systems, i.e., processes in which the spontaneous transitions do not take place within the two-level system but in which the electron is transferred to other atomic states. Hence, at least three levels must be taken into account.

The external field is assumed to be in resonance with any two levels of the three-level system. If the third level lies below the chosen pair (Fig.8.11a), a spontaneous transition can take place from the upper level of the two-level resonating system to this level. As a result a photon with a frequency higher

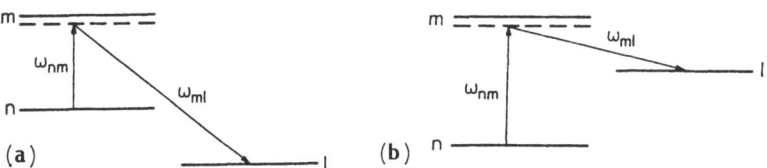

Fig.8.11a,b. The possible schemes of Raman scattering: (a) anti-Stokes scattering $(\omega_{m\ell} > \omega_{nm})$, and (b) Stokes scattering $(\omega_{m\ell} < \omega_{nm})$

than that of the incident radiation (anti-Stokes scattering) is observed.
If, however, the third level lies between the chosen pair, the frequency of
the emitted photon is obviously less than that of the incident light (Stokes
scattering). This case is illustrated in Fig.8.11b. Of course, when the incident light is scattered by atoms in their ground state, the result can only
be a Stokes photon.

The above process is known as Raman scattering. It is illustrated by the
following Feynman diagram (Sect.2.3):

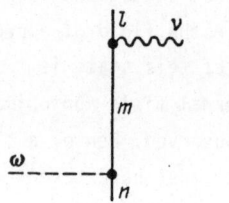

where, according to the law of energy conservation,

$$\nu = \omega - \omega_{\ell n} \, . \tag{8.68}$$

For Raman scattering to occur it is not important that there be an exact
resonance with the external field. In the absence of resonance, i.e., when
the detuning is comparable to the distance between the atomic levels, the
interaction is determined by the compound matrix element, which takes into
account any virtual transitions to intermediate atomic states. Since the
energy denominators are large, the nonresonant matrix element is always relatively small. In the event of resonance, i.e., when the energy denominator
is very small, all nonresonant terms can be neglected in comparison to the
resonant term.

Actually, the probability of Raman scattering was determined in Sect.8.1
both within and beyond the scope of perturbation theory. When perturbation
theory is applicable, the processes of resonance fluorescence and Raman scattering are qualitatively similar. In both cases the first transition is induced and the second is spontaneous. In resonance fluorescence the spontaneous
transition ends in the same state from which the electron was excited, while
in Raman scattering it ends in the third state. It is no wonder then that
the equations of perturbation theory in Sects.8.1,2 are similar. They are determined by the off-diagonal and the diagonal compound two-photon matrix element, respectively.

8.4.1 A Weak Perturbation

Let it be assumed that the amplitude of the external perturbation, $E \cos\omega t$, is so small that $|\mathbf{r}_{mn} \cdot \mathbf{E}| \ll \gamma_m$, where γ_m is the width of the spontaneous transition from the state m to the third state ℓ. If this condition is satisfied, the probability of a spontaneous transition from the state m to the state ℓ overshadows the probability of an induced transition from the state m to the state n, so that a resonance mixing of the states n and m does not occur. Then, as in resonance fluorescence in weak fields, the probability that the electron is transferred in unit time to the state ℓ with a spontaneous emission of a photon of frequency $\nu = \omega - \omega_{\ell n}$ is determined, in the framework of second-order perturbation theory, by the following equation:

$$dw_{n\ell} = \frac{\omega_{\ell m}^3}{8\pi c^3} \frac{|(\mathbf{r}_{mn} \cdot \mathbf{e})(\mathbf{r}_{\ell m} \cdot \mathbf{e}_\alpha'^{*})|^2}{(\omega_{mn} - \omega)^2 + \gamma_m^2/4} E^2 \, do \qquad . \qquad (8.69)$$

The cross section of this process, $d\sigma$, can be found by dividing (8.69) by the average flux density of the incident photons, $cE^2/8\pi\omega$. The function $dw_{n\ell}$, treated as a function of the detuning from resonance, $\Delta = \omega_{mn} - \omega$, (actually as a function of ω), exhibits a resonance peak with a width γ_{mk} equal to the width of the level m. The spectral distribution of the emitted photons, i.e., the dependence of the emission probability on the frequency of the emitted photons ν, is a delta function, just as in resonance fluorescence. However, contrary to resonance fluorescence, the middle of the peak is located at the combination frequency $\nu = \omega - \omega_{\ell n}$. In reality the line will be broadened due to the finite values of the widths γ_n and γ_ℓ.

The total probability per unit time can be found from (8.69) by integrating over the angle and the polarization of the spontaneously emitted photons (Sect.8.2). The concept of a probability per unit time is only valid when $t \ll 1/\gamma$. If $t \gg 1/\gamma$, the results of Sect.8.1.1 are valid.

Let us compare the probability of Raman scattering (8.69) with that of resonance fluorescence (8.9). The similarity of these two expressions results from the fact that the two processes are similar. The difference only lies in the final states. The transitions themselves are similar (either induced or spontaneous). It is quite natural then that the cross sections are of the same order of magnitude.

In addition to the cross section, the polarization of the spontaneously emitted photon is also an observable quantity. Let the incident wave be linearly polarized along the z axis, its direction of propagation be along the x axis, and the scattered photon be observed along the y axis (Fig.8.12).

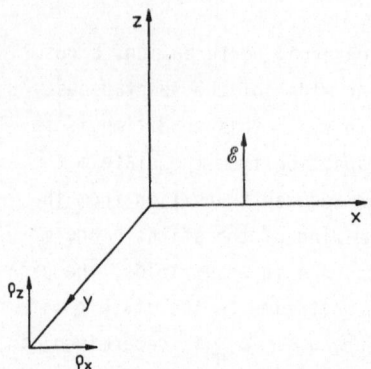

Fig.8.12. The depolarization in Raman scattering; the scattered light can have two polarizations, ρ_x and ρ_z

The scattered photon can be polarized along the x and z axes. The ratio of the respective scattering cross sections, σ_{zx}/σ_{zz}, is called the depolarization coefficient for Raman scattering. In the resonance case the depolarization coefficient is determined by the ratio of the matrix elements $|\mathbf{r}_{m\ell} \cdot \mathbf{e}_\alpha'^*|^2$ that enter (8.69) (for $\alpha = x$ and $\alpha = z$). It is a purely geometrical factor that depends on the orbital and the magnetic quantum numbers of the states. For nonresonant Raman scattering, because the energy differences do not cancel when σ_{zx} is divided by σ_{zz}, the summation is carried out over intermediate states m with different energies. As a result, the depolarization coefficient is a function of the frequency of the incident light. This enables the experimenter to determine the relative phases of the various dipole matrix elements [8.30,31].

If the limits of the validity of (8.69) are to be correctly determined, $dw_{n\ell}$ must be calculated using second-order perturbation theory. It is assumed that after the electron has gone over to the state m and has returned to the state n, it again goes over to the state m, and then spontaneously emits a photon to arrive at the state ℓ. The probability of such a process is $|\mathbf{r}_{mn} \cdot E|^2/[(\omega_{mn} - \omega)^2 + \gamma_m/4]$ times smaller than $dw_{n\ell}$.

Thus, perturbation theory can be used to describe Raman scattering when $|\mathbf{r}_{mn} \cdot E|^2/\gamma_m^2 \ll 1$ (the same criterion can be used for resonance fluorescence).

8.4.2 Raman Scattering Frequencies for a Strong Perturbation

What happens when $|\mathbf{r}_{mn} \cdot E| \gtrsim \gamma_m$, i.e., when perturbation theory cannot be applied? The interaction of a strong resonance field with a two-level system (n,m) gives rise to the appearance of quasi-levels. These are depicted in Fig.8.13. The two spontaneous transitions that end at the level ℓ originate from these quasi-levels. Let us first deal with the problem of finding the

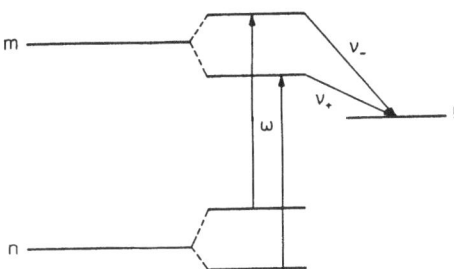

Fig.8.13. Raman scattering in a strong field of frequency ω

m ——————

v_-

v_+

ω

l

n ——————

frequencies of these transitions. In order to do this one must modify the theory in Sect.3.1 by introducing the spontaneous width γ_m of the level m into the equations for the resonance approximation. To this end one must substitute $E_m^{(0)} - i\gamma_m/2$ for $E_m^{(0)}$. Since this was already done in Sect.8.1, only the result of the substitution is noted (assuming for the sake of simplicity that $\gamma_n = \gamma_\ell = 0$). The difference between the two frequencies is

$$v_+ - v_- = \frac{1}{\sqrt{2}} \left\{ |\mathbf{r}_{mn} \cdot \mathbf{E}|^2 + \Delta^2 - \gamma_m^2/4 + \left[\gamma_m^2 \Delta^2 \right. \right.$$
$$\left. \left. + \left(|\mathbf{r}_{mn} \cdot \mathbf{E}|^2 + \Delta^2 - \gamma_m^2/4 \right)^2 \right]^{\frac{1}{2}} \right\}^{\frac{1}{2}} . \tag{8.70}$$

This relationship is of a general nature, since it includes the detuning from resonance, Δ, the spontaneous width of the excited state, γ_m, and the matrix element for the interaction of the resonating states, $|\mathbf{r}_{mn} \cdot \mathbf{E}|$. It follows from Sect.8.1 if one lets $\gamma_n = \gamma_\ell = 0$.

In the limit of a weak perturbation, i.e., when $|\mathbf{r}_{mn} \cdot \mathbf{E}| \ll \gamma_m$, $v_+ \rightarrow \omega_{m\ell}$ and $v_- \rightarrow \omega - \omega_{\ell n}$. As we saw in Sect.8.1, for a weak perturbation the emission of the photons with the combination frequency $\omega - \omega_{\ell n}$ prevails.

For a strong perturbation, i.e., when $|\mathbf{r}_{mn} \cdot \mathbf{E}| \gg \gamma_m$, the difference between the frequencies of the emitted photons (8.70) is equal to twice the Rabi frequency:

$$2\Omega = (\Delta^2 + |\mathbf{r}_{mn} \cdot \mathbf{E}|^2)^{\frac{1}{2}} .$$

8.4.3 The Probability of Raman Scattering Under the Influence of a Strong Perturbation

Let us recall the results for the probability of an electron spontaneously leaving the level m to arrive at the level ℓ (Sect.8.1).

Let us consider the limiting case for a very strong perturbation. From (8.70) one can see that the spectral distribution of the emitted photons exhibits two peaks at $v_\pm = \omega_{m\ell} - \Delta/2 \pm \Omega$ (separated by twice the Rabi frequency).

One can determine the population amplitude of the state ℓ (as was done in Sect.8.1 for the excited state m) by applying first-order perturbation theory in the vacuum field (perturbation V'):

$$a_\ell(\infty) = -i \int_0^\infty V'_{\ell m} a_m(t) \exp(i\omega_{\ell m}t - i\nu t)\, dt \quad , \tag{8.71}$$

where ν is the frequency of the spontaneously emitted photon. Substituting for $a_m(t)$ (Sect.8.1) and evaluating the integral, one finds that $|a_\ell(\infty)|^2$, as a function of ν, exhibits two Lorentzian peaks at ν_+ and ν_-. The widths of these peaks are

$$\Gamma_\pm = \frac{1}{2}\gamma_m \left[1 \mp |\Delta|\, (\Delta^2 + |\mathbf{r}_{mn} \cdot \mathbf{E}|^2)^{-1/2} \right] \quad . \tag{8.72}$$

The heights of these peaks are not the same. Equation (8.71) is obviously not only valid for very strong fields but also for the general case. However, this equation is too cumbersome.

In a very strong field the spectral widths are of the order of the natural width γ_m, although not equal to it or to each other. A similar situation was encountered in the case of resonance fluorescence, i.e., all of the spectral widths are of the order of γ_m. When $\Delta \ll |\mathbf{r}_{mn} \cdot \mathbf{E}|$, both the widths and the heights of the peaks are equal, and

$$\Gamma_+ = \Gamma_- = \frac{1}{2}\gamma_m \quad . \tag{8.73}$$

However, if $\Delta \gg |\mathbf{r}_{mn} \cdot \mathbf{E}|$, then (8.72) yields $\Gamma_+ = \gamma_m$ and $\Gamma_- = 0$. This was to be expected since a large detuning precludes a resonance mixing of the levels n and m. Thus, the situation becomes similar to the one encountered in a weak field, for which perturbation theory is valid.

Comparing (8.69), which describes the scattering process in a weak field, with (8.71-73), which describe the same process in a strong field, shows that the one maximum in the spectral distribution of the scattered light becomes two as the strength of the field increases (the distance between them is of the order of the Rabi frequency). Since the widths of the two peaks are of the same order of magnitude, the scattering probabilities differ very little. Therefore, the two cases can experimentally be distinguished only by the detailed structure of the spectrum of the scattered light, which must be measured with a resolution of the order of the natural widths of atomic levels. Unfortunately, all of the experimental data on Raman scattering has been obtained for non-monochromatic excited light under conditions in which the Doppler effect played a significant role and the scattering was usually stimulated. Under such conditions the only quantity that could be compared with the results of calculations was the scattering probability integrated over the frequencies of emitted photons. In all of the cases in which this quantity was measured, it was in good agreement with the theoretical results given above [8.32-36].

8.4.4 Multi-Photon Raman Scattering

Thus far, only processes in which the resonance between the initial level n and the intermediate level m is of a one-photon nature have been studied. However, — the shift in the frequency of the incident radiation an atom — may also be due to other processes in which the resonance is of a multi-photon nature.

In the event of a multi-photon resonance between the states n and m, all of the above relationships are still valid if the multi-photon matrix element $V_{mn}^{(K)}$ is substitued for the one-photon matrix element $(r_{mn} \cdot E)$ and the nonresonant shifts produced by the strong field are taken into account (Sect. 8.3). The magnitude of the nonresonant shift (including the Bloch-Siegert shift) in a one-photon resonance is small compared to the magnitude of the shift in a multi-photon resonance.

An experiment was described in [8.35] in which the 6S state of the potassium atom spontaneously decays after the two-photon excitation of the 4S state to the 6S state. Since the two-photon matrix element is proportional to the square of the strength of the external field, the yield of the scattered light is proportional to E^4. This process is described by the following diagram:

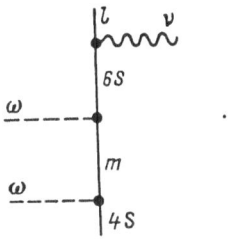

The main contribution to this diagram comes from the intermediate 5P state, although the energy of this state is not very close to $E_{4S}^{(0)} + \omega$ (the 4S → 6S transition is of a two-photon and not of a cascade nature).

The compound K-photon matrix element $V_{nm}^{(K)}$ for a K-photon resonance transition n → m contains a large number of energy denominators. If they are small, one must, according to the Breit-Wigner procedure, substitute $E_s^{(0)} - i\gamma_s/2$ for $E_s^{(0)}$, where γ_s is the spontaneous width of the level s. One must also take into account the nonresonant shifts of the intermediate states.

In switching over from a discussion of one-photon resonances to a discussion of multi-photon resonances, the terms detuning, Δ, and Rabi frequency must be replaced by the terms multi-photon detuning and multi-photon Rabi frequency, respectively (Chap.3). The corresponding criterion for the application of perturbation theory to the description of a K-photon resonance was given in Chap.3.

8.5 A Three-Level System in Two Resonance Fields

Due to the weakness of the vacuum field, Raman scattering has a low probability notwithstanding the presence of a resonance. The effect of this field can be considerably magnified if a second external field is applied to the transitions between one of the resonating levels and the third level. The third level may lie higher than the resonating levels. Of course, the frequency of the second field must be close to the frequency of the corresponding transition.

When intense electromagnetic radiation interacts with an extended atomic medium, not just with a single atom, the second field may be created by the large number of photons that are produced by the Raman scattering of the incident light. However, this process, i.e., stimulated Raman scattering, will not be discussed any further since it lies outside the scope of the phenomena involving an isolated atom [8.37]. Instead, the absorption of two photons by a three-level system will be considered. It will be assumed that a strong resonance field acts on the first transition between the levels n and m and that a weak resonance field acts on the second transiton between the levels m and ℓ.

8.5.1 The System of Equations

For the three-level system (n,m, and ℓ) the field $(1/2)(\mathbf{r}_{mn} \cdot \mathbf{E}) \cos\omega t$ acts on the transition between n and m, and the field $(1/2)(\mathbf{r}_{m\ell} \cdot \mathbf{E}') \cos\omega' t$ acts on the transition between m and ℓ. Let $\gamma_{n,m,\ell}$ be the spontaneous relaxation constants for the levels n,m,ℓ, respectively. In the resonance approximation the system of equations for the population amplitudes $a_{n,m,\ell}$ of the levels n,m, and ℓ are:

$$\dot{a}_\ell + \frac{1}{2}\gamma_\ell a_\ell = i(\mathbf{r}_{m\ell} \cdot \mathbf{E}')a_m \exp[i(\omega_{\ell m} - \omega')t] \quad ,$$

$$\dot{a}_n + \frac{1}{2}\gamma_n a_n = i(\mathbf{r}_{n\ell} \cdot \mathbf{E})a_m \exp[-i(\omega_{mn} - \omega)t] \quad ,$$

$$\dot{a}_m + \frac{1}{2}\gamma_m a_m = i(\mathbf{r}_{\ell m} \cdot \mathbf{E}')a_\ell \exp[-i(\omega_{\ell m} - \omega')t]$$

$$+ i(\mathbf{r}_{mn} \cdot \mathbf{E})a_n \exp[i(\omega_{mn} - \omega)t] \quad . \tag{8.74}$$

It is assumed that the transitions between n and m and between m and ℓ are dipole transitons. The initial conditions for finding the electron in the lowest level at $t = 0$ are

$$a_n(0) = 1 \quad , \quad a_m(0) = a_\ell(0) = 0 \quad . \tag{8.75}$$

The spontaneous widths are introduced into (8.74) using the Breit-Wigner procedure.

The expression

$$W_\ell = \gamma_\ell \int_0^\infty |a_\ell(t)|^2 \, dt \tag{8.76}$$

yields the total number of electrons that have left the state ℓ during the time $0 \leqslant t \leqslant \infty$. It can obviously also be interpreted as being equal to the probability of the absorption of two photons; one from the field E and the other from the field E'. The relative positions of the levels are not of any great importance. For example, if ℓ lies below m, one photon is absorbed from E while a photon with the frequency of E' is emitted. The latter process is called Raman scattering. In the following discussion no distinction will be made between the two types of processes.

The deviations of the frequencies of the strong field and the test field from the respective resonance frequencies are

$$\Delta = \omega - \omega_{mn} \quad , \quad \Delta' = \omega' - \omega_{\ell m} \quad . \tag{8.77}$$

8.5.2 Perturbation Theory

As in Sects.8.1,2,and 4, the discussion will be started with the case in which E is a weak field, i.e., when

$$|\mathbf{r}_{mn} \cdot \mathbf{E}| \ll \gamma_{n,m,\ell} \quad \text{or} \quad |\mathbf{r}_{mn} \cdot \mathbf{E}| \ll \Delta \quad . \tag{8.78}$$

If typical values for the atomic parameters are used, i.e., $|\mathbf{r}_{mn} \cdot \mathbf{e}| \sim 1$ Debye and $\gamma \sim 10^8 s^{-1}$, then the critical field strength for exact resonance lies between 10 and 100 V/cm. This is the lower estimate because both $|\mathbf{r}_{mn} \cdot \mathbf{e}|$ and γ can only be less than the above values. It is obvious that the critical field strength increases as the detuning from resonance increases. It may exceed the above value by many orders of magnitude.

If the criterion for the application of perturbation theory is met, the absorption of two photons is shown by the following diagram:

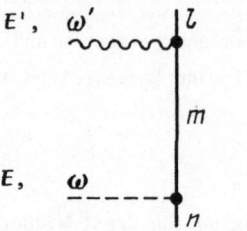

E', ω' l

m

E, ω n

If the field E is weak, one can solve (8.74) for the first order in E and in E'. A straightforward calculation yields

$$W_\ell = \frac{2|(\mathbf{r}_{mn} \cdot \mathbf{E})(\mathbf{r}_{m\ell} \cdot \mathbf{E}')|^2}{|\gamma_n - \gamma_m|^2/4 + \Delta^2}$$

$$\times \operatorname{Re} \left\{ \left(\frac{1}{\gamma_m} - \frac{1}{(\gamma_n + \gamma_m)/2 - i\Delta} \right) \frac{1}{(\gamma_\ell + \gamma_m)+2 + i\Delta'} \right.$$

$$\left. + \left(\frac{1}{\gamma_n} - \frac{1}{(\gamma_n + \gamma_m)/2 + i\Delta} \right) \times \frac{1}{(\gamma_\ell + \gamma_n)/2 + i(\Delta + \Delta')} \right\} \quad , \qquad (8.79)$$

where the terms

$$\frac{1}{(\gamma_n + \gamma_m)/2 \pm i\Delta}$$

reflect an interference between the two-stage and the two-photon transition $n \to m \to \ell$ (Sect.8.1). If

$$\Delta \gg \gamma_n , \gamma_m , \qquad (8.80)$$

the interference terms can be neglected. Equation (8.79) then yields

$$W_\ell = \frac{|(\mathbf{r}_{mn} \cdot \mathbf{E})(\mathbf{r}_{m\ell} \cdot \mathbf{E}')|^2}{\Delta^2}$$

$$\times \left\{ \frac{1}{\gamma_m} \frac{(\gamma_\ell + \gamma_m)/2}{(\gamma_m + \gamma_\ell)^2/4 + \Delta'^2} + \frac{1}{\gamma_n} \frac{(\gamma_\ell + \gamma_n)/2}{(\gamma_\ell + \gamma_n)/4 + (\Delta + \Delta')^2} \right\} \quad . \qquad (8.81)$$

The first term corresponds to a transition through a real intermediate state and describes the two-stage absorption. The second term corresponds to a transition through a virtual intermediate state and describes the two-photon absorption. In the present case one can speak of two-stage and two-photon absorption processes because the probability of the absorption of two photons is the sum of the two-stage and the two-photon transition probabilities. Depending on the value of Δ', either the two-stage or the two-photon absorption is dominant. When $\Delta' \approx -\Delta$, the second term in (8.81) is dominant. In other cases the two-stage absorption is dominant. One can therefore conclude that two-photon absorption places special requirements on Δ and Δ', i.e., the frequencies of the two fields must be such that

$$\omega + \omega' \approx \omega_{\ell n} \quad . \tag{8.82}$$

Each term in (8.81) represents a Lorentzian line with a corresponding width. With interference, the shape of the absorption line, i.e., the dependence of W_ℓ on ω', differs considerably from the Lorentzian. This is readily seen in Fig.8.14a, where W_ℓ is depicted as a function of the frequency ω' of the test field in terms of Δ'.

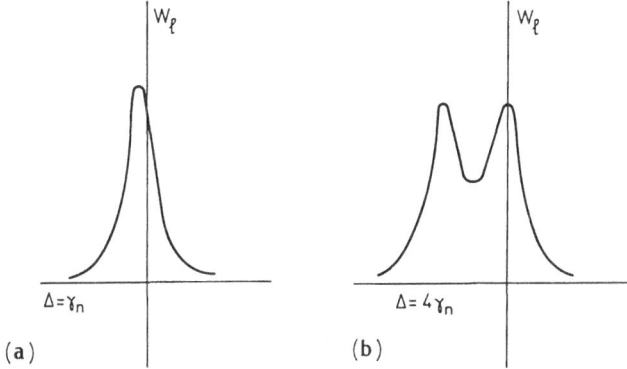

(a)   (b)

Fig.8.14a,b. The probability of the absorption of two photons as a function of the detuning Δ' when the two-stage and the two-photon processes interfere with one another

Figure 8.14b illustrates the situation for large Δ's, i.e., for conditions under which (8.81) is valid. The peak at $\Delta' = 0$ represents a two-stage transition, while the one at $\Delta' = -\Delta$ represents a two-photon transition. As Δ decreases, the interference between the two peaks increases, and at $\Delta = 0$ the two coincide.

If the attenuation rates γ_n and γ_m differ considerably, the main contribution to the absorption probability, according to (8.81), is made by the process with the lower attenuation rate. For example, if $\gamma_m \ll \gamma_n$ (a long-lived intermediate state), the absorption of two photons can be considered to be a two-stage process. For the opposite condition, i.e., $\gamma_m \gg \gamma_n$, the absorption is of a two-photon nature. Therefore, if the conditions $\gamma_m \ll \gamma_n$ or $\gamma_m \gg \gamma_n$ are met, one can be confident that the interference between the two-stage and two-photon absorption processes has been suppressed. The above-mentioned criterion $\Delta \gg \gamma_n$, γ_m can also be looked upon as a sufficient condition for the suppression of interference. If it is met, two-stage and two-photon absorption lie in different spectral regions, as shown in Fig. 8.14b.

In conclusion one can say that the above results do not refer only to the absorption of two photons, which was chosen as an example, but can also be applied to other processes involving two photons [8.38]. For instance, in the case of Raman scattering, a change in the shape of the spectral line due to interference has been observed in a number of experiments [8.39-41].

8.5.3 A Strong External Field

In Sect.8.5.2 the absorption of two photons was studied in the second order of E. Let us now assume that E is a strong field. The field E' of the frequency ω' is still a weak field, whose interaction with an atom will be considered in the framework of first-order perturbation theory. Under these conditions the solution to (8.74) can be found in the following way. The solution to the first two equations in (8.74), i.e., the equations for a_m and a_n, is similar to that of the Rabi problem for a two-level system in a strong resonance field (Sect.3.1). The only difference lies in the need to implement the Breit-Wigner procedure, since the energy terms take on complex values, i.e., $E_{n,m}^{(0)} \to E_{n,m}^{(0)} - i\gamma_{n,m}/2$. For example, the probability amplitude a_m for the intermediate state m is given by the following equation:

$$a_m(t) = A_1 e^{-\alpha_1 t} + A_2 e^{-\alpha_2 t} \quad , \qquad \text{where} \tag{8.83}$$

$$\alpha_{1,2} = \frac{1}{2}\left(\frac{1}{2}(\gamma_m + \gamma_n) + i\Delta \pm \left\{\left[\frac{1}{2}(\gamma_n - \gamma_m) + i\Delta\right]^2 - 4(\mathbf{r}_{nm} \cdot \mathbf{E})^2\right\}\right) \quad , \tag{8.84}$$

and the coefficients A_1 and A_2 are defined by the initial conditions given in Sect.8.5.1. The amplitude $a_m(t)$ is then subsituted into the first equation in (8.74) and $a_\ell(t)$ is determined for the first order in E. As a result, the following equation for the probability of the absorption of two photons is obtained:

$$W_\ell = \frac{2|(\mathbf{r}_{mn} \cdot \mathbf{E})(\mathbf{r}_{m\ell} \cdot \mathbf{E}')|^2}{|\alpha_1 - \alpha_2|^2}$$

$$\times \text{ Re } \left\{ \left(\frac{1}{\alpha_1 + \alpha_1^*} - \frac{1}{\alpha_1 + \alpha_2^*} \right) \frac{1}{\gamma_\ell/2 + \alpha_1 + i\Delta'} \right.$$

$$\left. \left(\frac{1}{\alpha_2 + \alpha_2^*} - \frac{1}{\alpha_2 + \alpha_1^*} \right) \frac{1}{\gamma_\ell/2 + \alpha_2 + i\Delta'} \right\} \quad . \tag{8.85}$$

In the limiting case of a very strong field E, i.e., for

$$(\mathbf{r}_{mn} \cdot \mathbf{E}) \gg \gamma_n, \gamma_m, \Delta,$$

(8.85) assumes a simpler form:

$$W_\ell = \frac{(\mathbf{r}_{m\ell} \cdot \mathbf{E'})^2}{\gamma_n + \gamma_m} \left(\frac{\gamma}{\gamma^2 + (\Delta' - |\mathbf{r}_{mn} \cdot \mathbf{E}|)^2} + \frac{\gamma}{\gamma^2 + (\Delta' + |\mathbf{r}_{mn} \cdot \mathbf{E}|)^2} \right) \quad , \tag{8.86}$$

where

$$\gamma = \frac{1}{2} \gamma_\ell + \frac{1}{4} (\gamma_m + \gamma_n) \quad . \tag{8.87}$$

Hence, as the strength of the external field increases, the distance between the absorption peaks can be expected to increase and finally reach a maximum equal to twice the Rabi frequency. As a result the role of the interference decreases. The quantity γ represents the widths of the peaks that correspond to the frequency ω. In this case this is obvious since one is dealing with transitions from the two quasi-energy levels of the state m, which differ by twice the Rabi frequency, to the state ℓ. It also follows from (8.86) that the amplitudes and the widths of the two peaks become equal as the intensity of the incident light increases.

The study of an atomic spectrum in the presence of a resonance with a strong external field has been carried out by a variety of methods, in all of which a very high spectral resolution of the order of the natural width was achieved. This makes it clearly possible to observe the perturbed spectrum. As an example one can take the D_2 line in sodium [8.20]. The frequency of the laser field (its strength being approximately 10 V/cm) was chosen so that there would be an exact resonance with the 3S → 4P transition. A test field of variable frequency was used to populate the 5S state from the 4P state (Fig.8.15a). The population of the 5S state was observed with the aid of the spontaneous cascade transition 5S → 4P → 3S. Figures 8.15b,c illustrate the probability of the two-photon absorption as a function of the frequency of the test field for two different values of the external field strength. In Fig.8.15b, as in Fig.8.14a, only one peak was observed in a weak external field. In a strong external field two peaks were observed (Fig.8.15c). This represents a splitting of the intermediate 4P level into two quasi-energy levels separated by twice the Rabi frequency (8.86).

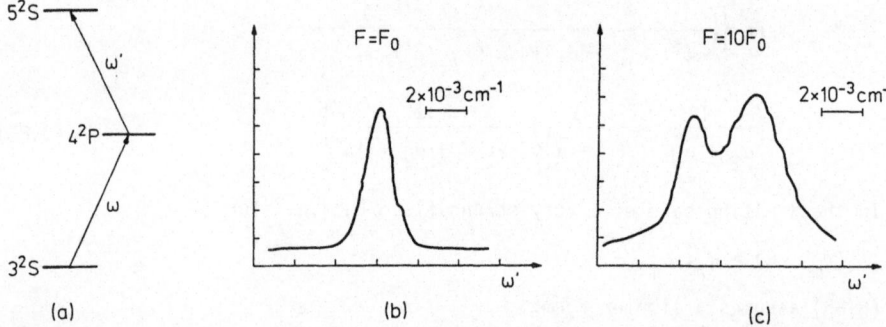

Fig.8.15a-c. The splitting of the D_2 line of a sodium atom: (a) the level diagram (ω and ω' are the frequencies of the strong field and the test field, respectively), (b) the population of the upper state in a weak field E (frequency ω), and (c) the population of the upper state in a strong field $3E$ (frequency ω)

This experiment shows that the distance between the peaks in Fig.8.15c is proportional to the field strength E, i.e., the effect is linear in the field, which agrees with (8.86).

8.5.4 Two Strong External Fields

Thus far the behavior of a three-level system under the influence of a strong external field and a weak test field was investigated. The case shall now be considered in which the transition between the levels n and m is acted on by a strong external resonance field $E \cos\omega t$ and the transition between the levels m and ℓ is acted on by another strong external resonance field $E' \cos\omega' t$. For equidistant levels, the two exciting photons may originate from the same external field.

This problem was studied in Chap.3 in the absence of relaxation processes. Therefore, the aim of this section is to investigate how spontaneous broadening influences the shape of the absorption lines. Such an influence happens to be significant if the field acts on the system for a long time, i.e., $t \geqslant 1/\gamma$.

The simplest case involves spontaneous widths which correspond to transitions from the levels of the three-level system to some other levels. In this case the results of Chap.3, which ignore the widths, can be used and the Breit-Wigner procedure applied, i.e., $E_{n,m,\ell}^{(0)} \rightarrow E_{n,m,\ell}^{(0)} - (i/2)\gamma_{n,m,\ell}$. The equations are cumbersome due to the Cardano formulas for quasi-energy levels (Chap.3) [8.42,43]. Qualitatively, one can say that each spectral term is split into three quasi-energy levels, and a spontaneous emission from any one of these levels yields three peaks (in contrast to the two peaks when only one field was strong). For very strong fields the distance between the peaks is proportional to the strengths of the external fields, i.e., the shift is linear in both fields.

The problem becomes more complex when spontaneous transitions take place within the three-level system. The Breit-Wigner procedure is no longer applicable so that the density matrix method must be applied. The fluorescence spectrum becomes much richer in detail. Indeed, since each level is split into three quasi-levels, the spontaneous transitions between pairs of levels should have resulted in nine peaks. According to the Floquet theorem three of these peaks should coincide, and the transition frequency corresponding to these peaks should coincide with that of the external field which is in resonance with this transition. Thus, each of the two fluorescence spectra generally have seven peaks [8.44]. The number of peaks may be less for various particular and limiting cases.

8.6 Resonance Ionization of Atoms

At the end of Chap.7 we considered non-resonance ionization of atoms, a process in which the energy of any integral number of quanta, $K' < K$ (K being the minimal number of quanta needed for ionization), differs considerably from the transition energies in the spectrum of the atom. This section is devoted to the opposite case, where the energy of a number of quanta, $K' < K$, proves to be close to the energy of transition from the initial state n into a certain excited bound electronic state m. In this case we speak of a K'-photon resonance, while the ionization process in the presence of such a K'-photon intermediate resonance is called *resonance ionization* (Fig.8.16).

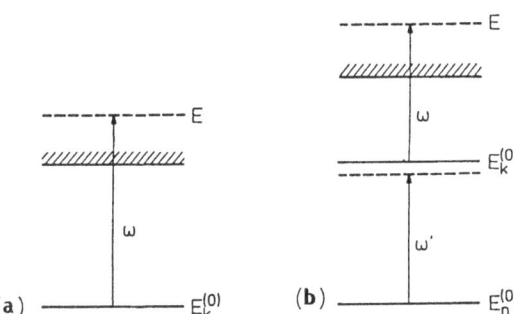

Fig.8.16. (a) Photoionization, and (b) resonance two-photon ionization

The role of the resonance depends on its detuning $\Delta = \omega_{mn} - K'\omega$. In practice, experimental observation of resonance ionization is limited by such Δ's that do not exceed the width γ_m of the mth resonance level, i.e., $\Delta \gtrsim \gamma_m$. One must bear in mind that when an atom is ionized by an intense light field, both the position of the levels (and therefore Δ) and their width may depend, to a great extent, on the external field strength E.

It is clear from general considerations how an intermediate resonance manifests itself in the simplest cases of atomic ionization. First, there is no doubt that at a fixed degree of nonlinearity the ionization probability in the presence of a resonance is many orders of magnitude higher than that of a direct ionization process (in the absence of resonance). In other words, there appears a peak in the curve representing the ionization probability as a function of frequency ω of the external field, known as a resonance peak. Thus, the resonance-ionization process determines the maximal nonlinear absorption of intense radiation by an atomic medium.

Second, another manifestation of resonance is the variation of the degree of nonlinearity d log(w)/d log(F), where w is the ionization rate, and F is the radiation intensity. In direct ionization this quantity is obviously equal to K, i.e., the minimum number of quanta whose absorption is necessary for ionization.

8.6.1 Resonance Ionization in a Weak Field

The nature of the ionization process in atoms depends largely on the specific atomic structure and the intensity of the external field. Two typical cases are those of a weak and a strong field. They differ in the extent to which the resonance states are perturbed by the external field. We shall call a field weak if in it the shift and the broadening of the resonance levels are smaller than the spontaneous widths of these levels.

The probability of resonance multi-photon ionization of an atom in a weak monochromatic field per unit time is given by the well-known Breit-Wigner formula [Ref.8.2, Sect.18]

$$w = \frac{\Gamma_f^2 \Gamma_i}{\Delta^2 + \gamma_m^2/4} \quad . \tag{8.88}$$

Here Γ_i is the ionization probability for the resonance level m defined as

$$\Gamma_i = 2\pi |V_{mE}^{(K-K')}|^2 \rho_E \tag{8.89}$$

(ρ_E is the density of the finite states of an electron with energy E, and $V_{mE}^{(K-K')}$ is the compound matrix element of the $(K-K')^{th}$ order; see Sect.2.4). The quantity Γ_f entering (8.88) is defined as

$$\Gamma_f = |V_{nm}^{(K')}| \tag{8.90}$$

and is called the field width. Equation (8.88) is a straightforward corollary of the expression for the multi-photon transition probability (Sect.2.4 and Chap.7) if in the multi-photon matrix element $V_{nE}^{(K)}$ one considers only one term corresponding to the mth level in the K'^{th} summation step. Moreover, in accordance with the Breit-Wigner procedure (see the beginning of this chapter), one must substitute $E_m^{(0)} - i\gamma_m/2$ for $E_m^{(0)}$ in the resonance denominator.

Equation (8.88) shows that the degree of nonlinearity remains equal to K notwithstanding the formation of an intermediate resonance.

Let us now investigate the validity regime of (8.88). First, it is valid for not too large periods of an applied external field, specifically, when wT is much less than unity (no saturation). Secondly, for (8.88) to be valid, $\delta E_{n,m}$ must be much less than γ_m, $\delta E_{n,m}$ being the Stark shifts of the levels n and m in a variable field (Chap.7). Thirdly, according to the results obtained at the beginning of this chapter, (8.88) is valid if Γ_f is much less than γ_m; in this case the splitting of the nth resonance level into quasi-energy states can be neglected.

The total neglect of broadening of a resonance in an external field presupposes not only a low intensity of the resonance but a high monochromaticity as well [8.45]. Precisely, for (8.88) to be valid one must also require that $\Delta\omega \ll \gamma_m$, $\Delta\omega$ being the spectral width of the radiation. This gives the lower limit for the pulse length: $T \gtrsim \Delta\omega^{-1}$.

If the above conditions are met, the resonance-ionization process is of no special interest, although it has several important fields of application. For example, it form the base for resonance-ionization spectroscopy of atoms and molecules.

In conclusion, we note that (8.88) does not require that the excitation and ionizing fields coincide; they may have different strengths and frequencies. The quantity defined in (8.89) is determined by the strength of the ionizing field, and that defined in (8.90) by the strength of the excitation field; the detuning from the resonance, Δ, in the denominator of (8.88) corresponds to the frequency of the excitation field.

8.6.2 Mechanisms of Resonance Ionization in a Strong Field

In relation to resonance ionization, the following three mechanisms of perturbation of an atomic spectrum under an external field can be observed [8.46,47].

1) Resonance mixing of the ground and the intermediate resonance levels. This process was discussed in great detail at the beginning of this chapter (see also Chap.3). Such mixing leads to a splitting of the resonance level into two quasi-energy states whose separation is determined by Γ_f. Note that in the case of a one-photon resonance (K' = 1) we have the following expression for the field width:

$$\Gamma_f = \frac{1}{2} |r_{mn} \cdot E_1| \quad , \tag{8.91}$$

where E_1 is the strength of the field exciting the intermediate resonance state m. It is easy to see that at K' = 1 the external field becomes intense ($\Gamma_f \gtrsim \gamma_m$) even for low field strengths E_1 (of the order of 10 to 10^2V/cm). As K' increases, the corresponding value of the field rapidly grows. At K' > 2 it is practically impossible to observe multi-photon resonance mixing in view of competing mechanisms for energy spectrum perturbation [8.48].

2) The shift of the ground and the intermediate resonance levels due to non-resonance interaction with the field through other atomic states (the quadratic dynamic Stark effect). This phenomenon leads to a variation in the transition energy for ω_{mn} by

$$\delta E_{mn} = \frac{1}{4} (\alpha_n - \alpha_m) E^2 \quad , \tag{8.92}$$

where $\alpha_{n,m}$ denotes the dynamic polarizabilities of states n and m. Both the strength of the excitation field E_1 and that of the ionizing field E_2 can be taken as E, depending on which of the two leads to a larger shift.

In the general case it is difficult to establish whether a field is strong, i.e., whether $\delta E_{nm} \gtrsim \gamma_m$. Indeed, as noted above (Chap.7), near a resonance the dynamic polarizability can vanish but it can also have very high values. If we take the value of α to be about 10 a.u., which is typical of interresonance gaps, the above criterion of a strong field is met for fields of about 10^4-10^5V/cm. Therefore, in this case the critical field strength proves to be much greater than the corresponding value necessary for field broadening (we found this value above). This comes as no surprise since the field width is proportional to E while the quadratic shift is proportional to E^2.

3) Broadening of the resonance level due to ionization. Quantitatively this effect is characterized by the ionization width Γ_i given by (8.89). We

shall estimate the strength at which the field may be considered strong, i.e., $\Gamma_i \gtrsim \gamma_m$, in the most common case $K - K' = 1$. The ionization width Γ_i is proportional to the photoionization cross section σ, whose order of magnitude is well known: $\sigma \approx 10^{-17} \text{cm}^2$. We can then easily find that the corresponding critical field strength is about 10^4-10^5 V/cm. It is natural then that the critical field strength for the quadratic Stark shift and for ionization broadening have the same order of magnitude, since both are proportional to E^2 (at $K - K' = 1$). However, we single out the case of ionization broadening because it influences the resonance ionization process quite differently from the Stark shift.

The critical field strength for multi-photon ionization broadening $(K - K' > 1)$ is naturally much higher than for one-photon ionization broadening.

The above estimates lead to a qualitative relationship between the nonlinearity parameters, K and K', in the resonance ionization, establishing the interrelationship between the various mechanisms by which an external field acts on resonance ionization. We assume that E_1 is approximately equal to E_2.

At $K' < 2(K - K')$ field broadening prevails over ionization broadening. For one, this is always the case at $K' = 1$, i.e., in one-photon resonance excitation.

At $K' > 2(K - K')$ ionization broadening is dominant. An important case exists for $K - K' = 1$, i.e., a one-photon transition from the resonance state to the continuum. Obviously, at $K' \geqslant 3$, i.e., in resonance excitation involving three and more photons and one-photon ionization from a resonance excitation state, ionization broadening is dominant, and this is accompanied by a non-resonance Stark shift of the levels.

At $K = 3$ and $K' = 2$ both mechanisms of broadening as well as the non-resonance shift of levels are of the same order of magnitude in their field strengths and may differ only in numerical factors of the multi-photon matrix elements.

The discussed mechanisms of resonances in the w versus ω curves do not exhaust all possibilities for real atoms. In the vicinity of a resonance level m there may be other atomic levels ℓ. Then, as soon as Γ_i, Γ_f, or δE_{mn} becomes comparable to the characteristic separation of levels m and ℓ, the above treatment of resonance ionization through an isolated intermediate level becomes invalid. In this case resonance or non-resonance mixing of levels m and ℓ becomes possible, and this complicates the picture.

8.6.3 Multi-Photon Ionization in a Strong Field

From general considerations it is clear that the width of the resonance curve reflecting the frequency (ω) dependence of the multi-photon ionization probability can be defined as the maximum of the quantities T^{-1}, γ_m, Γ_i, or Γ_f. Here T denotes the duration of the electromagnetic field.

We will describe the process of multi-photon resonance ionization in a strong field for $K' > 2(K - K')$, which is the most typical case from the experimenter's viewpoint. As discussed in Sect.8.6.2, in this case $\Gamma_i \gg \Gamma_f$, and there is ionization broadening of the resonance since ionization of the resonance state m occurs considerably faster than the transition from this state back to the ground state n. The ionization process can be described in terms of two successive transitions of the electron, nm and mE.

Under the conditions stated the ionization rate w can be obtained by substituting Γ_i for γ_m and $\tilde{\Delta} = \Delta + \delta E_{mn}$ for Δ in (8.88), which yields [8.49-51]

$$w = \frac{\Gamma_f^2 \Gamma_i}{\tilde{\Delta}^2 + \Gamma_i^2/4} \quad . \tag{8.93}$$

The condition wT < 1 imposes an upper limit on the duration of the field: $T \ll \Gamma_i \Gamma_f^{-2}$.

It is highly essential that we allow for level shifts due to the dynamic Stark effect, δE_{mn}, when studying the case of ionization broadening, since even when $K - K' = 1$, i.e., a one-photon escape to the continuum from a resonance state, the shift of level m is somewhat greater than the width of the level (the dependence on E is the same —both are proportional to E^2). But if $K - K' > 1$, the Stark shift is considerably greater than the ionization width.

Let us now turn to typical experimental data on multi-photon resonance ionization in a strong field. Figure 8.17 shows the yield of ions in 12-photon ionization of krypton atoms as a function of the field's frequency. The maxima are caused by 10-photon intermediate resonances with two discrete levels. The data was taken from [8.52]. The energy of transition to a resonance state differs considerably from the energies of transitions to the unperturbed spectrum because of the dynamic Stark effect [8.53,54].

Obviously, conditions of the experiment correspond to the mode of resonance-ionization broadening. However, the experiments are conducted in such a way that the total ionization time w^{-1} is of the same order of magnitude as the duration T of the exciting laser pulse. For this reason, the second transition (mE) is always realized in the saturation mode, which affects considerably the resonance width.

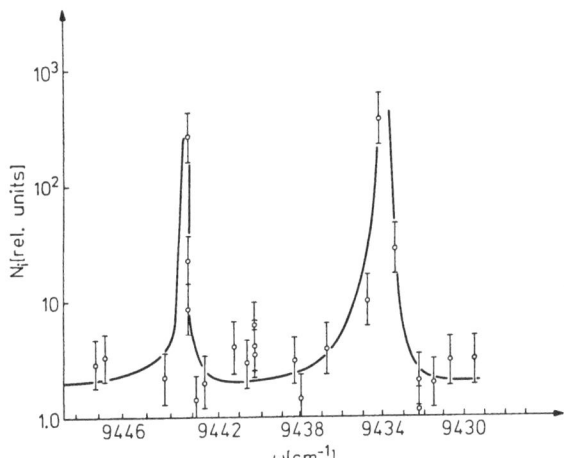

Fig.8.17. The resonance dependence of the ion yield of the twelve-photon ionization of a krypton atom on the frequency of the radiation

8.6.4 Resonance Ionization in the Adiabatic Inversion Mode

If the deviation Δ of the resonance from the intermediate state m is considerable, we arrive at a condition opposite to (3.8). The mode in which the external electromagnetic field is switched on or off is the adiabatic mode. In this process the time dependence of the positions of levels m and n in the field of varying amplitude, the variation being caused by the dynamic Stark effect, plays a considerable role in the formation of resonance peaks in the w versus ω dependence.

A drastically different mechanism of resonance broadening emerges when the Stark shift is so large that $\delta E_{mn} \gg \Gamma$, where Γ is the resonance width in the model with an instantaneous switch-on of the field (Sect.8.6.2).

When the field is introduced gradually (i.e., adiabatically), the population probability for state m is, according to the results of Sect.3.1, given by

$$|a_m(t)|^2 = \frac{1}{2}\left(1 - \frac{\Delta\Delta(t)/|\Delta|}{[\Delta^2(t) + \Gamma_f^2(t)]^{\frac{1}{2}}}\right) \tag{8.94}$$

where $\Delta(t)$ is the time-dependent detuning from resonance, namely

$$\Delta(t) = \omega_{mn} - \omega + \delta E_{mn}(t) \quad , \qquad \text{with} \tag{8.95}$$

$$\delta E_{mn} = \frac{1}{4}(\alpha_n - \alpha_m) E^2(t) \tag{8.96}$$

and $E(t)$ the amplitude of the external field introduced gradually, i.e.,

$$E(t) = Ef(t) \tag{8.97}$$

and $f(t) \to 0$ as $t \to \pm\infty$. The quantities α_n and α_m are the dynamic polarizabilities of the states n and m, respectively. The quantity $\Gamma_f(t)$ in (8.94) can be written as

$$\Gamma_f(t) = \Gamma_f f^{K'}(t) \quad , \tag{8.98}$$

where we have combined (8.97 and 90).

Multiplying the ionization rate of state m, which is Γ_i, by (8.94) and integrating over the duration of the laser pulse, we arrive at a formula for the ionization probability over a laser pulse:

$$W(\omega) = \frac{1}{2} \Gamma_i \int_{-\infty}^{+\infty} \left(1 - \frac{\Delta\Delta(t)/|\Delta|}{[\Delta^2(t) + \Gamma_f^2 f^{2K'}(t)]^{\frac{1}{2}}}\right) dt \quad . \tag{8.99}$$

What are the conditions for (8.99) to be valid? First, we must require that

$$\Gamma_i T \ll 1 \quad , \tag{8.100}$$

so that the total ionization probability over the duration of the field is small compared to unity: $W(\omega) \ll 1$ (no saturation). Second, we must require that

$$\Gamma_f T \gg 1 \quad , \tag{8.101}$$

so that in (8.99) the field width Γ_f is dominant in comparison with the width T^{-1}, the latter reflecting the fact that the radiation pulse is of finite duration. Third, we must require that the shift of levels is great, $\delta E_{mn} \gg \Gamma_f$, so that the effect is distinct (see below). Finally, we must require that $\delta E_{mn} \ll \Gamma_f^2 T$, so that in the range $\Delta(t)$ of small deviations from resonance, the adiabatic approximation remains valid, i.e., there should be no transitions between quasi-energy states.

The reader must note that the above four conditions can be met simultaneously only for special atoms and conditions of observation, namely, the field E must not be too strong but the values of the field width Γ_f must be very large (for numerical reasons).

Rapid variation of $|a_m(t)|^2$ from zero to unity over a short time interval t near the point where $\Delta(t) = 0$, a fact that follows from (8.99), corresponds to the well-known phenomenon of *adiabatic level inversion* [8.51]. In the standard adiabatic-inversion problem, total excitation of an atom is produced by changing the frequency ω of the external field. In the problem under consideration, however, the frequency ω of radiation is assumed constant, but the atomic level shifts in the field as its amplitude $E(t)$ increases (or decreases).

A qualitative feature of the resonance probability (8.99) is that, in contrast to the broadening mechanisms discussed above, the resonance is asymmetric in this case. This is due to the asymmetry in the dynamic polarizability: the Stark shift δE_{mn} has a definite sign. The resonance width is of the order of δE_{mn}. A detailed description of the behavior of W with frequency ω is given in [8.51,55,56].

8.6.5 The Polarization of Electrons and Nuclei in the Event of Resonance Ionization

The degree of the polarization, i.e., the preferred orientation of the spin, is one of the main characteristics of electrons and nuclei that determines their interaction with other atomic particles. A polarization of electrons and nuclei in the event of resonance ionization may be the result of a selective excitation of an atom by a circularly polarized field into a definite state. The essence of the phenomenon is the same in all cases: the absorption of a photon (or several photons) from the circularly polarized field changes the projection of the angular momentum of the atom and leads to a nonuniform (with respect to a projection of the total angular momentum) population of the excited states, i.e., to a definite degree of polarization. When a state of the fine structure (which is due to the coupling of the electron spin with the orbital angular momentum) is excited, the electron becomes polarized. When a state of the hyperfine structure (which is due to the coupling of the total angular momentum of the electron with the spin of the nucleus) is excited, both the electron and the nucleus become polarized. If the excited atom is subsequently ionized, the distribution of the electrons in the continuum and the distribution of the nuclei over the orientations of the spin angular momenta become nonuniform, i.e., the electrons and the nuclei in the final free state become polarized. Complete polarization is possible when the choice of the initial state permits the excitation of only one resonance state. This also determines a final state with a definite value of the projection of the spin of the particle in a specified direction.

Let us now consider the specific ways in which electrons can be polarized. Since the optimal method of obtaining electrons with a maximum polarization is to excite the states of the fine structure, there is no need to maximize the monochromaticity $\Delta\omega$ of the laser radiation or the duration τ of the laser pulses. The optimal values lie in the interval

$$\frac{1}{\Delta E_{hfs}} > \tau \sim \frac{1}{\Delta\omega} > \frac{1}{\Delta E_{fs}} \quad , \tag{8.102}$$

where ΔE_{hfs} and ΔE_{fs} are the hyperfine- and fine-structure constants, respectively [8.57,58]. The magnitude of the interval has been estimated: 10^{-8} s

$> \tau > 10^{-11}$s. Obvious difficulties arise when one tries to optimize the intensity of the external field [8.59]. On the one hand, the higher the intensity of the field the greater is the probability of transitions and the higher is the yield of polarized particles. On the other hand, as the strength of the field increases, the magnitude of the perturbation of the bound electronic states may become comparable to that of the fine-structure constant, which would lead to a breakdown in the conditions for selective excitation. For one-photon excitation, the optimal value for the strength of the external field is 10^5V/cm, which is equivalent to that of a pulsed nanosecond laser. Hence, the same field can be used to achieve excitation and ionization.

A high degree of polarization can be achieved with a transition of the type S → P → D, where S ($\ell = 0$), P ($\ell = 1$), and D ($\ell = 2$) are the initial, the resonance, and the final (free) states of the electron, respectively. The transition S → P is resonantly excited by a circularly polarized field [8.60, 61]. This was the scheme used for cesium atoms. The experiment provided a degree of polarization that was in good agreement with the above calculations [8.62]. The degree of polarization of the intermediate state $P_{1/2}$ was $(W_{1/2} - W_{-1/2})/(W_{1/2} + W_{-1/2}) \approx -60\%$, while that of the state $P_{3/2}$ was about +80%. (The plus sign signifies that the direction of the polarization is along the wave vector, the minus sign signifies an opposite direction, and W is the total probability that the electrons have a definite spin.)

Although a total polarization of electrons has not yet been achieved, this is undoubtedly possible.

Let us consider one of the possible methods of obtaining totally polarized electrons [8.63]. The method is based on the three-photon resonance ionization of alkali atoms (Fig.8.18). The first laser with an operating frequency ω_1 is tuned in resonance with the $|nS_{1/2}> \rightarrow |nP_{1/2}>$ transition. The intensity of this radiation is not assumed to be very high, so that a subsequent ionization from the level $|nP_{1/2}>$ is not important. The radiation of the first laser is assumed to be linearly polarized along the z axis. The wave vector of the light from the first laser is directed along the x axis. It is assumed that the wave vector of the second laser is directed along the z axis and that the radiation is circularly polarized (clockwise) in the xy plane. The operating frequency of the second laser is in resonance with the $|nP_{1/2}> \rightarrow |n'S_{1/2}>$ transition. The same radiation initiates a transition from the $|n'S_{1/2}>$ level to the continuum (ionization). Figure 8.18 contains the magnetic quantum numbers of those states that can absorb a photon ω_1 of the linearly polarized field and two photons ω_2 of the circularly polarized field. Due to the selection rules for the magnetic quantum number the state $|nS_{1/2}, M = 1/2>$ cannot be

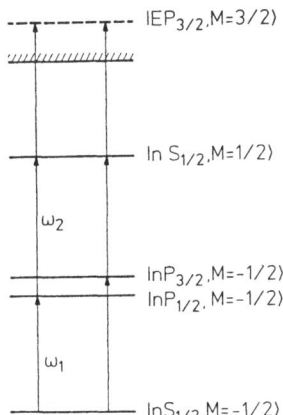

$|EP_{3/2}, M=3/2\rangle$

$|n\,S_{1/2}, M=1/2\rangle$

ω_2

$|nP_{3/2}, M=-1/2\rangle$
$|nP_{1/2}, M=-1/2\rangle$

ω_1

$|nS_{1/2}, M=-1/2\rangle$

Fig.8.18. One of the possible schemes for obtaining completely polarized electrons in the
event of a resonance ionization of an atom

resonantly excited and hence can be neglected. The final state $|EP_{1/2}\rangle$ does
not play a role either, since it cannot have M equal to 3/2.

The above ionization scheme produces completely polarized electrons. Similar results can also be obtained if the first laser is tuned in resonance
with the state $|nP_{3/2}\rangle$ that is closest to $|nP_{1/2}\rangle$, or if it is tuned to the
neighborhood of these states.

In addition, the light of both lasers can be directed along the z axis.
The light of the first must be linearly polarized along the x axis, while that
of the second must be circularly polarized in the xy plane. In this case, when
the first laser is tuned in resonance with the state $|nP_{1/2}\rangle$, a cascade tran-
sition occurs:

$$|nS_{1/2}, M = 1/2\rangle \xrightarrow{\omega_1} |nP_{1/2}, M = -1/2\rangle$$

$$\xrightarrow{\omega_2} |n'S_{1/2}, M = 1/2\rangle \xrightarrow{\omega_2} |EP_{3/2}, M = 3/2\rangle \quad . \tag{8.103}$$

In the first step, the transition operator

$$\mathbf{r} \cdot \mathbf{e} = x = r\,\sin\theta\cos\phi$$

$$= \frac{1}{2}\,r\,\sin\theta[\exp(i\phi) + \exp(-i\phi)] \tag{8.104}$$

initiates a transition in which $\Delta M = \pm 1$ and the state $|nS_{1/2}, M = -1/2\rangle$ trans-
forms into $|nP_{1/2}, M = 1/2\rangle$. The latter cannot be resonantly excited by the
circularly polarized field because a state $|EP_{3/2}\rangle$ with M = 5/2 does not exist.
Hence, this process also leads to a complete polarization of the electrons.

295

If the first laser is tuned in resonance with $|nP_{3/2}>$, then both final states $|EP_{1/2}>$ and $|EP_{3/2}>$ can reached:

$$|nS_{1/2}, M = 1/2> \xrightarrow{\omega_1} |nP_{3/2}, M = -1/2>$$

$$\xrightarrow{\omega_2} |n'S_{1/2}, M = 1/2>$$

$$\xrightarrow{\omega_2} |EP_{3/2}, M = 3/2> \quad ,$$

$$|nS_{1/2}, M = 1/2> \xrightarrow{\omega_1} |nP_{3/2}, M = 3/2>$$

$$\xrightarrow{\omega_2} |n'S_{1/2}, M = -1/2>$$

$$\begin{cases} \xrightarrow{\omega_2} |EP_{1/2}, M = 1/2> \\ \xrightarrow{\omega_2} |EP_{3/2}, M = 1/2> \end{cases} \qquad . \tag{8.105}$$

An analysis of the matrix elements of the individual transitions proves that the resulting degree of polarization is -80% [8.63].

As the frequency ω_1 is varied between the resonance frequencies for the $|nP_{1/2}>$ and $|nP_{3/2}>$ levels, the degree of polarization changes from 100 to -80% and vanishes for a certain frequency. Similar, yet simpler, schemes in which the last two transitions of the above three-photon ionization process were studied are illustrated in [8.64].

Let us now consider the polarization of nuclei. The first fact that one must bear in mind is that far from all states can have their hyperfine structure resolved because of the finite natural width of a state. Three cases can be distinguished:

1) The hyperfine structure is resolved in both the ground and excited states. This situation can be realized, for example, in alkali atoms.

2) Only the hyperfine structure of the ground state is resolved; e.g., in the hydrogen atom.

3) The hyperfine structure is only resolved in the excited states. This situation is realized with atoms in which the spin of the nucleus is equal to zero (e.g., the beryllium atom).

In all of these cases a certain degree of polarization can be obtained in the resonance ionization process. Obviously, the nuclei cannot be polarized in cases where the hyperfine structure of both states is not resolved.

The restrictions imposed on the radiation needed to excite the hyperfine structure states lie within the limits of modern lasers, which must operate in the single-frequency mode. The practical feasibility of such an experiment has been shown with cesium [8.65].

It is preferable to use two fields, an exciting and an ionizing field, in the polarization of nuclei. For a two-photon ionization in the presence of a one-photon resonance, the strength of the exciting field is less than 100 V/cm (this follows from the inequality $|\mathbf{r}_{kn} \cdot \mathbf{E}| < \gamma_k$, where γ_k is the natural width), while the strength of the ionizing field is less than 10^5V/cm.

The dependence of the degree of polarization of the nuclei on the various parameters that characterize the atomic system and the external fields has been calculated using the polarization density matrix [8.66]. Instead of discussing this method, because it cannot be illustrated graphically, let us study the S → $P_{1/2}$ → D transition in the hydrogen atom, in which the spin of the nucleus (a proton) is equal to 1/2. The scheme of such a transition induced by a right-handed circularly polarized field is shown in Fig.8.19. The frequency of the exciting field is equal to the frequency of the transition from the ground state $1S_{1/2}$ to the degenerate excited states $2P_{1/2}$. The latter are not resolved because their natural widths are greater than their hyperfine splitting. However, the $1S_{1/2}$ (F = 0) → $2P_{1/2}$ (F = 0) transition, where F is the total angular momentum of the atom, is forbidden by the selection rule $\Delta m_F = 1$. Thus, the only state that can be excited is the $2P_{1/2}$ (F = 1) state. Hence, the polarization of such a transition is complete. If the hydrogen atom is ionized after being excited into this state, the protons will be completely polarized.

High (although not 100%) degrees of polarization can also be obtained using other transitions in other atoms [8.67].

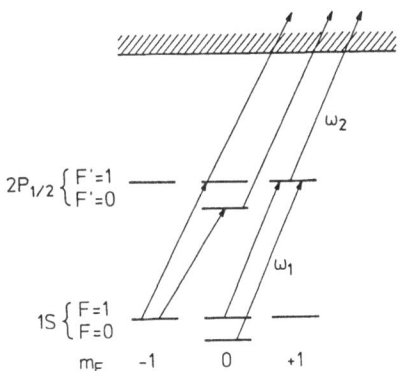

Fig.8.19. The transition diagram for the two-photon resonance ionization of a hydrogen atom

8.6.6 Angular Distribution of the Electrons in the Resonance Ionization Process

Let it be assumed that the ground state does not exhibit a preferred orientation, i.e., all values of the magnetic quantum number are equally probable. The various components of the intermediate state, which are characterized by different magnetic quantum numbers, are not uniformly populated. In a linearly polarized external field this nonuniformity is caused by the differences in the values of the corresponding matrix elements. In a circularly polarized field the nonuniformity is a result of the corresponding selection rules. The final states in the continuum are determined by the nonuniform distribution of electron in the intermediate state. Hence, the intermediate as well as the initial state determines the complex angular distribution of the electrons. An exception to the above statement is encountered in the transition $S_{1/2} \to P_{1/2}$, which takes place when a linearly polarized field is applied. In this case the intermediate state is obviously populated uniformly.

If one assumes a certain relaxation mechanism for the intermediate state, the distribution within this state becomes isotropic, and the angular distribution of the photoelectrons is only determined by the transition from the resonance state to the continuum (Chap.7). A relaxation process is needed that will mix the different values of the magnetic quantum number for the intermediate state. Such a mixing may be caused by a large ionization broadening of the resonance (8.98), when $|\omega_{kn} - \omega'| \ll w_E$, or by a strong mixing of the resonating states n and k, (8.99), when $|\omega_{kn} - \omega'| \ll |V_{kn}^{(K')}|$, [8.68].

Therefore, when the condition $|\omega_{kn} - \omega'| \ll \max(w_E, V_{kn}^{(K')})$ is fulfilled, the angular distribution of the ejected photoelectrons is only determined by the intermediate state. All of these assertions follow directly from (8.98,99) if the detuning from resonance is neglected.

The difference between the two cases stems from the difference between two-photon and cascade transitions (Sect.8.1). When a cascade transition takes place, the intermediate state k can be considered to be the initial state in which the electrons evenly populate the levels with different values of M.

The strong mixing of intermediate states can also be caused by a short perturbation pulse.

The results of the two-photon resonance ionization of the sodium atom in a linearly polarized field suggest an anisotropy in the angular distribution of the electrons from the $3S_{1/2}$ ground state to the $3P_{1/2}$ state [8.69]. Perturbation theory predicts that the angular distribution must be determined

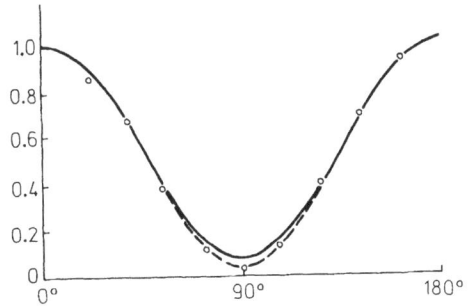

Fig.8.20. The number of electrons as a function of the scattering angle in the resonance ionization of a sodium atom; calculated (*solid line*) and experimental results (*dashed line*)

by the one-photon ionization from the state $3P_{1/2}$. The experimental data in Fig.8.20 is described by the well-known equation (7.75) in which $K = 1$.

9. Conclusion

This book has been devoted to a study of the behavior of an isolated atom in a monochromatic field whose amplitude is less than that of the atomic field. To a certain extent these boundaries are arbitrary.

Several problems that lie beyond the scope will be discussed below. However, the discussion will be kept brief. Only qualitative information will be supplied and for details the reader is referred to the sources quoted.

9.1 The Role of the Non-monochromaticity of Laser Radiation

In discussing various nonlinear phenomena in Chaps.7 and 8 it was assumed that a strong electromagnetic field is ideally monochromatic. However, this assumption is not valid for the majority of experiments. Laser radiation is only quasi-monochromatic (Sect.5.4). Of vital importance is the finite spectral line width of the radiation, $\Delta\omega$. Although small in comparison with the frequency of the field, ω, it may, in many cases, be greater than the natural line width of the atomic levels and the perturbation of these levels. Non-monochromatic laser radiation is characterized by fluctuations in the intensity of various modes. The duration of the fluctuation, τ_{cor}, is strictly correlated to the spectral width (in the multi-frequency mode, $\tau_{cor} \sim 1/\Delta\omega$).

Depending on the relative magnitude of the duration of the fluctuations and the response time of the atomic system, the observed effect is either a result of the average intensity of the fluctuating radiation, $<F>$ (for long response times), or of the instantaneous intensity $F(t)$ (for long fluctuation times). The response time of the atomic system is equal to the lifetime of the given state. The observation time is always assumed to be longer than the response time and τ_{cor}.

To characterize the radiation for an atomic system with long response times, it suffices to know $<F>$, which can be measured using standard photoelectronic methods. The main feature of radiation with long fluctuation times is the probability distribution of the intensity, $P(F)$, since the detection of instantaneous values of the intensity is a difficult problem to solve (Sect.5.1). The duration of the fluctuation, λ_{cor}, lies between 10^{-12}s (for neodymium glass lasers, $\Delta\omega \sim 10$ cm^{-1}) and 10^{-10}s (for ruby lasers, $\Delta\omega \sim 0.1$ cm^{-1}) (Sect.5.4). In practically all cases the observation time has been found to be longer than τ_{cor}, so that the results are averaged with respect to time.

As a result τ_{cor} is always smaller than the response time of an excited atomic state, provided that the latter is determined by the natural lifetime of this state. In this case the interaction of the atom with radiation is also determined by <F>. On the other hand, the lifetime of the nonresonant virtual states, determined by the time-energy uncertainty principle, is always much smaller than the fluctuation time. To characterize a K-photon radiation process it is sufficient to know the Kth moment of the intensity distribution. The role of the non-monochromaticity of the exciting radiation in the dependence of various phenomena on the intensity of the field must clearly by different for resonant and nonresonant processes.

This difference also becomes apparent if one considers the role of the finite spectral line width of the radiation and the frequency dependence of the various effects. Indeed, far from resonance the effective cross section of the excitation or ionization is a slowly varying function of the frequency. Thus, the frequency dependence can be neglected within the spectral width of the radiation, which therefore does not influence the probability of the process. On the other hand, the multi-photon cross sections in the neighborhood of a resonance are rapidly varying functions of the frequency. In this case, the parameters of the radiation must take into account the correlation function of the Kth order,

$$G_K(t) = <[E^*(t')E(t' + t)]^K> \quad , \tag{9.1}$$

of the amplitude of the field, E, or its Fourier transform, which is known as the effective spectrum of the Kth order:

$$S_K(\omega) = \int_{-\infty}^{\infty} G_K(t) \, e^{i\omega t} \, dt \quad . \tag{9.2}$$

A division into nonresonant and resonant processes is only advisable for weak, non-monochromatic fields, in which the perturbation of the atomic states can be neglected. An example of this is the nonlinear ionization of atoms and multi-photon excitation in a weak field, where the width of the excited state is determined by its spontaneous relaxation. When the perturbation of the

electronic states plays a significant role, the strength of the field determines the nature of the emerging effect. For example, a variation of the
field strength may increase the probability of resonance ionization from the
excited state, i.e., may change the width of this state. Therefore, the lifetime of this state may either be longer or shorter than the duration of the
fluctuations, depending on the strength of the field. Correspondingly, the effective cross section will either be determined by the average value of the intensity of the radiation, or by its instantaneous value. Quantitatively, the
same is true for level shifts. When $\delta E_n > \Delta\omega$ (slow fluctuations in the exciting field), the shift in the levels is determined by the instantaneous intensity of the radiation; otherwise for rapid fluctuations the shift is determined by the average intensity $<F>$.

If the different parameters for nonlinear, optical phenomena are compared
for a fixed value of the average intensity and for a frequency that corresponds to the maximum in the spectral distribution, ω, it can be seen that
the non-monochromaticity manifests itself in the average yield of nonlinear
processes, in the maximum value of the fluctuations of this yield, and in the
dependence of the yield on the frequency of the maximum of the distribution
and on the shape of the spectrum.

The role which the non-monochromaticity of the laser radiation plays in
some of the elementary, nonlinear, optical phenomena discussed in Chaps.7
and 8 will be considered. However, the conventional distinction between nonresonant and resonant processes will not be made [9.1].

9.1.1 Direct Multi-Photon Ionization

The probability of a direct K-photon ionization of an atom in a monochromatic
field is given by the well-known power law:

$$w(F) = \alpha_K(\omega,\rho)F^K \quad .\tag{9.3}$$

Far from resonance the multi-photon cross section $\alpha_K(\omega,\rho)$ is a slowly varying
function of the frequency of the field. Within the spectral width $\Delta\omega$ of the
non-monochromatic radiation this dependence can unquestionably be neglected.
Therefore, the difference between monochromatic and non-monochromatic radiation lies in the fact that the intensity of the former can be considered
to be a constant quantity while the intensity of the latter is a variable
quantity with a distribution $P(F)$. As a result, the ionization probability
in a non-monochromatic field can be considered to be equal to the ionization
probability in a monochromatic field with a slowly varying intensity:

$$w^*(<F>) = \int_0^\infty w(F)P(F) \, dF \quad , \tag{9.4}$$

where w^* is the ionization probability in a non-monochromatic field, and $<F>$ is a parameter of the distribution $P(F)$. In accordance with (9.3,4),

$$w^*(<F>) = \alpha_K \int_0^\infty F^K P(F) \, dF$$

$$= \alpha_K <F^K> \quad , \tag{9.5}$$

where the intensity of the radiation is averaged over the time it takes to measure the yield of the process (i.e., practically over an infinite time interval) and $<F^K>$ is the Kth moment of the distribution $P(F)$ and is equal to $G_K(0)$.

Comparing (9.5) and (9.3), one can see that, when the intensity of the monochromatic radiation F and the average intensity of the non-monochromatic radiation $<F>$ are equal, the probability of ionization in the non-monochromatic field is higher. Also the greater the degree of nonlinearity K the higher is the probability. For example, when $K = 2$ this is a result of the mathematical statement that the root-mean-square value is always greater than the arithmetic mean.

Quantitatively, the effect produced by the non-monochromaticity of the radiation when $<F> = F$ is determined by the statistical factor

$$g_K = \frac{w^*(<F>)}{w(F)} = \frac{<F^K>}{(<F>)^K} \quad 1 \quad . \tag{9.6}$$

If $P(F)$ is sufficiently accurately described by (5.3), which is characteristic of a source of heat, the statistical factor is equal to K!.

Strictly speaking, for laser radiation $P(F)$ is not described by the exponential relationship given in (5.3) but by the binomial distribution given in (5.2). Whether these two distributions are almost equal depends on the degree of the nonlinearity K and on the number of generated modes N. From (5.2), one can see that the dependence of g_K on K and N is as follows [9.2]:

$$g_K(N) = K!N^{(K)} \frac{(N - 1)!}{(N + K - 1)!} \quad , \tag{9.7}$$

which shows that as $N \to \infty$, the statistical factor increases in such a way that it becomes equal to K! Since N is proportional to $\Delta\omega$, the statistical factor also depends on $\Delta\omega$.

The last dependence of $g_K(N)$ on $\Delta\omega$ suggest that it may be feasible to use a variety of multi-frequency lasers to study various nonlinear phenomena. For example, the radiation from a ruby laser ($\Delta\omega \sim 0.1$ cm^{-1}, $N \sim 10^2$) may with sufficient accuracy be considered to be identical to the radiation of a heat source. Correspondingly, the equality $g_K = K!$ is only true for two- and three-photon processes, while the radiation of a neodymium laser ($\Delta\omega \sim 10$ cm^{-1}, $N \sim 5 \times 10^3$) is equivalent to that of a heat source with $K \leqslant 20$.

Equation (9.7) was experimentally verified for a two-photon process [9.3] and an eleven-photon process [9.4]. The validity of the limiting relationship $g_K = K!$ was verified for a two-photon process [9.3] and a five-photon process [9.5].

9.1.2 Tunneling Ionization in a Variable Field

Recalling the data on tunneling ionization in a monochromatic field (Chap.7), it becomes clear that qualitatively the non-monochromaticity of the radiation leads to the same effects observed in a direct ionization. The reasoning of Sect.9.1.1 makes it possible to use (9.4) for the probability of tunneling ionization in a non-monochromatic field. If it is assumed that P(F) and the ionization probability in a monochromatic field are defined by (5.3) and (7.78), respectively, then using (9.4) one arrives at an expression that describes the ionization probability in a non-monochromatic field to within exponential accuracy [9.6]:

$$w^* = \exp\left(-2\sqrt{3}\,\frac{E_n}{(<F>)^{1/3}}\right) , \qquad (9.8)$$

where E_n is the binding energy of the state from which the tunneling ionization starts.

Comparing (9.8) with (7.78), one can see that the probability of tunneling ionization in a non-monochromatic field is less dependent on the intensity of the field than in a monochromatic field.

Despite the different natures of the functions $w(F)$ and $w^*(F)$ one can introduce the statistical factor for equal values of <F> and F. This factor depends on F, which is not the case in the multi-photon limit.

The simultaneous decrease, predicted by theory and observed experimentally, in the statistical factor and the adiabaticity parameter γ (i.e., the transition from the multi-photon limit to tunneling ionization) is qualitatively quite obvious. As γ decreases from $\gamma \gg 1$ to $\gamma \sim 1$, i.e., in the multi-photon region, the dependence of w on F does not change, i.e., $w(F) \propto F^K$. The difference between the ionization probabilities in the non-monochromatic and the monochromatic field is accounted for by (9.7), which does not depend on γ (i.e., for a fixed frequency of the field, g_K does not depend on the intensity of the radiation). However, in the interval where $\gamma \sim 1$ and $\gamma \ll 1$ (the tunneling region), the dependence of w on F becomes increasingly weaker. Correspondingly, there is a decrease in the difference between the ionization probabilities in the non-monochromatic and the monochromatic fields. However, there is still a difference even as $\gamma \to 0$. Calculations of the statistical factor as a function of γ have been done for the tunneling and the intermediate regions ($\gamma \sim 1$) [9.7]. The experimental data for the intermediate region is in good agreement with these calculations [9.8].

9.1.3 Multi-Photon Excitation of Atoms

The multi-photon excitation of atoms in a non-monochromatic field is best studied separately for the case in which the perturbation of the resonance state is not essential and the case in which the perturbation is greater than the natural width of the state, i.e., where the perturbation substantially determines the characteristics of the excitation.

Let us start with the case in which the perturbation of the resonance state is not essential (a weak external field). In a monochromatic field the dependence of the probability of a K-photon excitation on the intensity of the radiation is given by the power law $w \propto F^K$. The frequency dependence of w is determined by the Lorentzian shape of the resonance absorption line. In a non-monochromatic field the excitation is determined by the relative magnitudes of the decay time for the correlation function $G_K(t)$ (which is proportional to τ_{cor} for values of K that are not too large) and the natural lifetime of a specific excited state (or in spectroscopic terms, the interrelation of the effective K^{th}-order spectrum $S_K(\omega)$ and the width of the excited state). Since the correlation time-interval for the Kth-order function is shorter than the interval for the first-order function, i.e., $\tau_{cor} \sim 1/\Delta\omega$, and the latter is always shorter than the natural lifetime of the excited state, the correlation time-interval is a deciding factor. One can also say that the spectral line width of the radiation, $\Delta\omega$, is always greater than the natural width of the atomic levels. Hence, the following power law is obtained: $w \propto (\langle F \rangle)^K$.

On the other hand, in the non-monochromatic case the width of the profile of $w(\omega)$ is determined by the effective K^{th}-order spectrum. If the fluctuations in the radiation intensity are Gaussian, then $S_K(\omega)$ is a K-fold convolution of the emission spectrum $F(\omega)$ [9.9]. If the spectrum itself is Gaussian, $S_K(\omega)$ is also Gaussian with a width $K^{\frac{1}{2}}$ times greater than that of $F(\omega)$, i.e., of the order $K^{\frac{1}{2}}\Delta\omega$.

In principle, the case in which the perturbation of the resonance state is essential can always be realized by increasing the intensity of the external field. The nature of the excitation process in a non-monochromatic field is determined by the relative magnitude of the perturbation of the resonance state (its shift or broadening) and the width of the emission spectrum. As long as the perturbation is smaller than $\Delta\omega$, the conclusions concerning the behavior of $w(F)$ and $w(\omega)$ that were made under the assumption that the perturbation of the resonance state is not essential remain valid. For higher intensities of the external field for which the perturbation becomes greater than $\Delta\omega$, the dependence of w on F does not follow the power law and is determined by the perturbed resonance state. This situation has only been encountered during the excitation of low-lying states. Indeed, when the perturbation is much greater than the spectral width of the multi-frequency laser radiation ($1-10$ cm^{-1}), it becomes difficult to single out a two-level system consisting of a ground state and a resonance state because of the mixing of other states.

This reasoning shows that it is only useful to compare the absolute magnitudes of the multi-photon excitation probabilities in non-monochromatic and monochromatic fields (or, simply, to calculate the statistical factor) when the resonance state is not perturbed or when the perturbation is much less than the spectral width. If the laser radiation contains a sufficiently large number of modes, then $g_K = K!$.

What has just been said about the multi-photon excitation of atoms is obviously also true for resonance ionization, since the resonance ionization probability is determined by the excitation of the resonance state (Sect.8.6). The experimental data on the resonance ionization in a non-monochromatic field is in good agreement with calculations [9.10].

Deviations from the power law $w \propto F^K$ are observed for even stronger fields (Sect.4.1). It turns out that the region in which perturbation theory can be used is reduced in changing over from a monochromatic to a stochastic field [9.11]. This is due to the considerable influence of the peaks in the transition probability in a stochastic field.

9.1.4 Perturbation of Atomic Levels

The nature of the perturbation of atomic levels is strongly dependent on the relative magnitude of the spectral width of the laser radiation and the amplitude perturbation (resonant or nonresonant).

Let us consider a nonresonant perturbation, specifically, the dynamic nonresonance polarizability of atomic levels in a non-monochromatic field.

The time-energy uncertainty principle shows that the atomic levels are shifted following the temporal variations in the intensity of the field when $\Delta\omega < \delta E_n$ (Sect.9.1). The energy shifts are determined by the instantaneous value of the field strength.

Over a time interval much longer than the correlation time, $\tau_{cor} \sim 1/\Delta\omega$, the amplitude of the field strength takes on almost all possible values continuously and adiabatically. The corresponding line shift also takes on almost all possible values, which can be observed as a nonuniform broadening. The instantaneous line shift is proportional to the square of the field strength. Hence, the magnitude of the broadening is proportional to <F>. In this case it is obviously not convenient to use the term "level shift", since the shift is small compared to the broadening and the level shift in a monochromatic field of the same intensity.

When the spectral width $\Delta\omega$ is greater than the level shift δE_n (weak field), the atom does not adiabatically follow the instantaneous value of the radiation intensity but only the average intensity. Hence here, as in a monochromatic field, one can speak of a level shift.

The line shift is equal to that in a monochromatic perturbing field of the same intensity. The linewidth is small. It is proportional to the square of the intensity of the light, that is, the linewidth is the next term in the expansion of the intensity. Such an expansion can be made because the field is weak.

These conclusions were reached for various stochastic distributions of the field strength [9.12] and have been experimentally verified [9.13].

A detailed theoretical description of a two-level system in a strong resonance field has also been made [9.12], but experimentally this has not yet been studied in detail [9.1].

Let us briefly examine the results of the theoretical description of a two-level system in a non-monochromatic field with a spectral width $\Delta\omega$. Each level of a two-level system in a monochromatic field is split into two quasi-energy levels. In a strong field the separation between the quasi-levels is equal to twice the Rabi frequency $(r_{mn} \cdot E)$. If weak auxiliary light is absorbed in the transition from one level of the two-level system to some third level, an absorption curve with two peaks is obtained. The distance between the peaks is equal to $(r_{mn} \cdot E)$, while the widths are of the order of γ_m.

If $\Delta\omega \ll |\boldsymbol{r}_{mn} \cdot \boldsymbol{E}|$, but $\Delta\omega > \gamma_m$, the absorption curve still exhibits two peaks with a dip in the center. Due to the fluctuations in E, the widths of the absorption peaks are of the order of $|\boldsymbol{r}_{mn} \cdot \boldsymbol{E}|$, i.e., they do not depend on the spectral width of the laser radiation.

A different situation is encountered for a wide spectrum, i.e., when $\Delta\omega \gg |\boldsymbol{r}_{mn} \cdot \boldsymbol{E}|$. Here, the two-peak structure is completely blurred, since the different frequencies $\omega \pm \Delta\omega$ exhibit entirely different Stark shifts. These shifts are determined by the quantity

$$[(\omega_{mn} - \omega \pm \Delta\omega)^2 + (\boldsymbol{r}_{mn} \cdot \boldsymbol{E})^2]^{\frac{1}{2}} \ .$$

If it is assumed that $\omega_{mn} - \omega \sim \Delta\omega$ when the external field is in resonance, the above quantity is approximately equal to

$$\Delta\omega + \frac{(\boldsymbol{r}_{mn} \cdot \boldsymbol{E})^2}{\Delta\omega} \ .$$

Due to the fluctuations in E, the second term represents the broadening of the line rather than its shift.

Therefore, in a wide spectrum, the absorption curve exhibits one very narrow peak, whose width is of the order of

$$\frac{(\boldsymbol{r}_{mn} \cdot \boldsymbol{E})^2}{\Delta\omega} \ .$$

This can be qualitatively explained by the fact that when the incident radiation has a wide spectrum, many frequencies are not in resonance with the transition frequency of the two-level system and the corresponding components of the field do not contribute anything to the absorption line.

A wide line for the non-monochromatic radiation (of the order of $\Delta\omega \gg |\boldsymbol{r}_{mm} \cdot \boldsymbol{E}|$) leads to a narrow absorption line (of the order of $|\boldsymbol{r}_{mn} \cdot \boldsymbol{E}|^2 / \Delta\omega \ll \Delta\omega$), whereas a narrow excitation line (of the order of $\Delta\omega \ll |\boldsymbol{r}_{mn} \cdot \boldsymbol{E}|$) leads to a wide absorption line (of the order of $|\boldsymbol{r}_{mn} \cdot \boldsymbol{E}| \gg \Delta\omega$).

A narrow spectrum was realized in an experiment with sodium [9.14]. While $\Delta\omega \sim 0.03$ cm^{-1}, the Rabi frequency was as high as 1 cm^{-1}, i.e., the condition for a narrow spectrum was fulfilled. When the detuning from resonance was equal to zero, the probability of the absorption of radiation from the test field for the $3P_{1/2} \rightarrow 4D_{3/2}$ transition as a function of the frequency is a symmetric curve with two peaks and a dip in the center. A strong resonance field acted on the $3S_{1/2} \rightarrow 3P_{1/2}$ transition and split the $3P_{1/2}$ state into two quasi-energy levels. The widths of the peaks are comparable to the distance between them, which is in agreement with the above theory.

The authors of [9.14] observed an additional effect associated with the non-monochromaticity of the strong field. When $\Delta = \omega - \omega_{mn}$ was small but finite (ω is the frequency of the strong field and ω_{mn} is the distance between the states $3P_{1/2}$ and $3S_{1/2}$), the left peak was higher than the right. As Δ increased, the two became equal in height. Finally, when Δ was increased even further, the right peak became much higher than the left; the latter is the result of the approximation of a strong monochromatic field. The peak inversion was due to the non-monochromaticity of the light, as a result of which the wings of the excitation band repopulated the quasi-energy levels of the $3P_{1/2}$ state.

A similar reasoning can be used to account for the influence of strong monochromatic light on the resonance fluorescence spectrum [9.15]. In a strong stochastic field with a broad spectrum, $\Delta\omega \gg |\boldsymbol{r}_{mn} \cdot \boldsymbol{E}|$, the three-peak structure of the fluorescence curve vanishes. The spectrum is represented by a one-peak curve, as in the absorption of radiation from a test field. Similar results have been obtained for resonance fluorescence with a narrow excitation spectrum.

9.2 Many-Electron Approximation

Up until now the behavior of an atom in a strong electromagnetic field has been studied using the single-electron approximation. Strictly speaking, this approximation is only valid for the hydrogen atom. The external electromagnetic field can also indirectly affect the valence electron through the dynamic polarizability of the atomic core. A number of complex atoms, e.g., those of the alkaline-earth metals and the lanthanides, possess bound two-electron and autoionization states, while the ions of these atoms contain three-electron states. In the event of resonance ionization, many-electron multi-photon transitions can take place. Let us consider the consequences of the fact that the atomic shell can contain more than one electron.

9.2.1 The Dynamic Polarization of the Atomic Core

For example, the atomic core of an alkali atom oscillates under the influence of a periodic external field. These oscillations are, of course, nonresonant, since the natural frequencies of the core are very high (of the order of Z in atomic units). They are simply forced dipole oscillations with the frequency of the applied field. In the Thomas-Fermi approximation the fluctuating dipole moment $d(t)$ is proportional to $Z^{-1}E \cos\omega t$, where Z is the atomic number [9.16].

The dipole moment can interact with the valence electron and excite the latter to other states. The corresponding interaction between the electron and the dipole is represented by the interaction potential $Z^{-1}E \cos\omega t$ (the distance between the valence electron and the dipole is approximately equal to one atomic unit). The perturbation of the electron due to the dynamic polarization is Z times less than that due to the field $E \cos\omega t$. Thus, the effect is small.

Let us now consider an atom whose outer shell contains more than one electron. In this case the Thomas-Fermi method cannot be used to describe the perturbation of one of the valence electrons by the other valence electrons. The most reasonable way to deal with the problem is to use the random-phase approximation [9.17,18].

The natural frequencies of these electrons are, undoubtedly, higher than the frequency of the perturbation, especially in the multi-photon case. Thus, the dipole oscillations in an electromagnetic field and the polarization perturbation of a valence electron are also nonresonant processes. Unfortunately, the present lack of studies in this field makes it impossible to draw any definite conclusions about the applicability of the single-electron approximation to atoms with several electrons in the valence shell.

In general one cannot account for the equivalent electrons in the outer shell by introducing a statistical factor that depends on the degeneracy of the level. The interaction between the electrons in this shell results in deviations from the single-electron approximation. For example, it is known that the interelectron interaction markedly influences the photoionization cross section [9.19]. Using the random-phase method to take into account the interelectron interaction in inert gas atoms makes it possible to obtain correct values for the photoionization cross section.

In most cases, however, the results obtained by employing the random-phase approximation appear to be almost identical to those obtained using lowest-order perturbation theory. The reason is that the principal part of the interelectron Coulomb interaction is included in the average Hartree-Fock self-consistent field. Hence, the single-electron states in such a field can always be used as a basis. The many-electron states are founded on this basis. The corrections to their energies due to the residual interelectron interaction are always small compared to the energies of the single-electron states.

In studying the valence shell it is not always clear what part of the Coulomb interaction between the electrons of this shell should be included in the average field and what part should be considered to be residual. The interelectron interaction should be observable in the multi-photon ionization of complex atoms. One should expect effects of the same order of magnitude as in photoionization (about 100 percent in the cross section), since in both cases the corrections are due to changes in the wave function of the ground states. A specific feature of the multi-photon case is the modification of the selection rules, which must lead to a change in the multi-photon matrix element. However, the experimental errors encountered in the measurement of the multi-photon cross sections are as high as one or even two orders of magnitude. For this reason the role of the residual interelectron interaction cannot be verified by such measurements.

9.2.2 Two-Electron Multi-Photon Ionization

The very first experiments involving the multi-photon ionization of a number of atoms having two electrons in the outer shell (alkaline-earth metals and the lanthanides) revealed the existence of singly and doubly charged ions [9.20]. Later studies revealed that in some cases (e.g., when a barium atom is ionized) the doubly charged ions are the result of the two-electron multi-photon ionization of the atom

$$A + k\hbar\omega = A^{2+} + 2e \tag{9.9}$$

and are not the result of a two-stage process [9.21]:

$$A + K_1\hbar\omega = A^+ + e \quad , \quad A^+ + K_2\hbar\omega = A^{2+} + e \quad . \tag{9.10}$$

This conclusion was drawn from the observed ratios of yields of singly charged and doubly charged ions, and from the different frequency dependences of the ion production probabilities. It must be assumed that the bound two-electron state or autoionization state experiences a multi-photon excitation which brings the two electrons into the continuum. The probability of such a process is of the same order of magnitude as the probability of the production of single charged ions [9.21]. This process plays a significant role in the interaction of intense radiation with complex atoms. Large cross sections for the multi-photon excitation of two-electron states are in full correspondence with the well-known data on the greater cross sections for one-photon excitation of two-electron states [9.22] and their excitation by electron impact [9.23]. In both cases these cross sections are of the same order of magnitude as the excitation cross sections for one-electron states. The production of 2^+, 3^+, and 4^+ charged ions was observed for krypton atoms [9.24]. A detailed theory to describe the features of many-electron multi-photon ionization has not yet been developed.

9.3 Ultrahigh Fields

Fields can be called ultrahigh if their intensity is greater than the characteristic intensity of the atomic field. The atomic field strength of the ground state of the hydrogen atom is about 5×10^9 V/cm (Chap.1). How does the intensity of the atomic field depend on the principal quantum number of a specific atomic state? For a qualitative analysis let us consider the states with large values of the principal quantum number, i.e., the Rydberg states. According to semi-classical arguments an atomic state is stable if the strength of the external field is smaller than [9.25]

$$E_{n,at} = \frac{5 \times 10^9 \, \omega^{2/3}}{7n^2} \text{ V/cm} \quad . \tag{9.11}$$

In the ordinary one-photon case ($E \sim E_{n,at}$) an atom in the nth state is ionized in $\tau_{at} = 2.4 \times 10^{-17} n^3$ s [9.26]. As the intensity of the field increases, the ionization proceeds with the electrons of the inner shells. This problem has been solved by applying the Thomas-Fermi model [9.27].

The multi-photon transition probabilities are still low in fields whose strengths are of the order of atomic fields (Sect.4.1). Field strengths of the order of KE_{at}, where K is the degree of the nonlinearity of the multi-photon transition, are needed for the transition time to become equal to the typical atomic period. This is not surprising because transitions are less probable when acted on by a multi-photon perturbation than when acted on by a one-photon perturbation. Since discrete atomic terms are not mixed, it would be interesting to conduct experimental and theoretical studies of the transitions and levels shifts in such fields.

At high frequencies, i.e., when $\omega \gg \omega_{mn}$, the electron undergoes classical oscillations with an amplitude of the order of E/ω^2. Even for fields of the order of E_{at} this amplitude may be smaller than the characteristic atomic dimensions, which leads to ionization times that are long compared to the atomic periods.

In principle, when the behavior of an atom in an ultrahigh field is studied, the atomic potential not the external field must be considered to be the perturbation. The Volkov wave function must be used for the solution (Sect.4.3). This makes it possible to obtain formulas for the transition probabilities. The problem of the ionization in a very strong electromagnetic field for the Coulomb atomic potential was considered along these lines [9.28]. It was found that Coulomb forces greatly increase the probability of multi-photon ionization. This probability becomes the factor $(E_{at}/E)^\lambda$, where $\lambda > 0$ depends on the adiabaticity parameter γ (Sect.4.3) and on the polarization of external field E. However, for the results to be valid, it must still be assumed that the external field is weak compared to the atomic field E_{at}.

The problem of multi-photon ionization of hydrogen by ultra-strong electromagnetic fields was solved in the limit where the field energy is greater than the Coulomb energy [9.29]. The transition rate tends to zero as the reciprocal of the field strength.

The calculations using one-dimensional delta-function atomic model potentials and an ultra-strong oscillating field was fulfilled by *Geltman* [9.30]. He found that the ionization probability becomes independent of the time. The physical nature of this result is not yet certain.

9.4 Highly Excited Atomic States in a Strong Electromagnetic Field

Highly excited atomic states exhibit two distinct features that enable a simplification of the problem of their interaction with a strong electromagnetic field to be made. First of all the highly excited states are similar to Rydberg states and their wave functions are close to Coulomb wave functions. This allows an analytical calculation of the various matrix elements that link the various highly excited states to be made. Most of the matrix elements have a comparatively simple form. In addition, the transitions between the Rydberg states can be described in classical rather than quantum mechanical terms. The transition to Newton's classical equations represents a significant simplification outside the limits of perturbation theory. The results obtained by taking into account these simplifying features are dealt with below.

9.4.1 Radiative Transitions Between Highly Excited Atomic States

To describe radiative transitions between highly excited states one has to estimate the radial dipole matrix elements, since the probabilities of induced and spontaneous radiative transitions are determined by the squares of the absolute values of these quantities.

The problem can be solved for low-lying excited states by dealing with approximate atomic wave functions [9.31]. However, the existing tables are insufficient for highly excited states. Hence, reliable analytical estimates of the radial dipole matrix elements are needed.

In highly excited states with principal quantum numbers much larger than unity, the electron is usually located far from the residual ion (at a distance of the order of n^2 in atomic units). Hence, the electronic wave functions can be considered to be hydrogen-like. The energies of these states are also close to those of hydrogen-like states because the potential of the residual ion only differs from the Coulomb potential at distances of the order of unity.

In studying highly excited atoms, one must differentiate between the two types of states: those with small values of the orbital quantum number ($\ell \sim 1$) and those with $\ell \gg 1$. States with $\ell \gg 1$ have a simpler structure. For all practical purposes the energies of these states coincide with those of hydrogen-like states. The centrifugal barrier keeps the electron away from the region where ℓ is of the order of unity, i.e., from the region where the residual ion potential differs greatly from the Coulomb potential. There is not such restriction for states with $\ell \sim 1$. Hence the levels become slightly elevated in energy compared to the hydrogen-like levels. The relative magnitude of this increase is of the order of n^{-1}. Although this quantity is small, it greatly influences the values of the matrix elements, as will be seen below.

Let us consider the case in which $\ell \gg 1$. This problem is completely reducible to the hydrogen-like problem. The exact values of the radial dipole matrix elements are expressed in terms of the combinations of hyper-geometric functions [9.32]. They are quite cumbersome, which hinders their practical use. For highly excited states, whose principal quantum numbers are large compared to unity, the expressions for the matrix elements can be considerably simplified. If $\Delta n = n' - n$ is of the order of unity, then the matrix elements are equal to the Fourier transforms of the corresponding classical quantity. Analytical formulas were obtained along these lines [9.33]:

$$R_{n,\ell}^{n-\Delta n,\ell-1} = \int_0^\infty R_{n\ell}(r) R_{n-\Delta n,\ell-1}(r)\, r^3 dr \qquad (9.12)$$

in terms of combinations of Bessel functions:

$$R_{n,\ell}^{n-\Delta n,\ell-1} = \frac{n^2}{Z|\Delta n|} \left\{ J'_{\Delta n}[\,|\Delta n|(1-x^2)^{\frac{1}{2}}] + \frac{x\ \text{sgn}\Delta n}{(1-x^2)^{\frac{1}{2}}} J_{\Delta n}[\,|\Delta n|(1-x^2)^{\frac{1}{2}}] \right\}\ , \qquad (9.13)$$

where $x \equiv \ell/n$. The same is true for the region in which $1 \ll |\Delta n| \ll n$.

A different estimate of the values of the matrix elements has been carried out for $\Delta n \sim n$, $n' \gg 1$ [9.34]. For $\ell \ll n$ we have

$$R_{n\ell}^{n',\ell\pm1} = \frac{2\ell^2\sqrt{nn'}}{\pi\sqrt{3}(n^2 - n'^2)} \; [K_{2/3}(\omega\ell^3/3) \mp K_{1/3}(\omega\ell^3/3)] \quad , \tag{9.14}$$

where $K_\nu(z)$ is a MacDonald function, and ω the atomic transition frequency, i.e.,

$$\omega = 1/2n^2 - 1/2n'^2 \quad . \tag{9.15}$$

The matrix elements have been estimated in [9.34] for the case where $\ell \sim n$, i.e., when they are exponentially small. Due to their complexity we shall not give them here.

There is an approximate selection rule based on the known numerical values for the dipole matrix elements [9.35], namely, when the orbital angular momentum ℓ changes in the course of a dipole transition, the principal quantum number n changes primarily in the same direction as ℓ. The above quasi-classical estimates enable us to obtain this selection rule theoretically. For one, the ratio of the matrix elements of "forbidden" and "allowed" transitions proves to be of the order of $(\Delta n)^{-1} \ll 1$ if $\ell \gg n^{2/3}$. This rule can be generalized as follows: in the course of a dipole transition the principal quantum number n changes in the same direction as ℓ and by the same amount. This rule is important when summing over intermediate states in multi-photon matrix elements (Chap.7).

For states of real atoms with $n \gg 1$ and $\ell \sim 1$ the quantum defect of Rydberg atomic energies is essential. Equations (9.13,14) are modified by substituting $n^* = n - \delta_\ell$ for n, with $\delta_\ell \sim 1$ the quantum defect for a definite value of ℓ. Here $\Delta n^* = \Delta n - \delta_\ell + \delta_{\ell'}$ is substituted for Δn and may differ substantially from it when Δn^* is approximately unity. Numerical values of the radial matrix elements have been given in [9.36,37]. A comparison of the results obtained with values found in the quantum defect approximation shows that the error of quasi-classical calculations does not exceed two percent for $n \geq 3$, so that the formulas have a very wide range of application.

9.4.2 The Dynamic Polarizability of Highly Excited Atomic States

In the case of highly excited atomic states one can obtain a fairly simple analytical expression for the dynamic polarizability. For the matrix elements $z_{n\ell m}^{n'\ell'm'}$ and the energy differences $\omega_{nn'}$ we can then use the known values since for $n \gg 1$ the effective atomic potential acting on an electron is a Coulomb potential. Here n, ℓ, and m are the principal, orbital, and magnetic quantum numbers. The value of $\alpha_{n\ell m}$ for an arbitrary ℓ is given in [9.38]. For the sake of simplicity, we shall restrict our discussion to the case

$\ell = 0$, which demonstrates all the characteristic features of α:

$$\alpha_{n00} = \frac{2}{3} n^6 \sum_{s=1}^{\infty} \frac{12s^2 - (\omega n^3)^2}{[s^2 - (\omega n^3)^2]^2} J_s'^2(s) \quad , \tag{9.16}$$

where $J_s(s)$ is a Bessel function. In the static limit $\omega = 0$ this yields

$$\alpha_{n00} \cong n^6 \quad , \tag{9.17}$$

while in the opposite limit of high frequencies we have

$$\alpha_{N00} = -1/\omega^2 \quad . \tag{9.18}$$

In the latter case the dependence on n vanishes and the expression corresponds to a free electron vibrating in the field of a plane electromagnetic wave.

Note that the dynamic polarizability $\alpha_{n\ell m}$ generally is of the order of n^6, i.e., it rapidly grows with the principal quantum number n. This estimate remains valid in the static limit.

The dynamic polarizability becomes infinite at $\omega = s/n^3$, s being an integer. Note that in contrast to ground and low-lying atomic states α does not change sign. This result follows from the following fact. For ground and low-lying states we encountered a fixed resonance with a definite discrete atomic level. For highly excited states, however, due to the equidistant separation of levels in the energy spectrum, there are two resonances in the vicinity of the nth level we are considering here simultaneously with a higher- and a lower-lying level (in relation to level n). They largely cancel each other, as a result of which α is of the order of n^6, which is n times smaller than separate terms in the entire sum for α. Note that in the vicinity of a resonance the dynamic polarizability α is always positive.

All these properties of α remain valid for $\ell \neq 0$, but a new feature is the dependence of α on m. When ℓ is much larger than unity, the dependence has the relatively simple form

$$\alpha \propto 1 - m^2/\ell^2 \quad . \tag{9.19}$$

As shown in [9.38], the sum of terms in the series for α whose numbers n' are far from n as well as the sum over the continuum provides only a small contribution to α due to the canceling of terms with n' > n and with n' < n. Although for terms with n' close to n such cancellation also takes place, these terms provide the main contribution due to the large quasi-classical matrix elements $z_{n\ell m}^{n'\ell'm'} \propto n^2$.

9.4.3 Multi-Photon Ionization of Highly Excited States

A specific feature of this problem is the considerable simplification of the formulas for the multi-photon matrix elements (2.22), which is due to the simplicity of the analytical expressions for the quasi-classical matrix elements (9.13,14) and the energy denominators in the multi-photon matrix elements.

All characteristic features of multi-photon ionization can clearly be seen in two-photon ionization, which we shall now consider. The probability of two-photon ionization is determined by the two-photon matrix element (2.13). It was shown in [9.39] that the sum over intermediate states in (2.13) in the quasi-classical approximation consists largely of terms in which the resonance denominator $\omega_{n'n} - \omega$ is approximately zero (here n and n' are the principal quantum numbers of the initial and intermediate states; both are much greater than unity). If the detuning from resonance with an intermediate state whose principal quantum number is n' is very small, we can restrict (2.13) to a few terms only.

Since n and n' differ greatly from the principal quantum number that characterizes the final state (this state can be described by the momentum $p \ll 1$ of the emitted photoelectron), the matrix elements entering into (2.13) are determined via (9.14). Here one must assume that $\ell \ll n$, since ionization from a state with $\ell \sim n$ has an exponentially low probability (Sect.9.4.1).

As a result, for the ionization cross section $\alpha_n^{(2)}$ summed over the orbital angular momenta of the final states and averaged over all the substates of the initial degenerate level with the principal quantum number n we have

$$\alpha_n^{(2)} = \frac{0.1}{\pi n^5 \omega^{22/3}} \cot^2 \frac{\pi n}{(1 - 2\omega n^2)^{1/2}} \quad . \tag{9.20}$$

A rough estimate of (9.20) for $\omega \sim n^{-2}$ has the form

$$\alpha_n^{(2)} \sim n^{29/3} \quad . \tag{9.21}$$

We note that just as in the one-photon case the ionization cross section does not depend on the momentum p of the emitted photoelectron explicitly. From (9.20) we see that when an intermediate resonance occurs, the argument of the cotangent becomes equal to an integer times π and the cross section becomes infinite.

In writing (9.20) we assumed that $\omega < 1/2n^2$, i.e., the intermediate resonance states lie in the highly excited part of the discrete spectrum. If $\omega > 1/2n^2$ and these states are in the continuous spectrum, then alongside with two-photon ionization there is one-photon ionization, whose cross section is given by the well-known Kramer's formulas. As for the two-photon matrix element (2.13), in contrast to the case where $\omega < 1/2n^2$, at $\omega > 1/2n^2$, it has an additional term (an imaginary part) since with the states p' in the continuum we must substitute $\omega_{p'n} - \omega - i\delta$ for $\omega_{p'n} - \omega$. The contribution of this term to the two-photon cross section can easily be calculated if one uses the quasi-classical matrix elements (9.14). As a result the imaginary part proves to be dominant in comparison with the real part (i.e., in comparison with the Cauchy principal value), with $\alpha_n^{(2)}$ defined by (9.20) where 1 has been substituted for $\cot^2(\dots)$ [9.40].

The imaginary part of the two-photon matrix element corresponds to the cascade transition $n \rightarrow p' \rightarrow p$ through a real (and not virtual) state p'. What we have said is proof that the probability of two-photon cascade ionization contributes the greatest part to the total probability.

9.4.4 Tunneling Ionization from Highly Excited States

In Chap.7 we studied tunneling ionization from the ground state of the hydrogen atom. Here we will consider the ionization of highly excited states by a low-frequency field. In a constant electric field the ionization probability for an atomic state with principal quantum number n and parabolic quantum numbers n_1 and n_2 is determined via the hydrogen-like approximation [9.41], which is valid for highly excited states, i.e., at $n \gg 1$:

$$w_{E=\text{const}} = \left(\frac{4}{n^3 E}\right)^{n-n_1+n_2} \frac{\exp[-2/3En^3 + 3(n_1 - n_2)]}{n^3 n_2!(n - n_1 - 1)!} . \qquad (9.22)$$

In the case of a monochromatic field with frequency ω so low that $\gamma^2 \ll 1$, i.e., $(\omega/nE)^2 \ll 1$, the tunneling rate can be found by a well-known method (Chap. 7 and [9.42]) and is

$$w = \left(\frac{3En^3}{\pi}\right)^{\frac{1}{2}} w_{E=\text{const}} . \qquad (9.23)$$

Here for the sake of definiteness we have put $n_2 > n_1$.

In reality, the result (9.23) holds, according to [9.43], for $\gamma^2 n \ll 1$, i.e., $(\omega/\sqrt{n}E)^2 \ll 1$, which is a condition more stringent than the one above. If $\gamma^2 n \gtrsim 1$ (but $\gamma^2 \ll 1$), splitting of the initial spectral term into quasi-energy levels becomes important. The result is then more cumbersome [9.43].

In the above reasoning we assumed that when the field is switched on the parabolic quantum numbers n_1 and n_2 remain "good" quantum numbers, notwithstanding the fact that the external field varies. The smallness of the mixing is due to the fact that the mixing occurs only if we allow for other principal shells (Chap.7). But this is unimportant if $E \ll E_{at,n}$, i.e., $E \ll n^{-4}$, a condition we assume to be valid from the start.

Note that under the given conditions, the frequency ω of the external field is very low compared with the classical frequency $1/n^3$ with which an electron revolves around the nucleus along a high orbit.

Let us now consider the case where the atomic states are characterized by an orbital angular momentum ℓ and its projection m. For highly excited atomic states this situation is realized when the level separation within a single multiplet but with different values of ℓ is large compared to the size of the perturbation, n^2E. According to [9.43], in a monochromatic field the tunneling ionization probability has the form

$$w = |C_{n\ell}|^2 \frac{(2\ell + 1)(\ell + |m|)!}{2^{|m|} \times 2n^2 \times (|m|)!(\ell - |m|)!}$$

$$\times \left(\frac{3n^3E}{\pi}\right)^{\frac{1}{2}} \left(\frac{2}{n^3E}\right)^{2n-|m|-1} \exp\left(-\frac{2}{3n^3E}\right) . \tag{9.24}$$

Here $C_{n\ell}$ is the coefficient in the asymptotic expansion of the unperturbed atomic wave function $\psi(\mathbf{r})$ at large distances r from the atom:

$$\psi(\mathbf{r}) = C_{n\ell} n^{-3/2} \left(\frac{r}{n}\right)^{n-1} e^{-r/n} Y_{\ell m}(\Omega) . \tag{9.25}$$
$$r \to \infty$$

Regardless of the dependence of the energy levels on ℓ due to the perturbation introduced by the atomic core potential, the wave function ψ of highly excited states are hydrogen-like with a high precision. According to [9.43],

$$C_{n\ell} = 2^n [n(n + \ell)!(n - \ell - 1)!]^{-\frac{1}{2}} . \tag{9.26}$$

Substituting (9.26) into (9.24), we find the tunneling ionization rate from a highly excited state $(n\ell m)$ in a monochromatic electromagnetic field:

$$w = (3/\pi)^{1/2} 2^{4n-2|m|-2} (2\ell + 1) n^{-6n+3|m|+3/2}$$

$$\times \frac{(\ell + |m|)! E^{-2n+|m|+3/2}}{(n + \ell)!(n - \ell - 1)!(|m|)!(\ell - |m|)!} \exp\left(-\frac{2}{3n^3E}\right) . \tag{9.27}$$

Note that in both (9.22 and 27) the exponent in the exponential function coincides with that for a short-range potential, but here there is a very large pre-exponential factor,

$$(n^3E)^{-n+n_1-n_2} \quad \text{or}$$

$$(n^3E)^{-2n+|m|} \quad ,$$

which results in a very high probability of tunneling. For this reason w becomes close to unity in considerably weaker fields than in the case of a short-range potential. Analyzing (9.22,27), one can easily see that the corresponding atomic field E_{cr} is proportional to E_n^2, while for a short-range potential (Chap.7) E_{cr} is proportional to $E_n^{3/2}$, where E_n is the energy of the level being ionized.

9.4.5 Stochastic Instability of the Classical Electron in a Variable Field and the Diffusion Ionization of Highly Excited Atoms

All the above reasoning was based on quantum mechanical laws. A specific feature of highly excited states is that laws of classical mechanics can be used to describe ionization, since $n \gg 1$.

Naturally, for a given value of n, ionization is most intensive when the external field frequency ω is close to, or an integral multiple of, the Kepler frequency $\Omega_n = 1/n^3$ of electron revolution along the classical orbit. Quantum-mechanically, this corresponds to resonance ionization.

For a weak field the probability of such ionization occurring is infinitely low because a large number of photons must be absorbed in the ionization process. As the intensity of the radiation is increased, there will be a point where the lowest order of perturbation theory for which energy conservation works ceases to be valid. We call such fields strong. In view of what has been said, the problem can be considered by classical means instead of quantum-mechanically.

In the field of an intense monochromatic (electromagnetic) wave, the nonlinear oscillations of an electron prove to be stochastic (i.e., the frequency is energy dependent), which leads to diffusion ionization of the atom [9.44]. The mechanism of this ionization works as follows. The electron gradually increases its energy, and the time τ_D it takes the electron to increase its energy from the initial value to $E = 0$ (which corresponds to ionization of the atom) is considerably greater than the time of revolution along an unperturbed orbit, which is n^3.

Both computer [9.45] and analytical [9.46] calculations provide estimates for the strength of the critical field necessary for stochastic instabilities to set in. Analytically this estimate has the same form as that for the critical field E_{cr} necessary for classical ionization by a constant electric field (i.e., above-the-barrier ionization):

$$E_{cr} = \frac{1}{C(\omega/\Omega_n)n^4} \quad . \tag{9.28}$$

The constant C depends on the frequency and type of polarization of the electromagnetic field.

In accordance with what has been said, C is minimum when ω is close to Ω_n and grows rapidly when ω deviates from the Kepler frequency. For instance, for a linearly polarized field, $C(1) \approx 25$ [9.45]. In contrast to the ionization mechanisms discussed earlier (multi-photon and tunneling), which are not threshold mechanisms, ionization due to stochastic instabilities does not occur below the threshold E_{cr}. When the ionizing field is higher than E_{cr}, it becomes necessary to calculate the probability for such an ionization process to occur.

As can any stochastic process, ionization under these conditions can be described by a diffusion equation if we consider only large times of field interaction or if we average over a large number of initial configurations of classical orbits of the atom in space in relation to the external field. We note once more that this diffusion process is classical in its physics. Within the scope of the quasi-classical approximation of quantum mechanics it was studied in [9.44]. When the total ionization probability W is much less than unity, it proves to be approximately $E^2 n^5$. This result has been verified by both numerical calculation [9.48] and analytical estimates [9.47]. Substituting (9.28), we find that $W \approx 10^{-3} n^{-3}$ (in atomic units).

At $\omega \ll \Omega_n$ the stochastic ionization mechanism is replaced basically by tunneling ionization considered in Chap.7. In fields of the order of (9.28), tunneling ionization is not exponentially small, and its probability must be determined via more complex formulas than those discussed in Chap.7. For highly excited states the tunneling probability has practically the same behavior as diffusion probability, i.e., it has a threshold and varies from zero to unity in a very narrow range of field strengths.

In the other limiting case, $\omega \gg \Omega_n$, stochastic ionization ceases, too. As in all situations discussed in this book, in this limiting case the conditions for using perturbation theory improve. In Sect.9.4.3 we discussed two-photon ionization from highly excited states. The multi-photon ionization probability from a highly excited state also has a threshold in the field strength; the corresponding constant in (9.28) was found in [9.25]. For one, it was found that at $\omega \approx \Omega_n = 1/n^3$ the probability of the nth level being ionized has the form

$$w_n \sim (E/E_{at})^{2K} \quad , \tag{9.29}$$

where, see (9.11),

$$E_{at} \cong 1/7n^4 \quad , \tag{9.30}$$

and $K = E_n/\omega = 1/2n^2$ is the number of photons necessary for ionization. Thus, in the region $1/80n^4 < E < 1/7n^4$ the mechanism of diffusion ionization is realized.

In conclusion, we note that classical diffusion ionization presupposes a quantum limit from below on the field strength. For the random walk of an electron to enclose many atomic levels (in other words, so that the discrete structure of the atomic spectrum does not manifest itself), E must be greater than n^{-6} [9.49]. Comparing this restriction with (9.28), we can see that the above conclusions concerning diffusion ionization require large values of n.

The ionization of states with $n \approx 60$ in a field of frequency $\omega = 9.9$ GHz $= 0.43 \, \Omega_n$ was experimentally observed and reported in [9.50]. The field strength in these experiments was 10 V/cm $= 1/25n^4$, which corresponds, with a fairly good accuracy, to the threshold of classical ionization (9.28), with the given values of ω and n.

Notation Index

Latin Symbols

A	Vector potential of the electromagnetic field
a_m	Probability amplitude of the state m
c	Speed of light
d	Dipole moment
$E_m^{(0)}$	Energy of the unperturbed state m
E_m	Energy of the perturbed state m
E_α	Quasi-energy of the state α
e	Polarization vector of light
E	Electric field strength
E_{at}	Atomic field strength
F	Radiation intensity; total angular momentum of an atom
G	Green function
G_K	Correlation function of the Kth order
g_K	Statistical factor
H	Hamiltonian
H	Magnetic field strength
J	Total angular momentum
K	Number of photons
k	Wave vectors
L, ℓ	Orbital angular momentum
M, m	Magnetic quantum number
n	Principal quantum number; number of photons
o	Solid angle
P	Power density
p	Momentum of the electron
R	Reflection coefficient
S	Spin angular momentum

T	Operation time of the radiation
t	Time
U	Atomic potential
V	Field perturbation
W	Transition probability
w	Transition probability per unit time
Z	Number of electrons in an atom
$z_{mn}^{(K)}$	K-photon dipole matrix element

Greek Symbols

α	Polarization of the radiation
α_m	Polarizability of the state m
Γ_m	Width of the state m
γ	Adiabaticity parameter
γ_m	Natural width of the state m
Δ	Detuning from resonance
Δ_K	Detuning from a K-photon resonance
$\Delta\omega$	Spectral linewidth of the radiation
δE_m	Shift in the energy of the state m
δT	Switch-on time
ε_N	Energy eigenvalues for the degenerate state N
λ	Wavelength
$\Delta\lambda$	Linewidth
μ_0	Bohr magneton
ν	Radiation frequency
π	Parity
ρ	Energy density
τ	Operation time of the radiation
τ_{cor}	Duration of the fluctuation peak
ϕ	Phase
χ	Susceptibility
Ψ	Wave function
ψ	Stationary wave function
$\psi(t)$	Pulse shape
ω	Laser frequency
ω_{mn}	Atomic transition frequency
Ω	Solid angle; Rabi frequency
Ω_K	K-photon Rabi frequency

References

Chapter 1

1.1 W. Heitler: *The Quantum Theory of Radiation*, 3rd ed. (Oxford University Press, London 1954)
1.2 I.I. Sobelman: *Atomic Spectra and Radiative Transition*, Springer Ser. Chem. Phys., Vol. 1 (Springer, Berlin, Heidelberg, New York 1979)
1.3 L.I. Schiff: *Quantum Mechanics*, 3rd ed. (McGraw-Hill, New York 1968)
1.4 N.H. March, W.H. Young, S. Sampanthar: *The Many-Body Problem in Quantum Mechanics* (Cambridge University Press, London 1967)
1.5 I. Lindgren, J. Morrison: *Atomic Many-Body Theory*, Springer Ser. Chem. Phys., Vol. 13 (Springer, Berlin, Heidelberg, New York 1982)
1.6 U. Fano, J. Cooper: Rev. Mod. Phys. **40**, 441 (1968)
1.7 B.A. Zon, B.G. Katsnel'son: Opt. Spektrosk. **40**, 952 (1976) [English transl.: Opt. Spectrosc. **40**, 544 (1976)]
1.8 M. Goeppert-Mayer: Naturwiss. **17**, 932 (1929)
1.9 F.V. Bunkin: Zh. Eksp. Teor. Fiz. **50**, 1685 (1966) [English transl.: Sov. Phys. JETP **23**, 1121 (1966)]
1.10 H. Bethe, R.W. Jackiw: *Intermediate Quantum Mechanics*, 2nd ed. (Benjamin, New York 1968)
1.11 M. Lax: *Fluctuation and Coherence Phenomena in Classical and Quantum Physics* (Gordon and Breach, New York 1968)
1.12 S. Stenholm: Phys. Repts. C**6**, 1 (1973)
1.13 F.V. Bunkin, A.M. Prokhorov: Zh. Eksp. Teor. Fiz. **46**, 1090 (1964) [English transl.: Sov. Phys. JETP **19**, 739 (1964)]
1.14 L. Allen, J.H. Eberly: *Optical Resonance and Two-Level Atoms* (Wiley, New York 1975)
1.15 E. Kamke: *Differentialgleichungen, Lösungsmethoden und Lösungen*, Vol. 1 (Akademische Verlagsgesellschaft Geest & Portig KG, Leipzig 1944)
1.16 Ya.B. Zel'dovich: Usp. Fiz. Nauk **110**, 139 (1973) [English transl.: Sov. Phys. Uspekhi **16**, 427 (1974)]
1.17 C. Cohen-Tannoudji, S. Haroche: In *Polarization, Matter and Radiation* (Jubilee volume in honor of Alfred Kastler) (Press Universitaires de France, Paris 1969)
1.18 C. Cohen-Tannoudji, S. Reynaud: In *Multiphoton Processes*, ed. by P. Lambropoulos, J.H. Eberly (Wiley, New York 1978) p. 103
1.19 H. Sambe: Phys. Rev. A**7**, 2203 (1973)
1.20 L.D. Landau, E.M. Lifshitz: *Quantum Mechanics: Non-Relativistic Theory*, 3rd ed. (Pergamon, Oxford 1977)
1.21 V.B. Berestetskii, E.M. Lifshitz, L.P. Pitaevskii: *Relativistic Quantum Theory*, Pt. 1 (Pergamon, Oxford 1971)
1.22 A.I. Baz', Ya.B. Zel'dovich, A.M. Perelomov: *Rasseyanie, reaktsii i raspady v nerelyativistskoi kvantovoi mekhanike* (Scattering, Reactions, and Decays in Nonrelativistic Quantum Mechanics), 2nd ed. (Nauka, Moscow 1971)

Chapter 2

2.1 S. Flügge: *Practical Quantum Mechanics*, P. II (Springer, Berlin, Heidelberg, New York 1971) Problem 180
2.2 L.D. Landau, E.M. Lifshitz: *Quantum Mechanics: Non-Relativistic Theory*, 3rd ed. (Pergamon, Oxford 1977) § 77
2.3 L.I. Schiff: *Quantum Mechanics*, 2nd ed. (McGraw-Hill, New York 1959) Chap. 7
2.4 N. Bloembergen: *Nonlinear Optics* (W.A. Benjamin, New York 1965) Chap. 2
2.5 D.C. Hanna, M.A. Yuratich, D. Cotter: *Nonlinear Optics of Free Atoms and Molecules*, Springer Ser. Opt. Sci., Vol. 17 (Springer, Berlin, Heidelberg, New York 1979)
2.6 W. Heitler: *The Quantum Theory of Radiation*, 3rd ed. (Oxford University Press, London 1954)
2.7 R. Loudon: *The Quantum Theory of Light* (Clarendon, Oxford 1973) Chaps. 8 and 12
2.8 I. Lindgren, J. Morrison:*Atomic Many-Body Theory*, Springer Ser. Chem. Phys., Vol. 13 (Springer, Berlin, Heidelberg, New York 1982)
2.9 N.H. March, W.H. Young, S. Sampanthar: *The Many-Body Problem in Quantum Mechanics* (Cambridge University Press, London 1967)
2.10 E.N. Economou: *Green's Functions in Quantum Physics*, 2nd ed., Springer Ser. Solid. State. Sci., Vol. 7 (Springer, Berlin, Heidelberg, New York 1983)
2.11 V.I. Ritus: Zh. Eksp. Teor. Fiz. **51**, 1544 (1966) [English transl.: Sov. Phys. JETP **24**, 1041 (1967)]
2.12 B.A. Zon, E.I. Sholokhov: Zh. Elsp. Teor. Fiz. **70**, 887 (1976) [English transl.: Sov. Phys. JETP **43**, 461 (1977)]
2.13 L. Hostler: J. Math. Phys. **5**, 591 (1964)
2.14 B.A. Zon, N.L. Manakov, L.P. Rapoport:*Teoriya mnogofotonnykh perekhodov v atomakh* (The Theory of Multiphoton Transitions in Atoms) (Atomizdat, Moscow 1978)
2.15 G. Simons: J. Chem. Phys. **55**, 756 (1971)
2.16 M. Seaton: Mon. Not. Royal Astron. Soc. **118**, 4504 (1958)

Chapter 3

3.1 L.D. Landau, E.M. Lifshitz: *Quantum Mechanics: Non=Relativistic Theory*, 3rd ed. (Pergamon, Oxford 1977), Problem to § 40
3.2 N.D. Sen Gupta: Phys. Lett. A**42**, 33 (1972)
3.3 E.T. Whittaker, G.N. Watson: *A Course of Modern Analysis*, 4th ed. (Cambridge University Press, London 1958)
3.4 L. Allen, J.H. Eberly: *Optical Resonance and Two-Level Atoms* (Wiley, New York 1975)
3.5 S.P. Goreslavskii, V.P. Yakovlev: Izvest. Akad. Nauk SSSR, seriya fizicheskaya **37**, 2211 (1973) [English transl.: Bull. Acad. Sci. USSR. Phys. Sci. **37**, 171 (1973)]
3.6 Ya.B. Zel'dovich: Usp. Fiz. Nauk **110**, 139 (1973) [English transl.: Sov. Phys. Uspekhi **16**, 427 (1974)]
3.7 N.B. Delone, V.P. Krainov, V.A. Khodovoi: Usp. Fiz. Nauk **117**, 189 (1975) [English transl.: Sov. Phys.=Uspekhi **18**, 750 (1975)]
3.8 B.A. Zon, B.G. Katsnel'son: Zh. Eksp. Teor. Fiz. **65**, 947 (1973) [English transl.: Sov. Phys. JETP **38**, 470 (1975)]
3.9 I.I. Sobelman, L.A. Vainshtein, E.A. Yukov: *Excitation of Atoms and Broadening of Spectral Lines*, Springer Ser. Chem. Phys., Vol. 7 (Springer, Berlin, Heidelberg, New York 1981)
3.10 L.A. Vainshtein, L.P. Presnyakov, I.I. Sobel'man: Zh. Eksp. Teor. Fiz. **43**, 518 (1962) [English transl.: Sov. Phys. JETP **16**, 370 (1963)1

Chapter 4

4.1 L.D. Landau, E.M. Lifshitz: *Quantum Mechanics: Non=Relativistic Theory*, 3rd ed. (Pergamon, Oxford 1977) Chap. 7

4.2 A.M. Dykhne: Zh. Eksp. Teor. Fiz. **41**, 1324 (1961) [English transl.: Sov. Phys. JETP **14**, 941 (1962)]
4.3 L.I. Schiff: *Quantum Mechanics*, 2nd ed. (McGraw-Hill, New York 1959) Sec.31
4.4 E. Jahnke, F. Emde, F. Lösch: *Tables of Higher Functions*, 6th ed. (McGraw-Hill, New York 1960)
4.5 R.P. Feynman, A.R. Hibbs: *Quantum Mechanics and Path Integrals* (McGraw-Hill, New York 1965)
4.6 I. Lindgren, J. Morrison: *Atomic Many-Body Theory*, Springer Ser. Chem. Phys., Vol. 13 (Springer, Berlin, Heidelberg, New York 1982)
4.7 D.F. Zaretskii, V.P. Krainov: Zh. Eksp. Teor. Fiz. **66**, 537 (1975) [English transl.: Sov. Phys. JETP **39**, 257 (1974)]
4.8 Yu.A. Bychkov, A.M. Dykhne: Zh. Eksp. Teor. Fiz. **58**, 1734 (1970) [English transl.: Sov. Phys. JETP **31**, 928 (1970)]
4.9 V.P. Krainov: Zh. Teor. Eksp. Fiz. **70**, 1197 (1976) [English transl.: Sov. Phys. JETP **43**, 622 (1977)]
4.10 D.F. Zaretskii, V.P. Krainov: Zh. Eksp. Teor. Fiz. **67**, 1301 (1974) [English transl.: Sov. Phys. JETP **40**, 647 (1975)]
4.11 J.V. Moloney, W.J. Meath: Molec. Phys. **31**, 1537 (1976)
4.12 V.P. Krainov, V.P. Yakovlev: Zh. Eksp. Teor. Fiz. **78**, 2204 (1980) [English transl.: Sov. Phys. JETP **51**, 1104 (1980)]
4.13 V.A. Kovarskii, S.A. Baranov: Zh. Eksp. Teor. Fiz. **71**, 2033 (1976) [English transl.: Sov. Phys. JETP **44**, 1067 (1976)]
4.14 L.V. Keldysh: Zh. Eksp. Teor. Fiz. **47**, 1945 (1964) [English transl.: Sov. Phys. JETP **20**, 1307 (1965)]
4.15 V.B. Berestetskii, E.M. Lifshitz, L.P. Pitaevskii: *Relativistic Quantum Theory*, Pt. 1 (Pergamon, Oxford 1971) § 32
4.16 A.I. Baz', Ya.B. Zel'dovich, A.M. Perelomov: *Rasseyanie, reaktsii i raspady v nerelyativistskoi kvantovoi mekhanike* (Scattering, Reactions, and Decays in Nonrelativistiv Quantum Mechanics) 2nd ed. (Nauka, Moscow 1971) Chap. 5, § 4

Chapter 5

5.1 H. Haken: Laser Theory (Springer, Berlin, Heidelberg, New York 1984)
5.2 H.K.V. Lotsch: Optik **26**, 112 (1967)
5.3 W. Koechner: *Solid-State Laser Engineering*, Springer Ser. Opt. Sci., Vol. 1 (Springer, Berlin, Heidelberg, New York 1976)
5.4 F.P. Schäfer (ed.): *Dye Lasers*, 2nd ed., Topics Appl. Phys. Vol. 1 (Springer, Berlin, Heidelberg, New York 1977)
5.5 Ch.K. Rhodes (ed.): *Eximer Lasers*, 2nd ed., Topics Appl. Phys., Vol. 30 (Springer, Berlin, Heidelberg, New York 1984)
5.6 P.G. Kryukov, V.S. Letokhov: Usp. Fiz. Nauk **99**, 169 (1969) [English transl.: Sov. Phys. Uspekhi **12**, 641 (1970)]
5.7 M. Young: *Optics and Lasers*, 2nd ed., Springer Ser. Opt. Sci., Vol. 5 (Springer, Berlin, Heidelberg, New York 1984)
5.8 M. Born, E. Wolf: *Principles of Optics*, 5th ed. (Pergamon, Oxford 1975)
5.9 R.H. Kingston: *Detection of Optical and Infrared Radiation*, Springer Ser. Opt. Sci., Vol. 10 (Springer, Berlin, Heidelberg, New York 1978)
5.10 R.J. Keyes (ed.): *Optical and Infrared Detectors*, 2nd ed., Topics Appl. Phys., Vol. 19 (Springer, Berlin, Heidelberg, New York 1980) Chap. 3
5.11 H.M. Smith (ed.): *Holographic Recording Materials*, Topics Appl. Phys., Vol. 20 (Springer, Berlin, Heidelberg, New York 1977) Chap. 2
5.12 S.L. Shapiro (ed.): *Ultrashort Light Pulses*, Topics Appl. Phys., Vol. 18 (Springer, Berlin, Heidelberg, New York 1977)
5.13 V. Korobkin, B. Stepanov, S. Fanchenko, M. Schelev: Opt. and Quant. Electr. **10**, 367 (1978)
5.14 L. Lompre, G. Mainfray, J. Thebault: Rev. Physique Appl. **17**, 21 (1982
5.15 B.Ya. Zel'dovich, T.N. Kuznetsova: Usp. Fiz. Nauk **106**, 47 (1972) [English transl.: Sov. Phys. Uspekhi **15**, 25 (1972)]
5.16 W. Demtröder: *Laser Spectroscopy*, Springer Ser. Chem. Phys., Vol. 5 (Springer, Berlin, Heidelberg, New York 1982)

5.17 L.D. Landau, E.M. Lifshitz: *The Classical Theory of Fields*, 4th ed. (Pergamon, Oxford 1975) § 50
5.18 W. Shurkliff: *Polarized Light* (Harvard University Press, Cambridge, Ma. 1962)
5.19 N.B. Delone, V.A. Kovarskii, A.V. Masalov, N.F. Perel'man: Trudy FIAN **115**, 140 (1980) [English transl.: Proc. P.N. Lebedev Physics Institute **115**, 140 (1981)]; also Usp. Fiz. Nauk **131**, 617 (1980) [English transl.: Sov. Phys. Uspekhi **23**, 472 (1980)]

Chapter 6

6.1 K. Smith: *Molecular Beams* (Wiley, New York 1955);
    N. Ramsey: *Molecular Beams* (Clarendon, Oxford 1956)
6.2 I. Delcroix, C. Mattos-Fereira, A. Richard: *Atomes metastables des les gaz ionises* (Acad. Press, Paris 1975)
6.3 L.V. Gurvich, G.V. Karachavtsev, V.N. Kondrat'ev, Yu.A. Lebedev, V.A. Medvedev, V.I. Potapov, Yu.S. Khodeev: *Energiya razryva khimicheskikh svyazei. Potentsialy ionizatsii i srodstvo k elektronu* (The Energy of Chemical Bond Breaking. Ionization Potentials and Electron Affinity) (Nauka, Moscow 1974)
6.4 S. Chin: Can. J. Phys. **48**, 1314 (1970)
6.5 A. Benninghoven (ed.): *Ion Formation from Organic Solids*, Springer Ser. Chem. Phys., Vol. 25 (Springer, Berlin, Heidelberg, New York 1983)
6.6 I.I. Sobel'man: *Atom Spectra and Radiative Transitions*, Springer Ser. Chem. Phys., Vol. 1 (Springer, Berlin, Heidelberg, New York 1979)
6.7 I.I. Sobel'man, L.A. Vainshtein, E.A. Yukov: *Excitation of Atoms and Broadening of Spectral Lines*, Springer Ser. Chem. Phys., Vol. 17 (Springer, Berlin, Heidelberg, New York 1981)
6.8 V.S. Lisitsa, S.I. Yakovlenko: Zh. Eksp. Teor. Fiz. **68**, 479)1975) [English transl.: Sov. Phys. JETP **41**, 233 (1975)]
6.9 S.I. Yakovlenko: Kvantovaya Elektron **5**, 259 (1978) [English transl.: Sov. J. Quantum Electron. **8**, 151 (1978)]
6.10 F.V. Bunkin, A.E. Kazakov, M.V. Fedorov: Usp. Fiz. Nauk **107**, 559 (1972) [English transl.: Sov. Phys. Uspekhi **15**, 416 (1973)]
6.11 P. Peyraud: J. Phys. **29**, 872 (1968)
6.12 J.B. Hasted: *Physics of Atomic Collisions* (Plenum Press, New York 1964) Chaps. 4-6
6.13 V.B. Berestetskii, E.M. Lifshitz, L.P. Pitaevskii: *Relativistic Quantum Theory*, Part 1 (Pergamon, Oxford 1971) § 90
6.14 Ya.B. Zel'dovich, Yu.P. Raizer: Zh. Eksp. Teor. Fiz. **47**, 1150 (1964) [English transl.: Sov. Phys. JETP **20**, 772 (1965)]
6.15 Yu.P. Raizer: *Laser-Induced Discharge Phenomena* (Consultants Bureau, New York 1977)
6.16 S.L. Shapiro (ed.): *Ultrashort Light Pulses*, Topics Appl. Phys., Vol. 18 (Springer, Berlin, Heidelberg, New York 1977)
6.17 N. Bloembergen, P.S. Pershan: Phys. Rev. **128**, 606 (1962)
6.18 S.A. Akhmanov, A.P. Sukhorukov, R.V. Khokhlov: Usp. Fiz. Nauk **93**, 19 (1967) [English transl.: Sov. Phys. Uspekhi **10**, 609 (1968)]
6.19 D.C. Hanna, M.A. Yuratich, D. Cotter: *Nonlinear Optics of Free Atoms and Molecules*, Springer Ser. Opt. Sci., Vol. 17 (Springer, Berlin, Heidelberg, New York 1979)
6.20 L.A. Ostrovskii: Pis'ma Zh. Eksp. Teor. Fiz. **6**, 807 (1967) [English transl.: JETP Lett. **6**, 260 (1967)]
6.21 A. Javan, P. Kelly: IEEE J. QE-**2**, 470 (1966)
6.22 V.S. Butylkin, A.E. Kaplan, Yu.G. Khronopulo: Zh. Eksp. Teor. Fiz. **61**, 520 (1971) [English transl.: Sov. Phys. JETP **34**, 276 (1972)]
6.23 D. Grischkowsky: Phys. Rev. Lett. **24**, 866 (1970)
6.24 S.A. Bakhramov, U.G. Gulyamov, K.N. Drabovich, Ya.Z. Faizulaev: Pis'ma Zh. Eksp. Teor. Fiz. **21**, 229 (1975) [English transl.: JETP Lett. **21**, 102 (1975)]
6.25 B.Ya. Zel'dovich, I.I. Sobel'man: Pis'ma Zh. Eksp. Teor. Fiz. **13**, 182 (1971) [English transl.: JETP Lett. **13**, 129 (1971)]
6.26 V.R. Nagibarov, A.V. Pirozhkov, V.V. Samartsev, R.G. Usmanov: Pis'ma Zh. Eksp. Teor. Fiz. **19**, 391 (1974) [English transl.: JETP Lett. **19**, 215 (1974)]

6.27 S. Edelstein, P. Lambropoulos, I. Duncanson, R. Berry: Phys. Rev. A**9**, 2459 (1974)
6.28 B. Cagnac, G. Grynberg, F. Biraben: J. Phys. **34**, 845 (1973)
6.29 K. Shimoda (ed.): *High-Resolution Laser Spectroscopy*, Topics Appl. Phys., Vol. 13 (Springer, Berlin, Heidelberg, New York 1976) Chap. 8
6.30 V.S. Letokhov, V.P. Chebotaev: *Nonlinear Laser Spectroscopy*, Springer Ser. Opt. Sci., Vol. 4 (Springer, Berlin, Heidelberg, New York 1977)
6.31 A.M. Bonch-Bruevich, V.A. Khodovoi: Usp. Fiz. Nauk **93**, 71 (1967) [English transl.: Sov. Phys. Uspekhi **10**, 637 (1968)]
6.32 N.B. Delone, B.A. Zon, V.P. Krainov, V.A. Khodovoi: Usp. Fiz. Nauk **120**, 3 (1976) [English transl.: Sov. Phys. Uspekhi **19**, 711 (1976)]
6.33 S. Curry, C. Sollins, M. Mizza, D. Popescu, J. Popescu: Opt. Commun. **16**, 252 (1976)
6.34 V.S. Letokhov: *Nonlinear Laser Chemistry*, Springer Ser. Chem. Phys., Vol. 22 (Springer, Berlin, Heidelberg, New York 1983)
6.35 A.M. Bonch-Bruevich, N.N. Kostin, V.A. Khodovoi, V.V. Khromov: Zh. Eksp. Teor. Fiz. **56**, 144 (1969) [English transl.: Sov. Phys. JETP **29**, 82 (1969)]
6.36 J. Bakos, N.B. Delone, A. Kiss, N.L. Manakov, M.L. Nagaeva: Zh. Eksp. Teor. Fiz. **71**, 520 (1976) [English transl.: Sov. Phys. JETP **44**, 268 (1976)]
6.37 K. Leung, J. Ward, B.J. Orr: Phys. Rev. A9, 2440 (1974)
6.38 T.U. Arslanbekov, V.A. Grinchuk, G.A. Delone, K.B. Petrosyan: Kr. soobshch. Fiz. (Brief Communications in Physics) No. 110, 33 (1975)
6.39 S. Chin, N.R. Isenor: Can. J. Phys. **48**, 1445 (1970); Y. Boulassier: Nouvelle Rev. Opt. **7**, 329 (1976)
6.40 G.A. Delone, N.L. Manakov, G.K. Piskova, L.P. Rapoport: Trudy FIAN **115**, 6 (1980) [English transl.: Proc. P.N. Lebedev Physics Institute **115**, 6 (Consultants Bureau, New York 1981)]
6.41 M.R. Cervenan, N.R. Isenor: Opt. Commun. **10**, 280 (1974)
6.42 N.B. Delone, V.A. Kovarskii, A.V. Masalov, N.F. Perel'man: Trudy FIAN **115**, 140 (1980) [English transl.: Proc. P.N. Lebedev Institute **115**, 140 (Consultants Bureau, New York 1981)]
6.43 N.B. Delone, V.A. Kovarskii, A.V. Masalov, N.F. Perel'man: Usp. Fiz. Nauk **131**, 677 (1980) [English transl.: Sov. Phys. Uspekhi **23**, 472 (1980)]

Chapter 7

7.1 V.B. Berestetskii, E.M. Lifshitz, L.P. Pitaevskii: *Relativistic Quantum Theory*, Pt. 1 (Pergamon, Oxford 1971) § 60
7.2 N. Bloembergen: *Nonlinear Optics* (Benjamin, New York 1965); D.C. Hanna, M.A. Yuratich, D. Cotter: *Nonlinear Optics of Free Atoms and Molecules*, Springer Ser. Opt. Sci., Vol. 17 (Springer, Berlin, Heidelberg, New York 1970)
7.3 N.B. Delone, B.A. Zon, V.P. Krainov, V.A. Khodovoi: Usp. Fiz. Nauk **120**, 3 (1976) [English transl.: Sov. Phys. Uspekhi **19**, 711 (1976)]
7.4 B.A. Zon: Opt. Spektrosk. **36**, 838 (1974) [English transl.: Opt. Spectrosc. **36**, 489 (1974)]
7.5 V.I. Kogan, V.M. Galitskii: *Problems in Quantum Mechanics* (Prentice-Hall, Englewood Cliffs, NJ 1963) Chap. 4, Problem 6
7.6 B.A. Zon, N.L. Manakov, L.P. Papoport: Opt. Spektrosk. **38**, 13 (1975) [English transl.: Opt. Spectrosc. **38**, 6 (1973)]
7.7 N.L. Manakov, M.A. Preobrazhenskii, L.P. Rapoport: Opt. Spektrosk. **35**, 24 (1973) [English transl.: Opt. Spectrosc. **35** (1975)]
7.8 B. Bederson, E. Robinson: Adv. Chem. Phys. **10**, 1 (1966)
7.9 B.A. Zon, N.L. Manakov, L.P. Rapoport: *Teoriya mnogofotonnykh perekhodov v atomakh* (The Theory of Multiphoton Transitions in Atoms) (Atomizdat, Moscow 1978)
7.10 V.A. Davydkin, B.A. Zon, N.L. Manakov, L.P. Rapoport: Zh. Eksp. Teor. Fiz. **60**, 124 (1971) [English transl.: Sov. Phys. JETP **33**, 70 (1971)]
7.11 M.Ya. Amus'ya, N.A. Cherepkov, S.G. Shapiro: Zh. Eksp. Teor. Fiz. **63**, 889 (1972) [English transl.: Sov. Phys. JETP **36**, 468 (1973)]
7.12 J.B. Hasted: *Physics of Atomic Collisions* (Plenum, New York 1964)

7.13 A.M. Bonch-Bruevich, N.N. Kostin, V.A. Khodovoi, V.V. Khromov: Zh. Eksp. Teor. Fiz. **56**, 144 (1969) [English transl.: Sov. Phys. JETP **29**, 82 (1969)]
7.14 J. Bjorkholm, P. Liao: Phys. Rev. Lett. **34**, 1 (1975)
7.15 P. Platz: J. Physique **32**, 773 (1971)
7.16 V.A. Grinchuk, G.A. Delone, K.B. Petrosyan: Fiz. Plazmy **1**, 320 (1975) [English transl.: Sov. J. Plasma Phys. **1**, 172 (1975)]
7.17 L.D. Landau, E.M. Lifshitz: *Quantum Mechanics: Non=Relativistic Theory*, 3rd ed. (Pergamon, Oxford 1977)
7.18 B.A. Zon, E.I. Sholokhov: Zh. Eksp. Teor. Fiz. **70**, 887 (1976) [English transl.: Sov. Phys. JETP **43**, 461 (1977)]
7.19 B.A. Zon, B.G. Katsnel'son: Zh. Eksp. Teor. Fiz. **65**, 947 (1973) [English transl.: Sov. Phys. JETP **38**, 470 (1974)]
7.20 G.A. Delone, B.A. Zon, K.B. Petrosyan: Pis'ma Zh. Eksp. Teor. Fiz. **22**, 519 (1975) [English transl.: JETP Lett. **22**, 253 (1975)]
7.21 I.I. Sobel'man: *Atomic Spectra and Radiative Transitions*, Springer Ser. Chem. Phys., Vol. 1 (Springer, Berlin, Heidelberg, New York 1979)
7.22 G.A. Delone, B.A. Zon, K.B. Petrosyan: Trudy FIAN **115**, 127 (1980) [English transl.: Proc. P.N. Lebedev Physics Institute **115**, 127 (Consultants Bureau, New York 1981)]
7.23 B.A. Zon: Opt. Spektrosk. **42**, 13 (1977) [English transl.: Opt. Spectrosc. **42**, 6 (1977)]
7.24 M. Gavrila: Phys. Rev. **163**, 147 (1967)
7.25 H. Bethe, R.W. Jackiw: *Intermediate Quantum Mechanics*, 2nd ed. (Benjamin, New York 1968)
7.26 B. Dubreuil, B. Ranson, J. Chapelle: Phys. Lett. A**42**, 323 (1972)
7.27 W. Heitler: *The Quantum Theory of Radiation*, 3rd ed. (Oxford University Press, London 1954)
7.28 H. Krüger, A. Oed: Phys. Lett. A**54**, 251 (1975)
7.29 P. Bräunlich, P. Lambropoulos: Phys. Rev. Lett. **25**, 135 (1970)
7.30 S. Yatsiv, M. Rokni, S. Barak: Phys. Rev. Lett. **20**, 1282 (1968)
7.31 M. Cardona, G. Güntherodt (eds.): *Light Scattering in Solids* II, Topics Appl. Phys., Vol. 50 (Springer, Berlin, Heidelberg, New York 1982) Chap. 4
7.32 P. Bräunlich, P. Lambropoulos: Phys. Rev. Lett. **25**, 986 (1970)
7.33 R. Miles, S. Harris: IEEE J. QE-**9**, 470 (1973)
7.34 H. Eicher: IEEE J. QE-**11**, 121 (1975)
7.35 J. Ward, G. New: Phys. Rev. **185**, 57 (1969)
7.36 K. Leung, J. Ward, B.J. Orr: Phys. Rev. A**9**, 2440 (1974)
7.37 I. Bigio, J. Ward: Phys. Rev. A**9**, 35 (1974)
7.38 N. Bloembergen: In *Spectroscopia non lineare* (Scuola internationale di Fisica "Enrico Fermi", LXIV Corso) (North-Holland, Amsterdam 1977)
7.39 R. Ditchburn: In *Atomic and Molecular Processes*, ed. by D.R. Bates (Academic, New York 1962) Chap. 3
7.40 G.V. Marr: *Photoionization Processes in Gases* (Academic, New York 1967)
7.41 Y. Samson: Phys. Rep. **28**, 303 (1976)
7.42 I. Lindgren, J. Morrison: *Atomic Many-Body Theory*, Springer Ser. Chem. Phys., Vol. 13 (Springer, Berlin, Heidelberg, New York 1982
7.43 P.A.M. Dirac: Proc. Roy. Soc. A**114**, 710 (1927)
7.44 H. Bethe, E.E. Salpeter: *Quantum Mechanics of One- and Two-Electron Atoms*, 2nd ed. (Rosetta, New York 1977)
7.45 L.V. Keldysh: Zh. Eksp. Teor. Fiz. **47**, 1945 (1964) [English transl.: Sov. Phys. JETP **20**, 1307 (1965)]
7.46 N.B. Delone: Usp. Fiz. Nauk **115**, 361 (1975) [English tranl.: Sov. Phys. Uspekhi **18**, 169 (1975)]
7.47 N.L. Manakov, L.P. Rapoport: Zh. Eksp. Teor. Fiz. **69**, 842 (1975) [English transl.: Sov. Phys. JETP **42**, 430 (1976)]
7.48 I.Y. Berson: J. Phys. B. **8**, 3078 (1975)
7.49 A.I. Baz', Ya.B. Zel'dovich, A.M. Perelomov: *Rasseyanie reaktsii i raspady v nerelyativistskoi kvantovoi mekhanike* (Scattering, Reactions, and Decays in nonrelativistiv Quantum Mechanics), 2nd ed. (Nauka, Moscow 1971) Chap. 1, § 3
7.50 E.N. Economon: *Green's Functions in Quantum Physics*, 2nd ed., Springer Ser. Solid-State Sci., Vol. 7 (Springer, Berlin, Heidelberg, New York 1983)

7.51 Yu.N. Demkov, V.N. Ostrovskii: *Metod potentsialov nulevogo radiusa v atomnoi fizike* (Zero-Radius Potential Method in Atomic Physics) (Leningrad State U.P., Leningrad 1975) Chap. 7, § 1
7.52 N.L. Manakov, A.G. Fainshtein: Doklady Akad. Nauk SSSR **244**, 567 (1979) [English transl.: Sov. Phys.-Doklady]
7.53 A.I. Nikishov, V.I. Ritus: Zh. Eksp. Teor. Fiz. **50**, 255 (1966) [English transl.: Sov. Phys. JETP **23**, 168 (1966)]
7.54 A.M. Perelomov, V.S. Popov, M.V. Terent'ev: Zh. Eksp. Teor. Fiz. **50**, 1393 (1966) [English transl.: Sov. Phys. JETP **23**, 924 (1966)]
7.55 A.M. Perelomov, V.S. Popov, M.V. Terent'ev: Zh. Eksp. Teor. Fiz. **51**, 309 (1966) [English transl.: Sov. Phys. JETP **24**, 207 (1967)]
7.56 S. Geltman: Phys. Lett. **4**, 168 (1963); and **19**, 616 (1965)
7.57 E. Robinson, S. Geltman: Phys. Rev. **153**, 4 (1967)
7.58 P.A. Golovinskii, B.A. Zon: Izv. Akad. Nauk SSSR, Ser. Fiz. **45**, 2305 (1981)
7.59 L.A. Lompre, G. Mainfray, C. Manus, J. Thebault: Phys. Rev. Lett. **36**, 949 (1976); Phys. Rev. A**15**, 1604 (1977)
7.60 J. Bakos, N.B. Delone, A. Kiss, N.L. Manakov, M.L. Nagaeva: Zh. Eksp. Teor. Fiz. **71**, 520 (1976) [English transl.: Sov. Phys. JETP **44**, 268 (1976)]
7.61 B.A. Zon, N.L. Manakov, L.P. Rapoport: Zh. Eksp. Teor. Fiz. **61**, 968 (1971) [English transl.: Sov. Phys. JETP **34**, 515 (1972)]
7.62 R.K. Sharma, K.C. Mathur: IEEE J. QE-**14**, 971 (1978)
7.63 E. Karule: In *Multiphoton Processes*, ed. by J.E. Eberly, P. Lambropoulos (Wiley, New York 1978) p. 259
7.64 G.A. Delone, N.L. Manakov, G.K. Piskova, L.P. Rapoport: Trudy FIAN **115**, 6 (1980) [English transl.: Proc. P.N. Lebedev Physics Institute **115**, 6 (Consultants Bureau, New York 1981)]
7.65 N. Isenor: In *Multiphoton Processes*, ed. by J.E. Eberly, P. Lambropoulos (Wiley, New York 1978) p. 179
7.66 D.T. Alimov, N.B. Delone, M.A. Preobrazhenskii, M.A. Tursunov: Pis'ma Zh. Tekh. Fiz. **6**, 1303 (1980) [English transl.: Sov. Tech. Phys. Lett.]
7.67 M. van der Wiel, E. Granneman: In *Multiphoton Processes*, ed. by J.E. Eberly, P. Lambropuolos (Wiley, New York 1978) p. 199
7.68 J.C. Tully, R.S. Berry, B.J. Dalton: Phys. Rev. **176**, 95 (1968)
7.69 Y. Gontier, N.K. Rahman, M. Trahin: J. Phys. B. **8**, L179 (1975)
7.70 Y. Gontier, N.K. Rahman, M. Trahin: Phys. Lett. A**53**, 83 (1975)
7.71 L.A. Lompre, G. Mainfray, C. Manus: J. Phys. B. **13**, 85 (1980)
7.72 S. Chin, G. Farkas, F. Yergeau: J. Phys. B. **16**, L223 (1983)

Chapter 8

8.1 M. Cardona, G. Güntherodt (eds.): *Light Scattering in Solids* II, Topics Appl. Phys., Vol. 50 (Springer, Berlin, Heidelberg, New York 1982) Chap. 2
8.2 W. Heitler: *The Quantum Theory of Radiation*, 3rd ed. (Oxford University Press, London 1954)
8.3 V.B. Berestetskii, E.M. Lifshitz, L.P. Pitaevskii: *Relativistic Quantum Theory*, P. 1 (Pergamon, Oxford 1971) § 63
8.4 L.D. Landau, E.M. Lifshitz: *Quantum Mechanics: Non-Relativistic Theory*, 3rd ed. (Pergamon, Oxford 1977) § 145
8.5 S. Rautian: In *Invited Papers 2nd Conf. Interaction of Electrons with Strong Electromagnetic Fields* (Central Research Institute for Physics, Budapest 1975)
8.6 R. Loudon: *The Quantum Theory of Light*, 2nd ed. (Clarendon, Oxford 1983)
8.7 W. Hartig, W. Rasmussen, K. Schieder, H. Walter: Z. Physik A**278**, 205 (1976)
8.8 H. Bethe, E.E. Salpeter: *Quantum Mechanics of One- and Two-Electron Atoms*, 2nd ed. (Rosetta, New York 1977)
8.9 I.I. Sobel'man: *Atomic Spectra and Radiative Transitions*, Springer Ser. Chem. Phys., Vol. 1 (Springer, Berlin, Heidelberg, New York 1979)
8.10 H. Torrey: Phys. Rev. **76**, 1059 (1949)
8.11 A.S. Davydov: *Quantum Mechanics*, 2nd ed. (Pergamon, Oxford 1976) § 14
8.12 A.A. Abrikosov, L.P. Gor'kov, I.E. Dzyaloshinskii: *Methods of Quantum Field Theory in Statistical Physics* (Prentice-Hall, Englewood Cliffs, NJ 1963)

8.13 M. Lax: Phys. Rev. **129**, 2342 (1963)
8.14 F. Schuda, C. Stroud, M. Hercher: J. Phys. B. **7**, 198 (1974)
8.15 B. Mollow: Phys. Rev. **188**, 1969 (1969)
8.16 F. Bloch, A. Siegert: Phys. Rev. **57**, 522 (1940)
8.17 E. Arimondo, G. Morucci: J. Phys. B. **6**, 2382 (1973)
8.18 B. Mollow: Phys. Rev. A**13**, 758 (1976)
8.19 P. Gahuzac, R. Vetter: Phys. Rev. A**14**, 270 (1976)
8.20 J. Picqué, Y. Pinard: J. Phys. B. **9**, L77 (1976)
8.21 C. Delsart, J. Keller: Opt. Commun. **18**, 231 (1976)
8.22 S. Moody, P. Lambropoulos: Phys. Rev. A**15**, 1497 (1977)
8.23 B.A. Zon, N.L. Manakov, L.P. Rapoport: Zh. Eksp. Teor. Fiz. **60**, 1264 (1971)
     [English transl.: Sov. Phys. JETP **33**, 683 (1971)]
8.24 J. Bjorkholm, P. Liao: Phys. Rev. Lett. **33**, 128 (1974); Phys. Rev. A**14**, 751
     (1976)
8.25 P. Lambropoulos, S. Moody, S. Smith, W. Lineberger: Phys. Rev. Lett. **35**, 159
     (1975)
8.26 C. Tull, M. Jackson, R. McFachran, M. Cohen: Can. J. Phys. **50**, 1169 (1972)
8.27 J. Ward, A. Smith: Phys. Rev. Lett. **35**, 653 (1975);
     C. Wang, L. Davis: Phys. Rev. Lett. **35**, 650 (1975)
8.28 M. Lipeles, R. Novick, N. Tolk: Phys. Rev. Lett. **15**, 690 (1965)
8.29 P. Bräunlich, P. Lambropoulos: Phys. Rev. Lett. **25**, 135 (1970)
8.30 L. Vriens: Opt. Commun. **11**, 396 (1974)
8.31 L. Vriens, M. Adriaancz: Opt. Commun. **11**, 402 (1974); J. Appl. Phys. **46**,
     3116 (1975)
8.32 M. Rokni, S. Yatsiv: Phys. Lett. A**24**, 266 (1967)
8.33 P. Platz: J. de phys. **32**, 773 (1971)
8.34 A. Flusberg, R. Weingarten, S. Hartmann: Phys. Lett. A**43**, 433 (1973)
8.35 S. Yatsiv, M. Rokni, S. Barak: IEEE J. QE-**4**, 900 (1968)
8.36 M. Jyumonji, T. Kobayasi, H. Inaba: Kvantovaya Elektron. **3**, 790 (1976)
     [English transl.: Sov. J. Quantum Electron. **6**, 430 (1976)]
8.37 M. Cardona (ed.): *Light Scattering in Solids* I, 2nd ed., Topics Appl. Phys.,
     Vol. 8 (Springer, Berlin, Heidelberg, New York 1983)
8.38 T. Hänsch, P. Toschek: Z. Physik **236**, 213 (1970)
8.39 Ya.S. Bobovich, A.V. Bortkevich: Usp. Fiz. Nauk **103**, 3 (1971) [English
     transl.: Sov. Phys. Uspekhi **14**, 1 (1971)]
8.40 M. Pfeifer, A. Lay, H. Weigmann, K. Lenz: Opt. Commun. **6**, 284 (1972)
8.41 A. Weber (ed.): *Raman Spectroscopy of Gases and Liquids*, Topics Current
     Phys., Vol. 11 (Springer, Berlin, Heidelberg, New York 1979)
8.42 V.P. Krainov: Zh. Eksp. Teor. Fiz. **70**, 1197 (1976) [English transl.: Sov.
     Phys. JETP **43**, 622 (1977)]
8.43 V.P. Krainov: Zh. Eksp. Teor. Fiz. **74**, 1616 (1978) [English transl.: Sov.
     Phys. JETP **47**, 565 (1978)]
8.44 R. Whitley, C. Stroud: Phys. Rev. A**14**, 1498 (1976)
8.45 I.I. Sobel'man, L.A. Vainshtein, E.A. Yukov: *Excitation of Atoms and
     Broadening of Spectral Lines*, Springer Ser. Chem. Phys., Vol. 7
     (Springer, Berlin, Heidelberg, New York 1981)
8.46 A.E. Kazakov, V.P. Makarov, M.V. Fedorov: Zh. Eksp. Teor. Fiz. **70**, 38 (1976)
     [English transl.: Sov. Phys. JETP **43**, 20 (1976)]
8.47 U. Fano: Phys. Rev. **124**, 1866 (1961)
8.48 L.P. Kotova, M.V. Terent'ev: Zh. Eksp. Teor. Fiz. **52**, 732 (1967) [English
     transl.: Sov. Phys. JETP **25**, 481 (1967)]
8.49 L.V. Keldysh: Zh. Eksp. Teor. Fiz. **47**, 1945 (1964) [English Transl.: Sov.
     Phys. JETP **20**, 1307 (1965)]
8.50 N.B. Delone: Usp. Fiz. Nauk **115**, 361 (1975) [English transl.: Sov. Phys.
     Uspekhi **18**, 169 (1975)]
8.51 N.B. Delone, M.V. Fedorov: Trudy FIAN **115**, 42 (1980) [English transl.:
     Proc. P.N. Lebedev Physics Institute **115**, 42 (Consultants Bureau, New
     York 1981)]
8.52 D.T. Alimov, N.B. Delone: Zh. Eksp. Teor. Fiz. **70**, 29 (1976) [English
     transl.: Sov. Phys. JETP **43**, 15 (1976)]
8.53 P. Agostini, C. Lecompte: Phys. Rev. Lett **36**, 1131 (1976)
8.54 L.A. Lompre, G. Mainfray, C. Manus: J. Phys. B. **13**, 85 (1980)

8.55 L. Armstrong, S. Feneuille: In *Multiphoton Processes*, ed. by J.E. Eberly, P. Lambropoulos (Wiley, New York 1978) p. 237
8.56 G. Mainfray, C. Manus: Appl. Opt. **19**, 3934 (1980)
8.57 I. Lindgren, J. Morrison: *Atomic Many-Body Theory*, Springer Ser. Chem. Phys., Vol. 13 (Springer, Berlin, Heidelberg, New York 1982)
8.58 J. Christiansen (ed.): *Hyperfine Interactions of Radioactive Nuclei*, Topics Current Phys., Vol. 31 (Springer, Berlin, Heidelberg, New York 1983)
8.59 N.B. Delone, M.V. Fedorov: Izvest. Akad. Nauk SSSR seriya fizicheskaya **43**, 428 (1978) [English transl.: Bull. Acad. Sci. USSR. Phys. Sci. **43**, 179 (1978)
8.60 H. Zeman: In *Electron and Photon Interactions with Atoms* (Festschrift for Professor Ugo Fano), ed. by A. Kleinpoppen and M.R.C. MacDowell (Plenum, New York 1976) p. 581
8.61 M. Teague, P. Lambropoulos, D. Goodmanson, D. Norcross: Phys. Rev. A**14**, 1057 (1977)
8.62 E. Granneman, M. Klewer, G. Nienhuis, M. van der Wiel: J. Phys. B. **10**, 1625 (1977)
8.63 P. Lambropoulos: J. Phys. B. **7**, L33 (1974)
8.64 P. Farago, D. Walker: J. Phys. B. **6**, L280 (1973)
8.65 E. Granneman, M. Klewer, K. Nygaard, M. van der Wiel: J. Phys. B. **9**, L87 (1976)
8.66 N.B. Delone, B.A. Zon, M.V. Fedorov: Zh. Eksp. Teor. Fiz. **76**, 505 (1979) [English transl.: Sov. Phys. JETP **49**, 255 (1979)]
8.67 N.B. Delone, B.A. Zon, M.V. Fedorov: Pis'ma Zh. Tekh. Fiz. **4**, 229 (1978) [English transl.: Sov. Tech. Phys. Lett. **4** (1978)]
8.68 P. Lambropoulos, R. Berry: Phys. Rev. A**8**, 855 (1978)
8.69 J. Duncanson, M. Stroud, A. Lindgard, R. Berry: Phys. Rev. Lett. **37**, 987 (1976)

Chapter 9

9.1 N.B. Delone, V.A. Kovarskii, A.V. Masalov, N.F. Perel'man: Usp. Fiz Nauk **131**, 617 (1980) [English transl.: Sov. Phys. Uspekhi **23**, 472 (1980)]
9.2 A.V. Masalov: Kvantovaya Elektron. **3**, 1667 (1976) [English transl.: Sov. J. Quantum Electron. **6**, 902 (1976)];
    J. Gersten, H. Mittleman: In *Electron and Photon Interactions with Atoms* (Festschrift for Professor Ugo Fano), ed. by A. Kleinpoppen and M.R.C. MacDowell (Plenum, New York 1976) p. 553
9.3 T.N. Smirnova, E.A. Tikhonov: Kvantovaya Elektron. **4**, 1105 (1977) [English transl.: Sov. J. Quantum Electron. **7**, 621 (1977)]
9.4 C. Lecompte, G. Mainfray, C. Manus, F. Sauchen: Phys. Rev. A**11**, 1009 (1975)
9.5 T.U. Arslanbekov: Kvantovaya Elektron. **3**, 213 (1976) [English transl.: Sov. J. Quantum Electron. **6**, 117 (1976)]
9.6 R.V. Karapetyan: Izvestiya Vysshikh Uchebnykh Zavedenii, radioelektronika **18**, 236 (1975) [English transl.: Radiophys. Quantum Electron. **18** (1975)]
9.7 V.P. Krainov, S.S. Todirashku: Zh. Eksp. Teor. Fiz. **78**, 26 (1980) [English transl.: Sov. Phys. JETP **52**, 34 (1980)]
9.8 T.U. Arslanbekov, N.B. Delone, A.V. Masalov, S.S. Todirashku, A.G. Fainshtein: Zh. Eksp. Teor. Fiz. **72**, 907 (1977) [English transl.: Sov. Phys. JETP **45**, 473 (1977)]
9.9 V.L. Strizhevskii: Opt. Spektrosk. **20**, 516 (1966) [English transl.: Opt. Spectrosc. **20** (1966)]
9.10 N.B. Delone, V.A. Kovarskii, A.V. Masalov, N.F. Perel'man: J. Phys. B. **13**, 4119 (1980)
9.11 S.A. Baranov: Izvestiya Akad. Nauk Moldavskoi SSR, seriya fiziko-tekhnicheskikh i matematicheskikh nauk No. 2, 83 (1978)
9.12 A.I. Burshtein, L.D. Zusman: Zh. Eksp. Teor. Fiz. **61**, 976 (1971) [English transl.: Sov. Phys. JETP **34**, 520 (1972)]

9.13 S.G. Przhibel'skii: Opt. Spektrosk. **43**, 542 (1973) [English transl.: Opt.
     Spectrosc. **43** (1973)];
     A.M. Bonch-Bruevich, S.G. Przhibel'skii, V.A. Khodovoi, P.A. Chigir': Zh.
     Eksp. Teor. Fiz. **70**, 445 (1976) [English transl.: Sov. Phys. JETP **43**, 230
     (1976)]
9.14 P.B. Hogan, S.J. Smith, A.T. Georges, P. Lambropoulos: Phys. Rev. Lett.
     41, 1609 (1979)
9.15 A.T. Georges, P. Lambropoulos, P. Zoller: Phys. Rev. Lett. **42**, 1609 (1979)
9.16 V.I. Kogan, V.M. Galitskii: *Problems in Quantum Mechanics* (Prentica-Hall,
     Englewood Cliffs, NJ 1963) Chap. 9, Probl. 5
9.17 N.H. March, W.H. Young, S. Sampanthar: *The Many-Body Problem in Quantum
     Mechanics* (Cambridge University Press, London 1967) Chap. 8
9.18 I. Lindgren, J. Morrison: *Atomic Many-Body Theory*, Springer Ser. Chem.
     Phys., Vol. 13 (Springer, Berlin, Heidelberg, New York 1982)
9.19 M.Ya. Amus'ya: Comments Atom. Molec. Phys. **10**, 155 (1981)
9.20 V.V. Suran, I.P. Zapesochnyi: Pis'ma Zh. Tekh. Fiz. **1**, 937 (1975)
     [English transl.: Sov. Tech. Phys. Lett. **1**, 420 (1975)]
9.21 I.S. Aleksakhin, N.B. Delone, I.P. Zapesochnyi, V.V. Suran: Zh. Eksp.
     Teor. Fiz. **76**, 887 (1979) [English transl.: Sov. Phys. JETP **49**, 447 (1979)];
     D. Feldmann, H. Krautwald, S. Chin, A. von Hellfeld, K. Welge: J. Phys. B.
     15, 1663 (1982);
     D. Feldmann, H. Krautwald, K. Welge: J. Phys. B. **15**, L529 (1982)
9.22 U. Brinkmann, W. Gartig, H. Telle, H. Walther: Appl. Phys. **5**, 109 (1974)
9.23 I.S. Aleksakhin, I.P. Garga, I.P. Zapesochnyi, V.P. Starodub: Opt. Spek-
     trosk. **38**, 228 (1975) [English transl.: Opt. Spectrosc. **38**, 126 (1975)]
9.24 A. L'Huiller, L. Lompre, C. Manus, G. Mainfray: Phys. Rev. Lett. **48**, 1814
     (1982)
9.25 N.B. Delone, M.Yu. Ivanov, V.P. Krainov: Preprint of FIAN (USSR) N 42, 1983
9.26 H. Bethe, E.E. Salpeter: *Quantum Mechanics of One- and Two-Electron Atoms*,
     2nd ed. (Rosetta, New York 1977)
9.27 V.P. Krainov, E.A. Manykin: Ukrainskii Fiz. Zhurnal **25**, 400 (1980)
9.28 A.I. Baz', Ya.B. Zeldovich, A.M. Perelomov: *Rasseyanie reaktsii i raspady
     v nerelyativistskoi kvantovoi mekhanike* (Scattering, Reactions, and Decays
     in Nonrelativistic Quantum Mechanics), 2nd ed. (Nauka, Moscow 1971)
9.29 M.H. Mittleman: Phys. Rev. Lett. **A47**, 55 (1974)
9.30 S. Geltman: J. Phys. B. **10**, 831 (1977)
9.31 I.I. Sobel'man: *Atomic Spectra and Radiative Transitions*, Springer Ser.
     Chem. Phys., Vol. 1 (Springer, Berlin, Heidelberg, New York 1979)
9.32 V.B. Berestetskii, E.M. Lifshitz, L.P. Pitaevskii: *Relativistic Quantum
     Theory*, P. 1 (Pergamon, Oxford 1971)
9.33 L.A. Bureeva: Astron. Zhur. **45**, 1512 (1968) [English transl.: Sov. Astron.
     -AJ]
9.34 S.P. Goreslavsky, N.B. Delone, V.P. Krainov: Zh. Eksp. Teor. Fiz. **82**, 1804
     (1982)
9.35 H. Bethe, R.W. Jackiw: *Intermediate Quantum Mechanics*, 2nd ed. (Benjamin,
     New York 1968)
9.36 V.A. Davydkin, N.B. Delone, B.A. Zon, V.P. Krainov: In *Doklady Vsesoyuznogo
     soveshaniya po kogerentnoi i nelineinoi optike* (Reports at the All-Union
     Conference on Coherent and Nonlinear Optics), Vol. I (Nauka, Novosobirsk
     1979) p. 175
9.37 J. Picart, A. Edmonds, N. Tran Minh: J. Phys. B. **11**, L651 (1978)
9.38 N.B. Delone, V.P. Krainov: Zh. Eksp. Teor. Fiz. **83**, 2016 (1982)
9.39 N.B. Delone, V.P. Krainov: Preprint of FIAN (USSR) N 15 (1983)
9.40 I.Ya. Bersons: Zh. Eksp. Teor. Fiz. **83**, 1276 (1982)
9.41 B.M. Smirnov, M.I. Chibisov: Zh. Eksp. Teor. Fiz. **49**, 841 (1965)
9.42 A.M. Perelomov, V.P. Kuznetsov, V.S. Popov: Zh. Eksp. Teor. Fiz. **53**,
     331 (1967) [English transl.: Sov. Phys. JETP **26**, 222 (1968)]
9.43 N.B. Delone, V.P. Krainov: Opt. Commun. **40**, 205 (1982)
9.44 N.B. Delone, B.A. Zon, V.P. Krainov: Zh. Eksp. Teor. Fiz. **75**, 445 (1978)
     [English transl.: Sov. Phys. JETP **48**, 223 (1978)]
9.45 J. Leopold, I. Percival: Phys. Rev. Lett. **41**, 944 (1978); and
     J. Phys. B. **12**, 709 (1979)

9.46 B.I. Meerson, E.A. Oks, P.V. Sasorov: Pis'ma Zh. Eksp. Teor. Fiz.
     29, 79 (1979) [English transl.: JETP Lett. **29**, 72 (1979)]
9.47 N.B. Delone, V.P. Krainov, D.L. Shepelyansky: Preprint of FIAN (USSR)
     N 16 (1983); Usp. Fiz. Nauk **140**, 355 (1983)
9.48 D. Jones, J. Leopold, I. Percival: J. Phys. B. **13**, 31 (1980)
9.49 E.V. Shuryak: Zh. Eksp. Teor. Fiz. **71**, 2039 (1976)
9.50 J.E. Bayfield, P.M. Koch: Phys. Rev. Lett. **33**, 258 (1974)

Subject Index

Absolute method 159

Absorption method 148,150

Adiabaticity parameter 117,209, 214,232

Adiabatic level inversion 292

Anti-resonance terms 86

Anti-Stokes scattering 202

Atomic field strength 5,164,196, 312

Atomic polarization 165,167

Atomic times 30

Bloch-Siegert shift 75,85,259,277

Born-Fock adiabatic approximation 95

Bound-bound transitions 96,265

Bound-free transitions 115

Breit-Wigner procedure 237,239,266, 277,279,282,284

Circularly polarized electromagnetic field 16,83,225

Classical action 97,104

Classical ionization 320

Continuous-wave laser 120

Correlation time 132

Counter-propagating beams 151,152

Degree of polarization 296

Density matrix 247

Depolarization coefficient 274

Detuning from resonance 67,69,82, 87,180,196,257,266,275

Diffusion ionization 320

Dipole-dipole interaction 62

Dipole interaction 10,13,183

Dipole moment 7,47,114,250

Dynamic polarizability 171,174,176, 181,192,218,288,315

Dynamic polarization 7,309

Elliptically polarized field 185

Energy denominators 22,51

Energy representation 60

Fermi rule 15

Feynman diagrams 38,40,195,201,203, 252,272

Feynman paths 103

Feynman principle 98

Field width 234,287

Floquet theorem 18,72,75,285

Free-particle propagator 40,42

Green's function 60,61

Heisenberg operator 255

Heisenberg uncertainty principle 161

Hill's equation 66

Hydrogen atom spectrum 186

Hyperfine structure 296

Hyper-polarizability 175,178
Hyper-Raman scattering 167,168,201, 207

Imaginary time method 118,220
Induced Compton effect 141
Induced width 243,297
Inverse bremsstrahlung 140,142

Kepler frequency 320,321
Kerr effect 144
Kramer's formula 318
Kramer's theorem 186

Lamb shift 238
Landau-Dykhne adiabatic approximation 94
Linearly polarized light 10,26,131, 171,223,228
Linear Stark effect 186,192
Linear susceptibility 163,164,170, 197,200
Longitudinal mode 124,128
Lorentz formula 8

Magnetic quantum numbers 10,11,84, 187,229,264,294
Markov approximation 253
Mathien's equation 66
Method of pumping 123
Method of Q-switching 123
Model potential 62,228
Multi-photon cross section 158,222, 223,301
Multi-photon emission 270
Multi-photon ionization method 149, 181,268
Multi-photon ionization probability 43
Multi-photon matrix element 41,77, 99,104,164,229,267,317
Multi-photon resonance fluorescence 263

Nonlinear susceptibility 163,167, 174,197,200,205

One-photon absorption 149
One-photon ionization probability 28,34
One-photon mixing 56
One-photon transition 25
Operation time 30
Optical theorem 256

Parabolic quantum numbers 318
Phase-diffusion model 134
Phase-plate method 131
Population inversion 91

Q-switched laser 120
Quantum defect 63,315
Quasi-classical approximation 100, 210,321
Quasi-energy 20,21,72,182

Rabi frequency 68,80,83,90,102, 109,237,241,252,275,279,283
Random-phase approximation 8,310
Rayleigh scattering 162,165,198, 203,249
Reference energy 23
Refraction index 144
Relative method 159,228
Resonance approximation 235,248
Resonance fluorescence 34,243,272, 309
Response time 300
Rotating wave approximation 18,20
Rydberg states 313

Saddlepoint method 95,215
Saturation 70,269
Selection rules 10
Self-energy part 74

Self-focussing 143

Self-frequency modulation 145

Self-induced distortion 135,143

Short-range potential 211

Spectral width 300

Spontaneous emission 236,239

Spontaneous Raman scattering 234, 271

Spontaneous width 148,234,235,242, 252,279

Static polarizability 173,176

Stimulated Raman scattering 89,234, 278

Stirling formula 99

Stochastic ionization 321

Sudden perturbation 37,70,71

Switch-on mode 37,71,112

Switch-on time 6,29,71,109

Test field 261

Total ionization probability 30

Transverse mode 124,128

Tunneling ionization 212,215,231, 304,318

Two-level approximation 233

Two-photon absorption 34,149,150, 206,208,281

Two-photon decay process 271

Two-photon ionization probability 34

Two-photon matrix element 32,50, 52,55,58,74,165,184,188,244,272 318

Two-photon mixing 53

Two-photon Rabi frequency 75

Two-photon transition 241,298

Two-stage process 240,281

Vibrational energy 225

Virtual transition 32,33,161,187, 221

Whittaker's equation 66

Wigner-von Neumann theorem 183

H. Haken, H. C. Wolf

Atomic and Quantum Physics

An Introduction to the Fundamentals of Experiment and Theory

Translated from the German by W. D. Brewer
1984. 254 figures. XIV, 394 pages. ISBN 3-540-13137-X

Contents: Introduction. – The Mass and Size of the Atom. –
Isotopes. – The Nucleus of the Atom. – The Photon. – The
Electron. – Some Basic Properties of Matter Waves. – Bohr's
Model of the Hydrogen Atom. – The Mathematical Framework of
Quantum Theory. – Quantum Mechanics of the Hydrogen Atom.
– Lifting of the Orbital Degeneracy in the Spectra of Alkali
Atoms. – Orbital and Spin Magnetism. Fine Structure. – Atoms
in a Magnetic Field: Experiments and Their Semiclassical Des-
cription. – Atoms in a Magnetic Field: Quantum Mechanical
Treatment. – Atoms in an Electric Field. – General Laws of
Optical Transitions. – Many-Electron Atoms. – X-Ray Spectra. –
Structure of the Periodic System. Ground States of the Elements.
– Hyperfine Structure. – The Laser. – Modern Methods of Optical
Spectroscopy. – Fundamentals of the Quantum Theory of
Chemical Bonding. – Appendix. – Bibliography. – Subject Index.

I. I. Sobelman

Atomic Spectra and
Radiative Transitions

1979. 21 figures, 46 tables. XII, 306 pages (Springer Series in
Chemical Physics, Volume 1). ISBN 3-540-09082-7

Contents: Elementary Information on Atomic Spectra: The
Hydrogen Spectrum. Systematics of the Spectra of Multielectron
Atoms. Spectra of Multielectron Atoms. – Theory of Atomic
Spectra: Angular Momenta. Systematics of the Levels of Multi-
electron Atoms. Hyperfine Strucutre of Spectral Lines. The Atom
in an External Electric Field. The Atom in an External Magnetic
Field. Radiative Transitions. – References. – List of Symbols. –
Subject Index.

This volume is continued by *Excitation of Atoms and Broadening
of Spectral Lines* published as Volume 7 in this series.

I. I. Sobelman, L. A. Vainshtein, E. A. Yukov

Excitation of Atoms and Broadening
of Spectral Lines

1981. 34 figures, 40 tables. X, 315 pages (Springer Series in
Chemical Physics, Volume 7). ISBN 3-540-09890-9

Contents: Elementary Processes Giving Rise to Spectra. – Theory
of Atomic Collisions. – Approximate Methods for Calculating
Cross Sections. – Collisions Between Heavy Particles. – Some
Problems of Excitation Kinetics. – Tables and Formulas for the
Estimation of Effective Cross Sections. – Broadening of Spectral
Lines. – References. – List of Symbols. – Subject Index. – Errata
for volume 1 of this series.

Springer-Verlag
Berlin
Heidelberg
New York
Tokyo

Ultrashort Light Pulses

Picosecond Techniques and Applications

Editor: **S.L.Shapiro**

2nd edition. 1984. 173 figures. XI, 435 pages (Topics in Applied Physics, Volume 18). ISBN 3-540-13493-X

Contents: *S.L.Shapiro:* Introduction – A Historical Overview. – *D.J.Bradley:* Methods of Generation. – *E.P.Ippen:* Techniques for Measurement. – *D.H.Auston:* Picosecond Nonlinear Optics. – D. von der Linde: Picosecond Interactions in Liquids and Solids. – *K.B.Eisenthal:* Picosecond Relaxation Processes in Chemistry. – *A.J.Campillo, S.L.Shapiro:* Picosecond Relaxation Measurements in Biology. – Bibliography. – Subject Index.

Excimer Lasers

Editor: **C.K.Rhodes**

2nd enlarged edition. 1984. 100 figures. XII, 271 pages (Topics in Applied Physics, Volume 30). ISBN 3-540-13013-6

Contents: *P.W.Hoff, C.K.Rhodes:* Introduction. – *M.Krauss, F.H.Mies:* Electronic Structure and Radiative Transitions of Excimer Systems. – *M.V.McCusker:* The Rare Gas Excimers. – *C.A.Brau:* Rare Gas Halogen Excimers. – *A.Gallagher:* Metal Vapor Excimers. – *D.L.Huestis, G.Marowsky, F.K.Tittel:* Triatomic Rare-Gas-Halide Excimers. – *H.Pummer, H.Egger, C.K.Rhodes:* High-Spectral-Brightness Excimer Systems. – *K.Hohla, H.Pummer, C.K.Rhodes:* Applications of Excimer Systems. – List of Figures. – List of Tables. – Subject Index.

Laser Spectroscopy VI

Proceedings of the Sixth International Conference, Interlaken, Switzerland, June 27–July 1, 1983

Editors: **H.P.Weber, W.Lüthy**

1983. 258 figures. XVII, 442 pages (Springer Series in Optical Sciences, Volume 40). ISBN 3-540-12957-X

Contents: Photons in Spectroscopy. – Spectroscopy of Elementary Systems. – Coherent Processes. – Novel Spectroscopy. – High Selectivity Spectroscopy. – High Resolution Spectroscopy. – Cooling and Trapping. – Collisions and Thermal Effects on Spectroscopy. – Atomic Spectroscopy. – Rydberg-State Spectroscopy. – Molecular Spectroscopy. – Transient Spectroscopy. – Surface Spectroscopy. – NL-Spectroscopy. – Raman and CARS. – Double Resonance and Multiphoton Processes. – XUV – VUV Generation. – New Laser Sources and Detectors. – Index of Contributors.

Springer-Verlag
Berlin
Heidelberg
NewYork
Tokyo